PLASTICS AND SUSTAINABILITY

PLASTICS AND SUSTAINABILITY
Practical Approaches

LEE TIN SIN
Faculty of Engineering and Science, Universiti Tunku Abdul Rahman (UTAR), Malaysia

BEE SOO TUEEN
Faculty of Engineering and Science, Universiti Tunku Abdul Rahman (UTAR), Malaysia

Elsevier
Radarweg 29, PO Box 211, 1000 AE Amsterdam, Netherlands
The Boulevard, Langford Lane, Kidlington, Oxford OX5 1GB, United Kingdom
50 Hampshire Street, 5th Floor, Cambridge, MA 02139, United States

Copyright © 2023 Elsevier Inc. All rights reserved.

No part of this publication may be reproduced or transmitted in any form or by any means, electronic or mechanical, including photocopying, recording, or any information storage and retrieval system, without permission in writing from the publisher. Details on how to seek permission, further information about the Publisher's permissions policies and our arrangements with organizations such as the Copyright Clearance Center and the Copyright Licensing Agency, can be found at our website: www.elsevier.com/permissions.

This book and the individual contributions contained in it are protected under copyright by the Publisher (other than as may be noted herein).

Notices
Knowledge and best practice in this field are constantly changing. As new research and experience broaden our understanding, changes in research methods, professional practices, or medical treatment may become necessary.

Practitioners and researchers must always rely on their own experience and knowledge in evaluating and using any information, methods, compounds, or experiments described herein. In using such information or methods they should be mindful of their own safety and the safety of others, including parties for whom they have a professional responsibility.

To the fullest extent of the law, neither the Publisher nor the authors, contributors, or editors, assume any liability for any injury and/or damage to persons or property as a matter of products liability, negligence or otherwise, or from any use or operation of any methods, products, instructions, or ideas contained in the material herein.

ISBN: 978-0-12-824489-0

For information on all Elsevier publications
visit our website at https://www.elsevier.com/books-and-journals

Publisher: Matthew Deans
Acquisitions Editor: Ana Claudia Garcia
Editorial Project Manager: John Leonard
Production Project Manager: Maria Bernard
Cover Designer: Vicky Pearson Esser

Typeset by STRAIVE, India

For our dearest children

Chen Chen and Yuan Yuan with love forever

Contents

About the Authors

Lee Tin Sin is a researcher, professional engineer and associate professor. He graduated with Bachelor of Engineering (Chemical-Polymer) First Class Honours and PhD in Polymer Engineering from Universiti Teknologi Malaysia. Dr. Lee has been involved in rubber processing, biopolymers, nanocomposites and polymer synthesis, with more than 100 publications, including journal papers, book chapters and conferences. He was the recipient of the Society of Chemical Engineers Japan Award for Outstanding Asian Researcher and Engineer 2018 for his contribution to polymer research. He also received a Meritorious Service Medal from the Sultan of Selangor, Malaysia in 2019. He coauthored *Polylactic acid* (Elsevier), the first and second editions of which were published in 2012 and 2019, respectively.

Bee Soo Tueen is a researcher, professional engineer and associate professor. She graduated with Bachelor of Engineering (Chemical-Polymer), Master of Engineering (Polymer Engineering), and PhD in Polymer Engineering from Universiti Teknologi Malaysia. Dr Bee has published more than 70 journal papers on nanocomposites, flame retardants and biopolymers. She coauthored *Polylactic acid* (Elsevier), the second edition of which was published in 2019.

Preface

It is an indisputable fact that we use plastics every day of our lives. Plastics have brought uncountable benefits, but improper disposal of plastics has caused serious environmental issues. On 2 March 2022, heads of state, environmental ministers and other representatives from 175 nations endorsed a historic resolution at the United Nations Environment Assembly in Nairobi to stop plastic pollution by forging an international and legally binding agreement by the end of 2024. A majority of global nations have recognized that reasonable measures need to be taken urgently to manage plastic consumption. Otherwise, our planet is going to be overwhelmed by nondegradable plastic wastes. In this book, sustainable plastics usage, handling of plastic wastes, recycling technology, biopolymers and policies related to managing of plastics are discussed. Although a variety of approaches are being engaged to tackle the plastic pollution problem, education remains the most important element driving the successful implementation of all the policies and regulations. Communities are being educated on how to reduce, reuse and recycle plastics (3R). In fact, many of the policies and regulations have been formulated based on 3R. The authors hope that this book can provide comprehensive information to policymakers, researchers, educators and environmentalists on the basic knowledge related to responsible plastic consumption and sustainable plastic applications from a long-term perspective. Hopefully, we can achieve a cleaner and better environment while not sacrificing the conveniences and benefits that plastics provide to us.

Lee Tin Sin
Bee Soo Tueen

1

Plastics and environmental sustainability issues

1.1 Introduction

Since their invention plastics, or polymers, have become one of the most important materials used by humans, often replacing metals and wood. Usually plastics replace metals due to their lighter weight, and they replace wooden products mainly because of durability and antitermite characteristics (while minimizing deforestation). In fact, plastics applications are an indication of modernization, with plastics bringing many conveniences to human life, having applications in household use, transportation, packaging, electronics and healthcare, as well as the military. Modern plastics, including polyethylene, polypropylene, polystyrene, polyvinyl chloride and polyethylene terephthalate, have taken less than 100 years to achieve production and use on a gigantic scale. For instance, plastics production was merely 1.5 million metric tons in 1950 and almost 250 times that, at 359 million metric tons, in 2018. According to interesting data published by Tiseo (2020) for Statista, shown in Table 1.1, plastic production and wastes are significantly larger than generally expected. Environmental pollution caused by plastics is a harsh issue worldwide. When the global plastic consumption volume is as high as 12 billion metric tons, as projected for the year 2050, humans may be overwhelmed by plastic wastes. This is a crucial problem, because plastic wastes can remain for hundreds of years in landfills. Until now, efforts to reduce, reuse, and recycle (3R) plastics have been the best approach we have to minimize environmental impacts. Nevertheless, prior to embracing 3R, it is best to review the kinds of plastics in our surroundings so that the best approach can be engaged when managing plastic wastes, other than merely sending them to landfills, when recovery and second life/usages might

Plastics and Sustainability. https://doi.org/10.1016/B978-0-12-824489-0.00006-4
Copyright © 2023 Elsevier Inc. All rights reserved.

Table 1.1 Interesting global plastics statistics (Tiseo, 2020).

Estimated volume of plastic waste since 1950	5 billion metric tons
Country generating largest amount of domestic plastic waste	China
Plastic waste created by PepsiCo in India	20,213 metric tons
Average number of plastic fibres in United States tap water	4.8 per 500 mL
Global plastic waste volume by year 2050	12 billion metric tons

be practical. In the next section, common types of plastic demand and supply are discussed so that readers can understand the recyclability of the different types.

1.2 Types of plastics and demand

In general, plastics can be categorized as either thermoplastics or thermoset plastics. The term thermoplastic can be separated into 'thermo' and 'plastic', which means a material that can be deformed when subjected to thermal effects, indicating that these polymeric materials can be softened at high temperature, followed by reforming into new shapes or structures. Thermoplastics are polymeric materials that can be easily recycled using injection moulding, extrusion, blow moulding, blown film or most of the hot melt processing technologies. Although the melting and reshaping process of thermoplastics seemingly can be done infinitely, the recycling of thermoplastics is still limited by thermal degradation that leads to chain scissioning. The most common observations of thermal degradation of recycled polymers are brittleness and yellowish colour. This indicates that the plastic materials are reaching their end of service life before going to the landfill or being incinerated to transform waste into energy. Most of the commodity plastics like polyethylene, polypropylene, polystyrene, polyvinyl chloride, polyethylene terephthalate, acrylonitrile-butadiene-styrene (ABS) and nylon are examples of thermoplastics that can easily undergo a thermal recycling process.

On the other hand, the term thermoset, as implied by the word 'set', is a kind of polymer that is fixed or crosslinked after the first thermal processing. A thermoset plastic is usually limited to a once-only thermal process and, upon end of service life, the thermoset is discarded and seldom recycled. Examples of thermoset plastics include epoxy, unsaturated polyesters, vulcanized rubber/elastomers and polyurethane. Commonly, engineering products made of glass fibre, carbon fibre and vulcanized rubbers, such as tyres, hoses and tubes, and antiseismic elastomers, are thermoset plastics. They are unable to be recycled, yet they still possess secondary applications: for instance, tyres can undergo a further conversion process to produce reclaimed rubber. The main reason why thermoset plastics cannot be recycled is that the crosslinked structure restricts molten deformation. Moreover, the crosslinked structure tends to degrade easily when subjected to high temperature, subsequently generating a pungent odour and turning into char, being unusable thereafter. In short, consumers need to be well guided on the recyclability of plastics products in order to avoid being misled by participating in recycling programs in which nonrecyclable plastic wastes are transported to other countries for disposal, causing serious environmental pollution. A well-known case of this type occurred (Fig. 1.1) when a plastics waste shipment from Australia was

Fig. 1.1 Malaysian Minister of Energy, Science, Technology, Environment and Climate Change Yeo Bee Yin shows samples of plastics waste shipment from Australia on May 28, 2019 (CNN Report, 2019).

exported to Malaysia. Subsequently, the Malaysian government returned the containers of plastic waste to the origin countries.

Commodity thermoplastics account for the majority of plastics usage globally. In general, commodity thermoplastics means polyolefin (polyethylene and polypropylene), polyvinyl chloride, polystyrene and polyethylene terephthalate (PET). As reported by market research platform Markets and Markets (2020), the plastic market size at year 2020 was $US 468.3 billion with a compound annual growth rate (CARG) of 6%. COVID-19 has seriously impacted the global economy, and in particular restricted travelling has led to a large drop in food-serving utensils such as those used in airlines, restaurants and hawker foods and drinks. However, the decline of demand in this tourism-related industry has actually been taken up by the medical supplies industry and the electrical/electronics industry. In fact, the medical devices and equipment supply industry saw tremendous growth in 2020: for instance, as reported by Financial Times (2020a, 2020b), China exported 3.1 million patient monitors in the first quarter of 2020, four times more than in the same period of time in 2019. Also the electronics industry demand has rocketed due to implementation of working from home and learning-from-home activities by governments (Nikkei Asia, 2020). Thus it can be expected that plastics and electronic wastes will flood the landfills in coming years. Therefore knowledge of plastics recycling remains relevant for the near future, to protect the environment for future generations. Although there are hundreds of types of plastics available on the market, this chapter focuses on commonly found plastics in consumer products, with Resin Identification Codes as shown in Table 1.2.

Table 1.2 Resin identification code for recyclable types of plastics.

Code	Plastic type
1	Polyethylene terephthalate
2	High-density polyethylene
3	Polyvinyl chloride
4	Low-density polyethylene
5	Polypropylene
6	Polystyrene
7	Others

1.3 Demand and supply for common types of plastics

1.3.1 Polyolefin

Generally, polyolefins include polyethylene and polypropylene. There are many grades of polyethylene and polypropylene available on the market. For instance, polyethylene can be divided into high-density, low-density, linear low-density, metallocene and ultra-high-molecular weight polyethylene. Meanwhile, there is only one type of polypropylene. During the grade selection of polyethylene and polypropylene for production, one of the main considerations is the melt flow index (MFI). This is a very important indicator that measures the weight (in grams) of polymer extruded over a specified period of time and temperature, through an extrusion die (also known as an orifice) using a plastometer consisting of a piston and weight fitted on top. In general, low-density polyethylene tends to possess high MFI. High MFI (3–6 g/10 min or higher) is likely to be used for injection moulding, compared to extrusion, which requires lower MFI (0.1–0.5 g/10 min) (Krevelen & Nijenhuis, 2009). None of the polyethylene and polypropylene is inherently biodegradable or degradable in general. Nevertheless, there is a product on the market known as oxo-degradable polyethylene, which exhibits accelerated degradation induced by thermooxidative metal complexes, which cause chain scissioning activity to occur in the presence of oxygen, heat and ultraviolet rays. A study by Abed et al. (2020) of oxo-degradable polyethylene in the marine environment is shown in Fig. 1.2; specific activity of bacterial genera *Alteromonas* and *Zoogloea* was found on the oxo-degradable polyethylene, suggesting that both species are possibly involved in mediating biotic degradation. Although oxo-degradable polyethylene can be transformed into fragments in a shorter period of time, concerns still arise about whether the fragmentation process tends to cause other serious problems of microplastic exposure to the environment. A recent publication from Shruti et al. (2020) studied the disposal of oxo-degradable plastic in Mexico City. The study found that the single-use plastic composed of 85% high-density polyethylene and 15% low-density polyethylene with oxo-degradable additives mostly exceeded the standard international norms of heavy metals for Cu, Cr, Mo, Zn, Fe and Pb, with maximum concentration of 1898 mg/kg, 1586 mg/kg, 95 mg/kg, 1492 mg/kg, 1900 mg/kg and 7528 mg/kg, respectively. This indicates that, when the oxo-degradable plastics are degraded, there is possible leaching of heavy metals into soils, causing water

Fig. 1.2 Development of biofouling communities after 20 and 80 days on films of OXO-PE (oxo-degradable polyethylene), PE (neat polyethylene) and PET (net polyethylene terephthalate). Reproduced from Abed, R. M. M., Muthukrishnn, T., Al Khaburi, M., Al-Senafi, F., Munam, A., Mahmoud, H. (2020). Degradability and biofouling of oxo-biodegradable polyethylene in planktonic and benthic zones of the Arabian Gulf. *Marine Pollution Bulletin, 150*, 110639. Reproduced with permission of Elsevier.

source pollution. Hence, the safety and degradability of oxo-degradable plastics remains ambiguous, with additional investigation being recommended into solutions for managing polyolefin wastes.

Polyolefins are thermoplastics that can easily undergo a recycling process. The recycling code for high-density polyethylene is 2, low-density polyethylene is 4 and polypropylene is 5 (refer to Fig. 1.3). High-density polyethylene and low-density polyethylene must be segregated during the recycling process, because any mixing of high-density and low-density polyethylene can lead to deterioration of properties. Typically, low-density polyethylene with lower molecular weight, upon mixing with high-density

Fig. 1.3 HDPE, LDPE and PP recycling codes.

polyethylene, causes the mechanical strength of the recycled polymer to be weakened; in addition the processability behaviour changes, such that the MFI becomes higher, resulting from the low molecular weight LDPE promoting chain motions at the molten state.

As mentioned earlier, the recycling of HDPE, LDPE and PP requires the first steps of segregation. Completion of segregation is followed by a process of grinding and crushing into pellet size, prior to supplying to the manufacturer. Commonly, the recycled pellets of plastics are added in the amount of 10 wt.% to 50 wt.% to the virgin resins. The combination with virgin resins is crucial because recycled resins have variations in quality that can affect the processability and mechanical properties. In fact, the recycling process, which involves mechanical forces and thermal effects, can lead to depolymerization reactions that cut down the molecular chain length. It is also possible that recycled HDPE, LDPE and PP are from different grades having different properties, even though they are similar types. In other words, this action aims at minimizing the effects of recycled resins on the product quality by reducing variation impacts, while enhancing processability and quality control.

In addition, recycling of plastics from a well-defined source can involve the addition of recycled plastics to virgin plastics up to 80 wt.% or even higher. Well-defined sources of recycled plastics are actually obtained from factories that produced rejected plastics: for instance, blow mouldings of beverage bottles that used virgin HDPE resin are collected and reused for production bottles to pack engine oil. Most of the time, the plastic recycler tends to utilize such sources of rejected plastics as the input to blend with virgin plastics, because the quality is more reliable and cleaner, compared to sources like domestic recycling centres. In general, the demand for and uses of polyethylene remain strong, while proper recycling needs to be used to minimize environment impacts.

1.3.2 Styrenic polymers

Styrenic polymers are commonly known as polystyrene, or styrofoam for expanded polystyrene. Expanded polystyrene has been an important material for decades, used primarily for packaging and insulation. Expanded polystyrene possesses a closed cell and rigid structure produced by polystyrene beads impregnated with the foaming agent pentane. It has a density of 11–32 kg/m^3, which is very light and thus easily floats on the surface of water. This is why expanded polystyrene rubbish is found

Table 1.3 Polystyrene and acrylonitrile-butadiene-styrene copolymer main producers.

Styrenic polymer	Region	Producers
GP-PS and HIPS	Americas	Alpek
		Americas Styrenics
		Cellofoam North America
		Dow Chemicals
		Nova Chemicals
		Styrotech Inc.
		Trinseo
		Resirene
	Europe	BASF
		Dell Polymers
		ENI S.p.A and Versalis
		INEOS
		Total Petrochemicals
	Asia	China National Petroleum Corporation
		Samsung Cheil Industries
		Formosa Plastics
		LG Chem
		Lotte Advance Material
		Sabic
		Saudi Polymers
		Idemitsu
ABS	Worldwide	Denka
		Ineos
		Sabic
		Toray
		Trinseo

virtually everywhere, particularly in rivers, lakes, drains and beaches, causing severe clogging of sewage systems, subsequently causing flash floods and breeding of mosquitoes.

Styrenic plastics commonly consist of general-purpose polystyrene (GP-PS), high-impact polystyrene (HIPS) and acrylonitrile-butadiene-styrene copolymer (ABS). Table 1.3 shows the main producers of polystyrene and ABS, respectively. The general properties of styrenic plastics are listed in Table 1.4. Observing the data as tabulated, GP-PS shows the highest tensile strength

Table 1.4 Properties of general-purpose polystyrene (GP-PS), high-impact polystyrene (HIPS) and acrylonitrile-butadiene-styrene copolymer (ABS) (Fried, 2003).

Property	Specific gravity	Tensile strength (MPa)	Tensile modulus (GPa)	Elongation at break (%)	Impact strength, notched Izod (J/m)	Heat deflection temperature (°C) at 455 kPa
ASTM Standard	D792	D638	D638	D638	D256	D648
GP-PS	1.04–1.06	36.6–54.5	2.41–3.38	1–2	13.3–21.4	75–100
HIPS	1.03–1.06	22.1–33.8	1.79–3.24	13–50	26.7–587	75–95
ABS	1.03–1.58	41.5–51.7	2.07–2.76	5–25	160–320	102–107

and modulus as compared to HIPS. GP-PS possesses a strong and rigid structure, mainly contributed by the large benzene ring structure, which resists movement of polymer chains when subjected to external forces. Nonetheless, the result of the rigid structure of GP-PS is that it is poor in elongation and impact strength. On the other hand, HIPS is produced to overcome this poor elongation of GP-PS, by incorporating ground rubber into the styrene stream followed by peroxide-initiated polymerization of styrene in a rubber-styrene solution (Polimeri Europa, 2009). By incorporation of a rubber component (i.e. polybutadiene) in the microstructure of HIPS, under the observation of transmission electron microscopy (see Fig. 1.4), it was found that the polybutadiene particles are dispersed in the matrix of polystyrene (Alfarraj & Nauman, 2004). The dispersed phase acts to disrupt the compact structure of neat polystyrene and promotes the toughness (ability to absorb shock energy) by easing chain sliding when deformation occurs.

On the other hand, acrylonitrile-butadiene-styrene (ABS) copolymer is commonly categorized as an engineering plastic. ABS is produced by grafting a styrene-acrylonitrile copolymer onto partially crosslinked particles of polybutadiene. With a similar function to polybutadiene in HIPS, the polybutadiene increases impact strength over styrene-acrylonitrile copolymer alone, as shown in the electron micrograph of a section cut parallel to the surface of a deformed sample of ABS, as reported by Matsuo (1969) in Fig. 1.5. When the deformation was applied in the direction shown by the arrow, crazes were observed to form perpendicular to the deformation direction and pass through

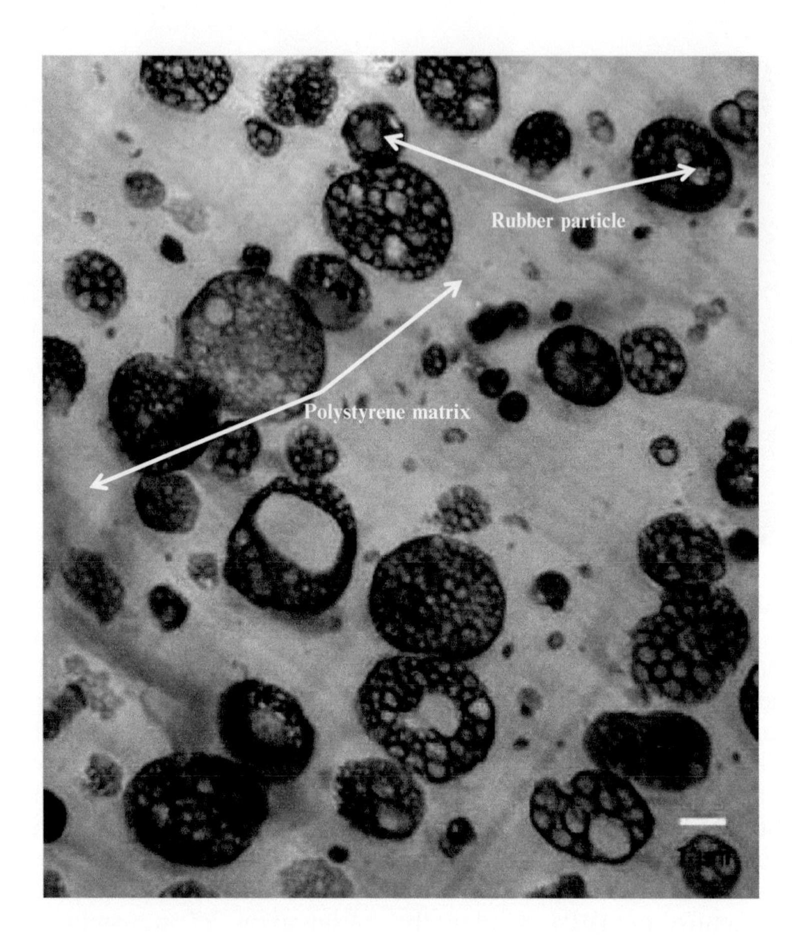

Fig. 1.4 Micrograph of transmission electron microscopy of HIPS produced by SABIC grade LADEN PS330 (content 23 volume % rubber with average rubber particle diameter 1.25 μm). Reproduced from Alfarraj, A., Nauman, E. B. (2004). Super HIPS: Improved high impact polystyrene with two sources of rubber particles. *Polymer, 45*, 8435–8442. Reproduced with permission of Elsevier.

the polybutadiene particles. Meanwhile, the presence of acrylonitrile provides better chemical and weathering resistance due to the fact that polystyrene and polybutadiene are susceptible to sunlight damage.

Styrofoam, or expanded polystyrene, is commonly used for packaging and insulation (as shown in Fig. 1.6). Expanded polystyrene is produced from the slid beads of polystyrene, which consists of polystyrene forming the cellular structure and pentane serving as the blowing agent. During the fabrication process of expanded polystyrene in final products such as clamshell food boxes, steam is used to boil the beads; subsequently the beads expand 40–50 times their original volume, followed by moulding into shapes. Expanded polystyrene is very lightweight and has low thermal conductivity due to its closed cell structure. Expanded polystyrene is suitable for packing electronic components due to its low moisture absorption and excellent cushioning properties. Although the low thermal conductivity of expanded

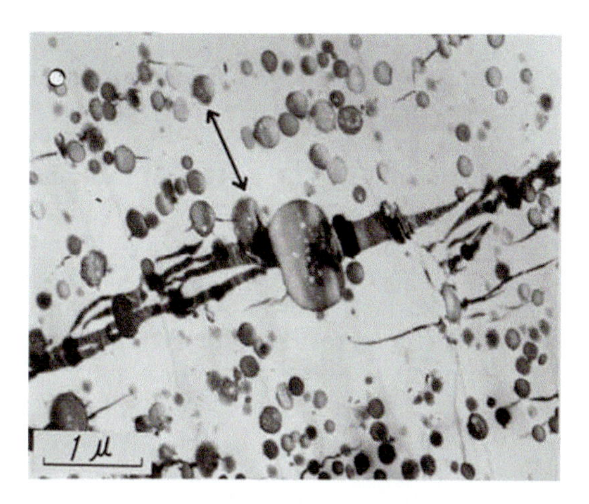

Fig. 1.5 Electron micrographs of commercial ABS when load applied in the direction of *arrow*. Crazes (*dark* region) can be observed running from left to right in the micrographs. Reproduced from Matsuo, M. (1969). Fine structures and facture processes in plastic/rubber two-phase polymer systems. II. Observation of crazing behaviours under the electron microscope. *Polymer Engineering and Science*, 9, 206–212. Reproduced with permission of John Wiley & Sons.

Fig. 1.6 Examples of common EPS products.

polystyrene makes it suitable for use in building insulation, the drawback is that expanded polystyrene is highly flammable and emits toxic gases when it catches fire. As a result, blending of polystyrene with flame retardants is crucial to safeguard users as well as to fulfil the stringent requirements for fire safety.

Despite expandable polystyrene providing many benefits to a variety of industries, many states have banned the use of expandable polystyrene. This ban is not due to its being hazardous or nonrecyclable. Indeed, there are many myths about expandable polystyrene being harmful to users. However, the fact is that expandable polystyrene is as stable as its originated polystyrene for food serving and packing. Consumers like to use expandable polystyrene due to its light weight, low cost, insulating characteristics, and wide availability, and they tend to exploit the convenience of expanded polystyrene by overusage to pack a variety of content, especially as single-use plastics. Commonly, after a single use, the expandable polystyrene is discarded without proper waste management, resulting in accumulation in the surrounding environment. Similar to other synthetic plastics, expandable polystyrene takes hundreds of year to degrade. Moreover, the lightweight/low density characteristic of expandable polystyrene means that it can float on the surface of water, and subsequently can cause substantial environmental pollution. Table 1.5 summarizes the answers to frequently asked questions about expandable polystyrene. Overall, polystyrene has brought many conveniences to humans; however, humans have overexploited it, causing environmental pollution. Overall, the uses of polystyrene are unavoidable, and responsible usage is the only way to minimize its impact on the environment.

1.3.3 Polyethylene terephthalate

Polyethylene terephthalate, commonly known as PET, is the most important polyester. PET has the characteristic of being a transparent, amorphous thermoplastic when it is subjected to rapid cooling, while it behaves as a semicrystalline plastic when cooled slowly or when cold-drawn. PET is produced from the polycondensation of ethylene glycol and terephthalic acid. PET can be processed using a common moulding method like injection moulding, blown moulding and extrusion. It is also suitable to be used to fabricate thin layer products like stretched film and thermoforming. PET is widely used to fabricate carbonated beverage bottles because it has high strength and toughness, good abrasion and heat resistance, low creep at elevated temperatures, good chemical resistance and excellent dimensional stability. Another interesting application of PET is to make artificial fibres for textiles (commonly found in the clothes tag, written as Polyester). Fibres made from PET have outstanding wear resistance, low moisture absorption and are very durable. Textile applications include blankets, bed sheets, comforters, carpets, cushioning in pillows, upholstery padding and upholstered furniture. According to the data source Tiseo (2021), the total production capacity of PET was 30.3 million tons in 2017, with the major producers as listed in Table 1.6.

Table 1.5 Frequently asked questions about expandable polystyrene.

What is expandable polystyrene?

Expandable polystyrene, commonly known as EPS or styrofoam, originates from polystyrene.

Where can we find expandable polystyrene and how can it be identified as expandable polystyrene?

Expandable polystyrene can be easily found in food and beverage serving utensils, with examples as shown in Fig. 1.6. Commonly the colour of EPS is white, yet it can be available in a variety of colours. It is very lightweight and is able to provide insulation effects to keep foods and beverages in warm or cool conditions for a longer time. In fact, EPS is widely used on the market to keep raw meat, seafood, etc. at low temperature for ease of transportation. On the other hand, EPS is also widely used for building insulation as well as for safe storing or cushioning of equipment from damage, because EPS can absorb limited impacts. Its light weight can reduce transportation costs as well.

Why is expandable polystyrene (EPS) banned?

Expandable polystyrene is banned not because it possesses health risks, but because of irresponsible disposal of EPS wastes causing environmental problems.

Is expandable polystyrene (EPS) recyclable?

Since expandable polystyrene (EPS) originates from polystyrene and polystyrene is classified as thermoplastic, EPS can undergo a thermal melting process and be moulded into other plastic products. In other words, the recycling of EPS is almost identical to the polyolefin recycling process.

Since expandable polystyrene can undergo a recycling process, why do most community recycling centres reject segregated expandable polystyrene goods?

Expandable polystyrene is a very lightweight plastic material. It has a volume of 98% air, with the remainder as plastic content. Expandable polystyrene is unfavourable to undergo a recycling process because it is bulky and requires huge space, effort, and transportation costs to ship the items to factories for recycling. For instance, when a 40-lorry load of expandable polystyrene is transported to the factory for recycling, eventually the 40-lorry volume of expandable polystyrene can only produce a volume of 1 lorry of solid polystyrene. In short, this makes the recycling unprofitable and thus no recycler is interested in pursuing such business.

Since it is not profitable to recycle expandable polystyrene, are there any solutions for managing the waste?

Expandable polystyrene is very bulky and it can cause overloading if dumped in landfills. Moreover, when expandable polystyrene waste is simply disposed of at illegal dump sites, there is the possibility of it being washed down to drains, rivers, lakes and seas. The light weight of expandable polystyrene can cause it to float on the surface of water, causing clogging of drainage and leading to flash floods. The floating of expandable polystyrene on the water surface can also cause blockage of sunlight under the sea, which affects photosynthesis. Importantly, marine life may accidentally swallow the waste, subsequently causing suffocation.

One of the best approaches to treat expandable polystyrene is through an incineration process, provided the emissions are monitored thoroughly, since combustion of polystyrene can generate harmful substances.

Can consumers totally avoid usage of expandable polystyrene since it is banned?

The EPS applications are too wide and difficult to avoid, particularly for food packaging with insulation functions. Consumers are urged to reduce their use of EPS for single-use applications, such as carry-out foods and beverages. For cushioning purposes, there are alternatives such as using a polyethylene air cushion/column method to replace EPS, as shown in Fig. 1.7. Such polyethylene air cushion can be easily inflated by releasing air (i.e. punching holes) and the inflated sheet can be accumulated in bulk and sent for recycling.

Table 1.6 Major producers of polyethylene terephthalate (PET).

Region	Producers
Americas	Alfa Alpek
	Celanese
	Amco Polymers
	DAK Americas
	Dupont
	Invista
Europe	BASF
	Lyondell Basell
Asia	Indorama Ventures
	Teijin
	Samyang
	Sabic Petrochemical
	Mitsui Chemicals
	Kolon Plastics

Fig. 1.7 Polyethylene air cushion/column to protect content during transportation.

As mentioned earlier, PET plays an important role in carbonated beverage packaging because it has lower gas permeation compared to other polymer materials. First, by referring to Fig. 1.8 to compare low-density polyethylene (LDPE), polystyrene (PS), polylactic acid (PLA) and PET, it is obvious that the permeation properties of PET are lowest among the polymeric materials. When the polymeric materials have low permeability, this brings improvements in product shelf life, where the pressurized carbon dioxide in the carbonated beverage is unlikely to penetrate through the PET wall and escape, which would cause the drink to no longer be in a pressurized condition. Second, the external atmospheric gases, particularly oxygen, will not penetrate through the PET wall into the beverage, causing the taste of the beverage to change or even spoiling the contents. Hence, one

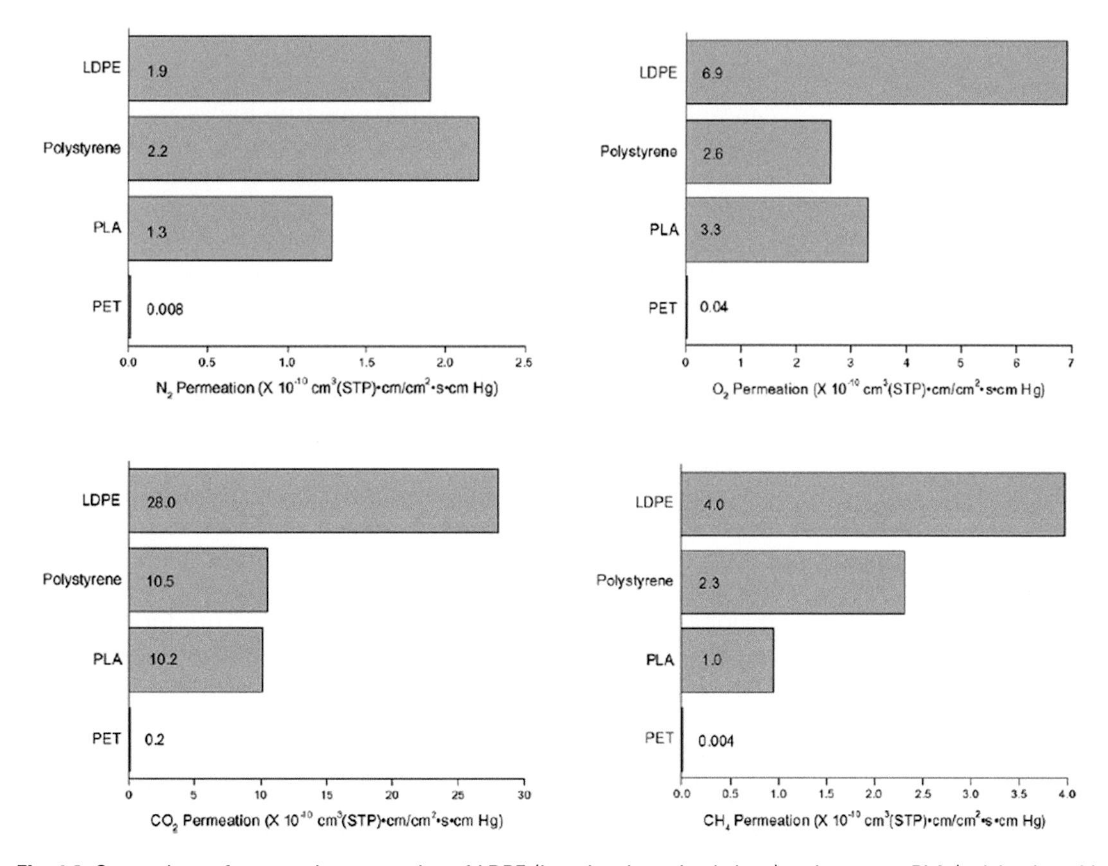

Fig. 1.8 Comparison of permeation properties of LDPE (low-density polyethylene), polystyrene, PLA (polylactic acid with isomer L: isomer D at 96:4) and PET (polyethylene terephthalate). Reproduced from Lehermeier, J. H., Dorgan, J. R., Way J. D. (2001). Gas permeation properties of poly(lactic acid). *Journal of Membrane Science, 190,* 243–251. Reproduced with permission of Elsevier.

Fig. 1.9 Paper carton box for beverage with similar low gas permeation properties to PET.

can well notice that, even though some fruit juices and fresh milk are packed in paper box cartons as shown in Fig. 1.9, such paper box cartons actually consist of multilayers of film to protect the contents from external gases penetrating and reacting with the contents. Hence, PET is undeniably helping to reduce food wastage and safeguard food quality for consumers.

In addition, PET fibre also plays an important role in synthetic textile applications. One of the common PET fibre applications is in sportswear, such as shirts (jersey) and trousers. Compared to cotton, PET polyester is durable, dries quickly, and is wrinklefree; however, the main drawbacks are its poor breathability and lack of moisture absorption. Thus it is commonly blended with both cotton and PET polyesters to overcome these drawbacks. PET polyester is also widely used to manufacturer fibres for cushioning purposes, such as pillows, upholstery, and carpet. Such PET fibres are commonly sourced from recycled PET bottles. For instance, major apparel companies like Nike and Uniqlo produce products using >38% of recycled PET. A study done by Kruger, Kauertiz, and Detzel (2009) reported that the environmental impact of recycled PET is less than that of virgin PET as well as biopolymer-polylactic acid. Table 1.7 shows the extracted data from the report. Obviously, recycling is a better approach to saving the environment.

1.3.4 Polyvinyl chloride

Polyvinyl chloride (PVC) is on the market as a low-cost plastic with very wide applications, including packaging, containers, toys, medical and building materials. PVC can be commonly

Table 1.7 Comparison of polylactic acid (PLA), virgin PET (vPET) and recycled PET (rPET) environmental impacts based on European Union framework for clamshell container.

	Landfill			Incineration		
Treatment approach	*PLA*	*vPET*	*rPET*	*PLA*	*vPET*	*rPET*
Renewable primary energy (GJ)	0.53	0.02	0.02	0.52	0.01	0.02
Nonrenewable primary energy (GJ)	1.22	1.70	1.04	0.96	1.37	0.88
Aquatic eutrophication (g PO_4)	9.73	3.81	2.20	6.61	0.68	0.62
Acidification (kg SO_2)	0.52	0.34	0.20	0.49	0.33	0.19
Climate change (kg CO_2)	60.6	77.8	49.4	81.8	104	62.7
Fossil resources (kg crude oil)	13.5	26.0	14.6	9.9	21.4	12.3

Data extracted from (Kruger et al., 2009) and republished in Lee and Bee (2019).

found in both rigid and flexible forms. Rigid PVC is used as sheets, pipes and window profiles, among others, while the flexible form of PVC includes toys, stationery, and a variety of medical parts like tubes, blood bags and fittings. Most of the commercial-grade PVC is produced by a suspension polymerization process, although emulsion polymerization of PVC is also being used to produce a high grade of PVC for a glove manufacturing process, where smaller particle size is needed for better dispersion in the solution process. PVC has a high glass transition temperature (T_g), which limits its applications in cold climate regions. T_g is the temperature range where the polymer structure changes from a rigid glass material to a soft material (but not melted). Most of the common PVCs have a T_g in a range from 60°C to 80°C. This means that when the temperature of an application is lower than T_g, especially in winter when the temperature drops below freezing, PVC is very brittle and tends to be broken easily, even when subjected to small loads. In order to use PVC at low temperatures or to transform rigid PVC into flexible PVC, the addition of plasticizer is crucial to lubricate the motion of PVC chains, so that the chains can slide between each other when deformation occurs.

There are many producers of PVC worldwide, including Shin Etsu (Japan), Kaneka (Japan), Formosa (Taiwan), Sinopec (China), Solvay (Belgium) and others. China is the largest PVC producer in the world, with total production capacity exceeding 25 million tons/year. Meanwhile, as mentioned earlier, the processing of PVC into consumer end products includes blending together with plasticizer. The most common PVC plasticizers available on the market have been phthalates for the past several decades. However, the application of certain types of phthalates has been banned in European Union countries and the United States in certain consumer products, such as toys in childcare products: for instance, Commission Regulation (EU) 2018/2005 of 17 December 2018 amending Annex XVII to Regulation (EC) No 1907/2006 of the European Parliament and of the Council concerning the Registration, Evaluation, Authorisation and Restriction of Chemicals (REACH) as regards bis(2-ethylhexyl) phthalate (DEHP), dibutyl phthalate (DBP), benzyl butyl phthalate (BBP) and diisobutyl phthalate (DIBP). This regulation restricts DEHP, DBP, BBP and DIBP from being added to any toys and childcare articles. For the United States, the Consumer Product Safety Improvement Act of 2008 (CPSIA) banned the use of three phthalates in toys and childcare articles at concentrations greater than 0.1%: di-2-ethylhexyl phthalate (DEHP), dibutyl phthalate (DBP) and butyl benzyl phthalate (BBzP). CPSIA also restricts the use of di-isononyl phthalate (DINP), di-isodecyl phthalate (DIDP) and di-noctyl phthalate (DnOP) in toys that can be mouthed and childcare articles. In replacing the banned plasticizers, the PVC product producers have shifted to nonphthalate plasticizers such as trimellitates, adipates, sebacates, benzoates, citrates and phosphate plasticizers. Selection of the appropriate plasticizer is important not only to ensure flexibility of PVC; while plasticizers are also able to act as flame retardants, such as phosphate-type plasticizers, substantial addition of a plasticizer can actually cause the PVC products to more easily ignite when subjected to a fire source.

As mentioned earlier, PVC can be used to fabricate a variety of products including toys, building materials, artificial leathers, and tubing/bags for medical applications. Table 1.8 lists the common applications of PVC with their safety concerns. Although PVC is a thermoplastic, the recycling of PVC is not promising due to the types of plasticizer added, the nature of its applications and processing ability. For instance, the mixing of PVC containing different types of plasticizers can cause exudation (plasticizer migration to the surface) resulting from the incompatibility of the plasticizers. An example is PVC containing chlorinated polyethylene (CPE), which cannot be recycled and mixed with PVC products for use in high temperature applications. This can cause the PVC products to harden as the result of migration of chlorinated polyethylene to the surface, while a wet surface can be found due to the accumulation of migrated plasticizer. Besides, the processing of PVC requires attention particularly on the processing temperature, because when PVC is processed at high temperature, it has a tendency to release hydrogen chloride gases that

Table 1.8 Common applications of PVC with safety concerns.

Application	Products	Safety concerns
Toys and childcare	 	Concern about types of plasticizers to be added. Avoid adding DEHP, DBP, BBP, DIBP, BBzP, DINP, DIDP, DnOP and any phthalates banned by regulations
Building materials	 	Concern about the types of plasticizers being used. In addition, the PVC products need to be fire resistant and preferably low smoke when they catch fire. Efforts need to be spent on the flame-retardant package to be applied
Artificial leather		Concerns about the types of plasticizers being used. Besides, the PVC products need to be fire resistant and preferably low smoke when they catch fire. Efforts need to be spent on the flame-retardant package to be applied

Continued

Table 1.8 Common applications of PVC with safety concerns—cont'd

Application	Products	Safety concerns
Gloves	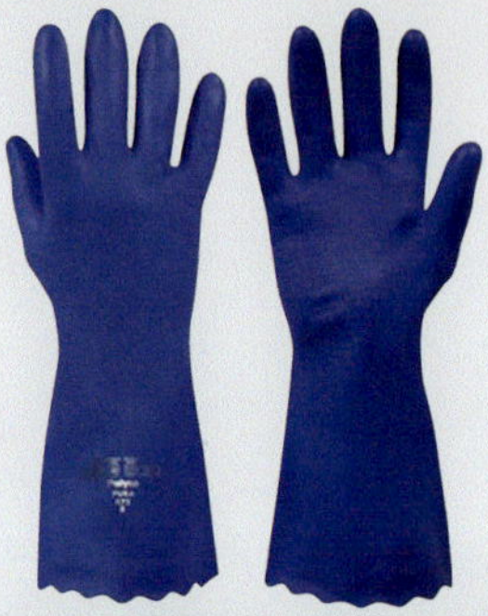	Concerns about the dispersion of PVC particles to achieve homogeneity and less tendency to puncture
Medical applications	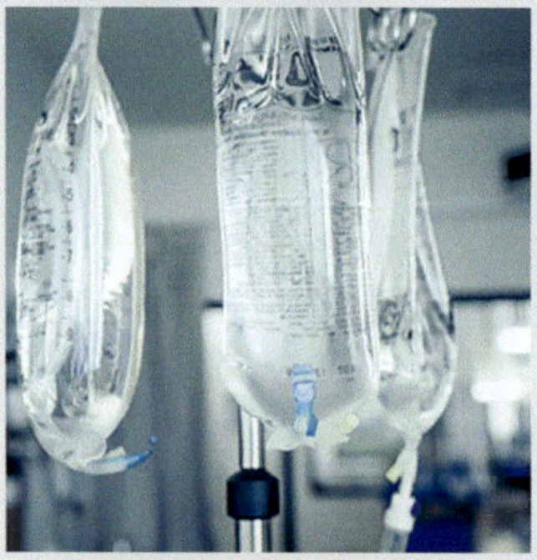	Concerns about the types of plasticizers used. Avoid banned phthalate plasticizers
Packaging		Concerns about the types of plasticizers used. Avoid banned phthalate plasticizers

not only corrode the equipment, but also release strong radicals that can attack the PVC chains and cause degradation. The signs of degradation can be visualized when the products become yellowish or, worse, brownish, indicating degradation has occurred and the resulting products are expected to have poorer mechanical properties. Finally, the artificial leather, glove and medical devices made of PVC are not suitable to be recycled. The simple reason is because such PVC products are contaminated after use. Generally, PVC is an important plastic material for which proper recycling techniques are required to ensure the efficiency of PVC recycling.

1.4 Plastic wastes and impacts—Soil and aquatic

Plastic waste issues have been gaining worldwide attention for decades. While many efforts have been spent to educate users to reduce, reuse, and recycle plastic products, the pollution caused by plastic is worsening and drastic action needs to be taken. During the COVID-19 pandemic, the handling of single-use disposable personal protective equipment (PPE), face masks, medicine packaging, etc. became serious, in order to safeguard the surroundings from contamination while preventing the spread of the virus (Parasha & Hait, 2021). Indeed, not only have medical wastes increased, but also bottled water, disposable wipes, hand sanitizers, and cleaning agent packaging have increased tremendously and these can be found almost anywhere on the streets (Picheta, 2020). Meanwhile, many studies have been carried out to evaluate the impact of the COVID-19 pandemic on plastic waste generation. Benson et al. (2021) estimated that in June 2020 the total plastic waste generated was 1.6 million tonnes, with a daily facemask disposal of about 3.4 billion pieces, as shown in Table 1.9. Based on the estimated amount of disposed plastic wastes generated daily, when 20%, or 116 million tonnes, of the wastes are mishandled and end up in the ocean, this can produce a disaster for marine life. At the time of the writing of this chapter, many countries are facing the second and third spikes of COVID-19 infection rates and the future is still difficult to anticipate; consequently the plastics pollution could be even worse and the impacts could spread to future generations.

Commonly, pollution is divided into three major environmental categories: terrestrial (soil), aquatic (water) and atmospheric (air). Plastic wastes are usually accumulated extensively in terrestrial and aquatic ecosystems. For atmospheric pollution, the source is open burning of plastic wastes leading to smoke and fly ash, but it is comparatively shorter lived than plastic pollution accumulation in terrestrial and aquatic elements. A report by Pietrelli, Poeta, Bttisti, and Sighicelli (2019) stated that polyolefin remains the most commonly found plastic waste and it is the main source of macro- and microplastic particles contaminating

Table 1.9 Estimated plastic waste generation corresponding to region.

Region	Africa	Asia	Europe	South America	North America	Oceania	Total
Population[a]	1,340,598,147	4,641,054,775	747,636,026	653,952,454	368,869,647	42,677,813	–
Total COVID-19 cases[a]	212,271	1,470,640	2,149,248	1,267,248	2,361,458	8,896	–
Facemask acceptance rate[b] by population (%)	70	80	80	75	80	75	–
Average facemask/person/day	1	1	1	1	1	1	–
Estimated daily facemask disposed	411,814,854	1,875,181,681	445,022,934	380,414,703	244,335,150	21,682,379	3,378,451,702
Estimated plastic waste generated (tonnes)	100,544,861	348,079,108	56,072,702	49,046,434	27,665,223	3,200,836	584,609,165
Estimate plastic waste generated per day (tonnes)	275,465	953,641	153,623	134,373	75,795	8769	1,601,666

[a]Data of population and total COVID cases were obtained from Worldometers in June 2020.
[b]Data of facemask acceptance rate are based on estimated value.
Data from Benson, N. U., Bassey, D. E., Palanisami, T. (2021). COVID pollution: Impact of COVID-19 pandemic on global plastic waste footprint. Heliyon, *7, e06343.*

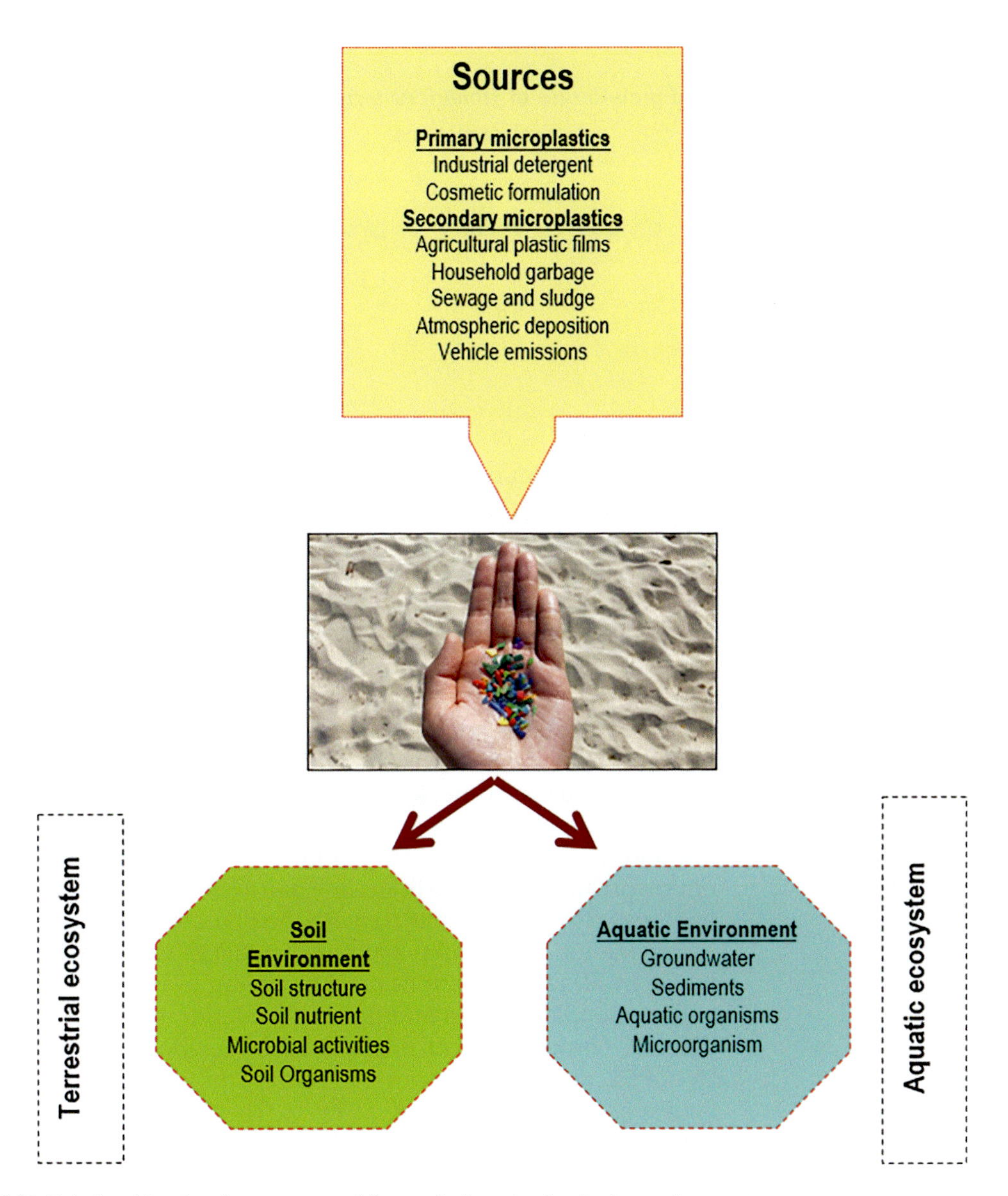

Fig. 1.10 Relationship of main sources and fates of microplastics in the environment.

land and water sources. Qi, Fu, Wang, Gao, and Peng (2020) also reviewed the sources of microplastics and summarized the linkages of terrestrial and aquatic pollution as shown in Fig. 1.10. According to Qi et al. (2020), soil is not only the sink of microplastics, but also is the main contributor to microplastics in groundwater and aquatic environments. For instance, after the degradation of plastics within the terrestrial boundary,

Table 1.10 Mortality and growth rate of *Lumbricus terrestris* earthworm in polyethylene microplastics.

	0% MP	7% MP	28% MP	45% MP	60% MP
Mortality	0	0	8.3	8.3	25
Growth rate (mg/worm day)	10.3	5.8	0.5	1.4	−0.5

MP denotes polyethylene microplastics.

microplastics can be further washed out to drainage in lakes and rivers and, finally, to the marine environment. Moreover, there are also sources of microplastics from agricultural soils; for example, usage of mulch film subsequently leads to contamination in topsoils. Liu et al. (2018) studied microplastics (sizes of 20 μm to 5 mm) and mesoplastics (5 mm to 2 cm) in farmland soils for 20 vegetable fields around the suburbs of Shanghai, China. They found the abundance of microplastics was 78.00 ± 12.91 and 62.50 ± 12.97 items per kg in shallow and deep soils, respectively. Also, mesoplastics were found, with an abundance of 6.75 ± 1.51 and 3.25 ± 1.04 items per kg in shallow and deep soils. Out of these plastics particles, 48.79% and 59.81% were of sizes <1 mm in shallow and deep soils, consisting of fibre, fragment and film. It was not surprising to observe that more particles with larger size could be found closer to the surface with the content make-up of polypropylene (50.51%) and polyethylene (43.43%). In another study by Du et al. (2020), with the soil sampling to include industrial and residential areas in addition to farmland, the results showed that the composition of microplastics included polypropylene, PVC, PET, and nylon 6, with PET and nylon 6 accounting for the largest proportion, with both at 30.2%. This shows that microplastics composition strongly depends on location, and industrial area content is mostly made up of engineering plastics.

Such microparticles of plastics can migrate from one location to larger coverage through a variety of mechanisms, including bioturbation, tillage operation and water infiltration (Qi et al., 2020). Animals may be attached to traces of particles either physically on the surface or through ingestion and defecation by livestock of worms, insects and plants. Subsequent washouts by rainwater of microplastics-containing soil can cause the microplastics to spread to wider areas and eventually contaminate the aquatic ecosystem. Although the migration of microplastics to groundwater can be less likely to happen due to the depth of soil, a study from Panno et al. (2019) reported that contamination was found with a maximum concentration of 15.2 particles/L in springs and wells from two karst aquifers in the state of Illinois in the United States.

Many studies have been conducted to examine the impacts of microplastics on soil and aquatic entities, summarized as follows.

1.4.1 Soil impacts

(a) Soil structure—When large quantities of plastic particles are present in the soils, it can be catastrophic, because this reduces the water retention ability of rainwater due to the hydrophobic characteristics of most plastics. Subsequently, this can cause irrigation times to be higher. In addition, the soil aeration and water permeability are lower when soil is mixed with plastic traces, with the result being reduction of root growth and productivity.

(b) Soil physicochemical structure—The presence of plastic particles can predominantly change the chemical behaviour of soil, as it becomes more hydrophobic. This can increase the soil's tendency to attract chemical substances. For instance, pesticides tend to accumulate in the soil, causing toxicity in the soil habitat. In addition, the equilibrium of carbon and nitrogen contents can be significantly decreased with an increase in plastic residue. This leads to poor growth of plants. Moreover, the leaching of mulch film containing additives such as plasticizers can lead to transfer into the soil. For example, PVC mulch film is more durable then polyethylene mulch film, being more resistant to weathering for prolonged use. However, the fabrication of PVC film requires phthalate-type plasticizers. The typical phthalate plasticizers are endocrine-disrupting chemicals, also known as 'environmental hormones'. Excessive exposure to plasticizers can affect mammalian reproduction as well as having carcinogenic and other toxic effects. A study by Shi et al. (2019) found that phthalate esters in grains from mulching plots range from 4.1 to 12.6 mg/kg. Di-n-butyl phthalate and di-(2-ethylhexyl) phthalate were found in soil and grain samples. Although phthalate is a noncarcinogenic compound, phthalates remain a risk to younger groups of people. In short, degradation of plastics into the soils can pose a major threat to ecosystem and human health.

(c) Soil organisms and plants—A study by Lwanga et al. (2016) focused on microplastics in the earthworm *Lumbricus* *terrestris* species when exposed to polyethylene particles at a size of <150 μm. They found that when the composition of microplastics was higher, the mortality and growth rate of earthworms were higher and lower, respectively, as shown in Table 1.10. Later Cao et al. (2017) studied polystyrene microplastics (58 μm) on *Eisenia foetida* species earthworms. When the amount of polystyrene particles was <0.5 w/w%, there was an insignificant effect. However, when the concentration increased to >1%, there was significant inhibition of growth of earthworms, with higher mortality rate. This phenomenon was explained by Bandopadhyay, Martin-Closas, Pelacho, and DeBruyn (2019) as follows: plastic mulches can affect soil microbial activity in relation to changing the soil microclimate and soil physical structure by enabling contaminants to adhere to the film fragments.

1.4.2 Aquatic impacts

In recent years, many reviews have been published on the aquatic impacts of microplastics. For instance, a review by Rebelein et al. (2021) summarized the microplastics concentrations of various aquatic environments globally, as shown in Table 1.11. The sources of microplastics in aquatic environments are usually (1) Plastic wastes directly discharged into water bodies by daily human use, such as microbeads in the textile, cosmetic and washing industry and personal care products; (2) Plastic wastes discharged from domestic and industrial sewage treatment plants; and (3) Plastic wastes in ship navigation and fisheries such as aquaculture-required ropes and floating drilling rigs.

(a) Combination of microplastics and antibiotics: A review by Tang et al. (2021) of several studies suggested that there is a possibility of microplastics being loaded with antibiotics and swallowed by aquatic organisms. As revealed in a study by Li et al. (2018), different surface characteristics and crystallographic junctions of microplastics can induce adsorption effects on antibiotics. Li et al. (2018) reported that nylon (also known as polyamide) has the strongest adsorption capability, with distribution coefficient values of 7.36 ± 0.257 to 756 ± 48.0 L/kg in the freshwater system. The main reason for polyamide possessing such a good adsorption characteristic is its ability to form hydrogen bonding. Meanwhile, polyethylene, polystyrene, polypropylene and polyvinyl chloride have lower antibiotic adsorption. Overall, the adsorption capability of different types of antibiotics on microplastics can be ordered as

Table 1.11 Amount of microplastics in various locations globally.

Area	Range (mean) [Microplastic particles per m^3]
Open and coastal ocean	
Atlantic Ocean	Range of means (0.01–2.4)
Northeast Atlantic Ocean	0–22.5
Atlantic Ocean transect	13–501
Pacific Ocean	Range of means (0.017–7.25)
Northeast Pacific Ocean	8–9180 (2080)
North Pacific, offshore	0.43–2.23
North Pacific, inshore	5–7.25
South Africa	258–1215
Mariana Trench, deep sea	2060–13,510
Arctic water + Arctic sea ice	0–18
Enclosed ocean areas	0–650,000 (88,000)
Jade system, North Sea	
Baltic Sea	0–0.8
Black Sea	(1100)
Qatar's Exclusive Economic Zone	0–3 (0.7)
Gulf of Mexico	4.8–18.4
Estuaries	
Changjiang Estuary, China	(157)
East China Sea	(113)
Estuaries worldwide	Range of means (0.1–100)
Freshwater	
Neckar, Germany	8–59.3
Rhine, Germany	2.9–214.2
Donau, Germany	9.8–150.8
Rhine River	0.1–141.6
Inland freshwaters, China (rivers and lakes)	1660–8925
Three Georges Reservoir, China	1597–12,611 (4703)
Saigon River, Vietnam	10–519,000
Great Lakes, United States	0.05–32
Karst water system, Illinois, United States	0–15,200

Modified from Rebelein, A., Int-Veen, I., Kammann, U., Scharsack, J. P. (2021). Microplastic fibers—Underestimated threat to aquatic organisms. Science of the Total Environment, *777, 146045, recommended further reading.*

ciprofloxacin > amoxicillin > trimethoprim > sulfadiazine >tetracycline. Other relevant research (Dan et al., 2019; Ninwiwek et al., 2019) on the antibiotic adsorption of microplastics has also been reported, with identical observations.

(b) Combination of microplastics and heavy metals: The combination of heavy metals with microplastics is usually related to coatings such as antirust paint for ports, boats, ships, etc. Other sources include fisheries and marine equipment. In a study by Brennecke, Buarte, Paiva, Cacador, and Canning-Clode (2016) it was found that the heavy metals copper and zinc were leached from antifouling paint from polystyrene beads and polyvinyl chloride particles in seawater. Such heavy metals not only accumulate in the bodies of aquatic animals, but the heavy metal composition can accumulate in the food chain, causing severe damage to the ecosystem. This is the effect called biomagnification, meaning a buildup of toxins in the food chain. For instance, if heavy metal concentration is initially in parts per million, as the trophic level increases in a food chain, the amount of toxic buildup increases. Toxins build up in organisms' fat and tissue. Predators (such as carnivores and humans) accumulate higher toxins than prey (such as herbivores and plankton).

(c) Combination of microplastics and organic pollutants: It is well known that almost all types of plastics have additives for ease of processing, colouring, reinforcement, antioxidants, flame retardants, stabilizers, etc. Such additives are derived from organic compounds or heavy metal complexes, which possess risks when leached out from the microplastics. Compared to the original plastic articles, when the plastic is fragmentized into microplastics, the surface area of contact increases tremendously, subsequently exposing the additives to react with the external environment. Many researchers have raised concerns about endocrine types of additives in plastics, such as brominated types of flame retardant, e.g. polybromiated dipheyl ethers (PBDEs) and phathalate type plasticizers. These compounds, when released to the environment, can accumulate in aquatic life, subsequently causing bioaccumulation and health risks.

(d) Physical interaction of macro-microplastics with aquatic life: Although the preceding paragraphs mention impacts that seem to be not so 'obvious and direct', the most important impacts of plastics result from large articles that cause aquatic life to suffer physically, leading to death. The photos

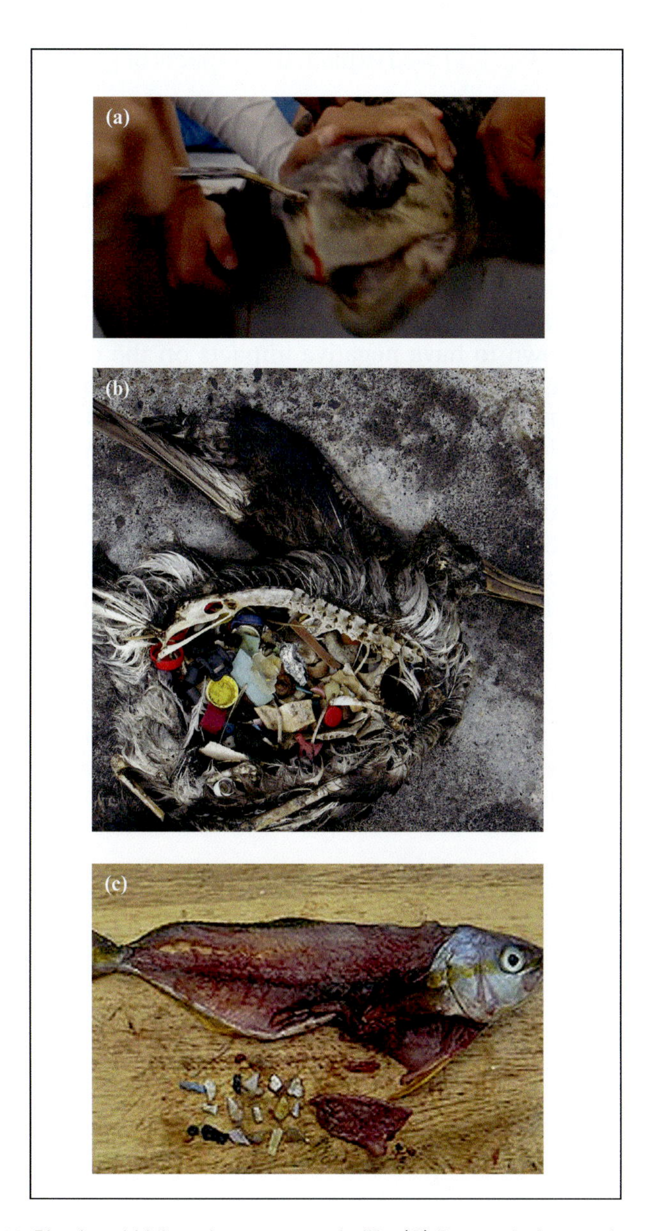

Fig. 1.11 Plastic rubbish endangers aquatic life. (A) Removal of straw from turtle nose; (B) Plastic trash in the stomach of seagull; (C) Plastic fragments found in the stomach of fish. (B) Photo credit: Dan Clark/USFWS/AP; (C) Photo credit: Gyres Institute.

shown in Fig. 1.11 well explain the impacts of plastic trash on marine life.

1.5 Concept of sustainability for plastics and packaging

Plastics and packaging are common components in almost every product that we purchase daily. Although plastics and packaging are known to have negative impacts on the environment, proper management of plastics and packaging can help to minimize these impacts. Through the concept of sustainability, the impacts of plastics and packaging can be reformed with innovative packaging designs. Sustainable packaging innovations can be divided into four main initiatives (refer to Fig. 1.12):

(1) Application initiative: In general, the application approach indicates that the innovator wants to minimize the waste or turn the waste into something useful for another industry. For instance, the packaging is designed so that, when discarded, it can be further used as an input for another industry, such as PET beverage bottles used to produce polyester fibre for textiles (i.e. curtain fabric). After the textiles have aged, they can be further transformed into rugs, before finally being incinerated for energy recovery. The product

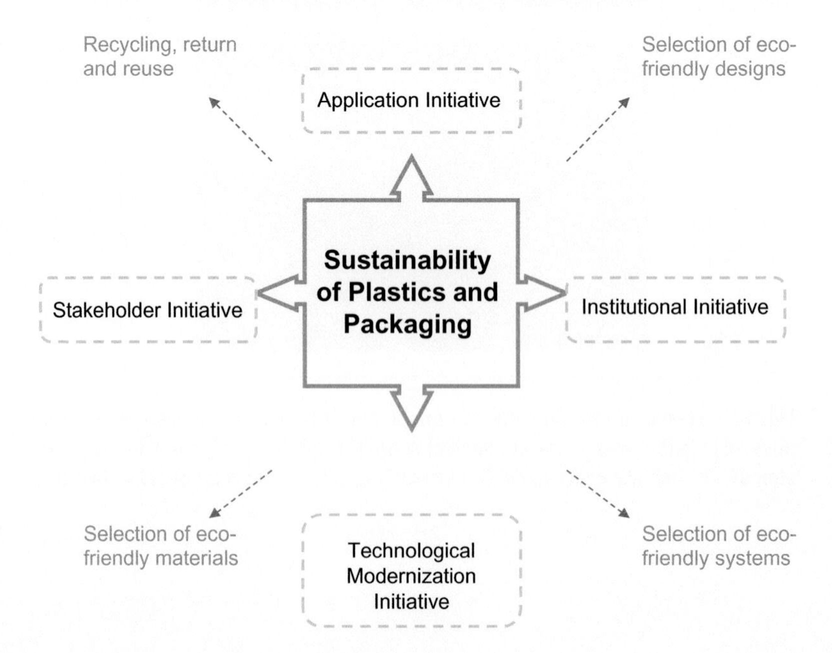

Fig. 1.12 Sustainability initiatives and approaches for plastics and packaging.

fabricators at each stage are actually thinking of producing something which, at the end of its service life, can still be used as an input for products. This is referred to as a 'circular economy', with the energy recovery from the incineration process used by household consumers for lighting or other purposes. The circular economy is a process to recover as many benefits as possible by considering the applications of each stage, so that it can generate as much value as possible, while minimizing the impacts. The designers have the important role of ensuring when products are introduced to the market that they are able to fulfil the circular movement from beginning to end, so that environmental impacts are lower.

(2) Institutional initiative: This approach addresses how the legislation, laws, regulations, and policies affect the decision-making of the packaging producers as well as product manufacturers when introducing products into the consumer market. For instance, countries of the European Union have imposed bans and levies on the use of unsustainable packaging. In order to meet the legal requirements to penetrate the markets of different countries, the producers will take the initiative to improve their plastic packaging. Implementation of laws and regulations are considered an instant approach that can affect the operation of industrial players in a short time; however, the long-term outcomes still depend on the effectiveness of the implementation.

(3) Stakeholders initiative: The stakeholders are any parties with an interest that can affect the decision-making of producers, such as investors, bankers, customers, suppliers, material suppliers, brand owners, recyclers, etc. For instance, investors and bankers as the capital providers can state their intentions to finance a better business prospect that includes sustainable packaging. In addition, the brand owners can express their intention to improve a product image through a sustainable packaging design to meet the expectations of consumers over the long run. All these initiatives from stakeholders will indirectly pressure the producers to produce more environmentally friendly products.

(4) Technological modernization initiative: Some of the advantages that technology advancement brings to the industry are improvements in competitiveness, cost, and resource savings. The pushing factors that enable technological modernization to successfully achieve sustainability mainly depend on government encouragement, while growth of business competition motivates industry players to make major changes in order to maintain and expand their local and global market presence. For instance, the plastics industry players can select more environmentally friendly types of biopolymers as packaging materials for their products. Also, they can take the initiative to obtain the packaging material from closer sources (i.e., from a local manufacturing plant instead of importing from other countries, because long-distance transportation can increase the carbon footprint). In addition, the plastic industrial players can adopt changes with material savings and innovative designs to minimize material usage and requirements: for instance, a honeycomb or polyethylene air cushion to protect the contents during transportation (refer to Fig. 1.7). Many times, the innovations are readily available, but industry players such as food and beverage, fast moving consumer goods (FMCG), and services (e.g. aviation catering services) require initiatives to select those more beneficial to their business that also make the environment cleaner and better.

There are some common practices to address sustainable plastics use, including (1) recycling, return, and reuse; (2) selection of ecofriendly designs; (3) selection of ecofriendly materials; and (4) selection of ecofriendly systems. Recycling, return, and reuse practices have been well known in our daily lives for some time. Most of the time, the success of recycling, return, and reuse practices depends on personal willingness to carry them out. In other words, the community should take these practices as norms of behaviour, not only for plastics but for other materials as well. Recycling, return, and reuse should be introduced to children at preschool age. The Japanese are a successful example, as they have educated their citizens at a young age on recycling of plastic trash. The result of this education can be well observed, as Japanese rivers and drainages are the cleanest in the world, with no plastics rubbish found in the drainages.

On the other hand, the selection of ecofriendly designs, materials, and systems are considered to be more complicated practices to meet sustainability requirements of plastic packaging. These practices are usually considered as a higher level of innovation to be performed by industrial players. To a certain extent, the brand owners and consumers with better awareness can insist on products with sustainability features. For instance, consumers can reject products with excessive packaging, such as electrical and electronic products with multiboxes or unnecessary free gifts. For single-use plastic articles, such as drink cups, spoons, plates and bowls, consumers can select thinner thermoforming products compared to injection moulded plastic products. Commonly, thermoforming containers have lighter weight compared to injection moulded containers. Consumers can also opt for biodegradable plastics with a renewable feature to substitute for fossil-type plastics. In addition, manufacturers or brand owners can opt to use recycled plastic materials for their products: for instance, instead of packing engine oil using HDPE bottles, 60–80 wt.% recycled HDPE can be used. All these remain the efforts and initiatives of consumers and manufacturers to achieve sustainability features in plastic materials.

1.6 Methods of recycling, opportunities and challenges

In general, there are two types of recycling methods for plastics: physical recycling (also known as melt recycling) and chemical recycling. Between both recycling methods, the physical recycling method remains the most straightforward and it can be performed after segregation of plastics according to their types, as listed in Table 1.2. On the other hand, the chemical recycling method is complicated, as it engages advanced technology to transform waste plastics to monomers or pyrolysis oils before undergoing further processing to be turned into other valuable substances. Many times, recycling of plastic wastes is known as valorization of waste to wealth, while reducing environmental pollution. Table 1.12 compares the advantages and disadvantages of physical and chemical recycling of plastics.

1.6.1 General operation principle of physical recycling

Physical recycling of plastics is also known as the melt recycling method. As mentioned in Table 1.12, physical recycling is a straightforward recycling method. The first step of physical recycling is the segregation of plastic wastes according to the Resin Identification Code, as shown in Table 1.2. This is followed by cleaning of the collected items. However, most of the time the cleaning of collected plastic items is a major challenge, particularly when they are sourced from domestic trash with contamination from leftover food, liquids and soils. The most dangerous are unidentified chemical compounds. At this point, both the plastic collectors and processors should possess the wisdom to reject highly contaminated plastic trash from unidentified sources, so that the contamination will not affect the cleanliness of the recycled products. Meanwhile, the washing/cleaning of the

Table 1.12 Comparison of physical and chemical recycling of plastics.

Recycling method	Advantages	Disadvantages
Physical	– Well-established technology with acceptable cost of capital (starting cost can be as low as USD 100,000 with a single screw extruder) – The process can be easily adopted in any existing plastic manufacturing factory to recycle from self-produced reject products – Able to handle small quantities of wastes, i.e. <200 kg/day – Moderate knowledge is needed through practitioners experienced on plastic materials – Acceptable quality of recycled plastics depends on the quality of segregated plastics	– Limited to recycling well-known types of plastics only. Unable to recycle plastics with different composition of additives – Preferable to recycle virgin plastics – Plastic segregation is needed according to types of plastics. Poor quality of segregation can cause low quality of produced recycled plastics – Operators can lack awareness of appropriate operation of factories, which can lead to soil, sound, air and water pollution – The prices of recycled plastics are low and the recycled plastics usually can only be used to blend with similar type of plastics
Chemical	– Most of the plastics are able to undergo a pyrolysis process to break down at high temperature into monomer or pyrolysis oil – Acceptable quality of pyrolysis oil depends on the quality of segregated plastics – Pyrolysis products, i.e., low molecular weight chemicals, can be used for wider industries to produce new/virgin materials – Better operation unlikely to cause substantial pollution	– Requires high technical knowledge to manage the pyrolysis process – Very complicated and costly set-up process with estimated capital cost >USD 1 million – Requires consistency of raw material quality – Process/technology is set up for limited types of input – Requires more attention on the emission-handling process to avoid air pollution

collected plastic items requires huge quantities of water and the wastewater also requires a proper treatment process before being released to rivers. Nevertheless, the cleaning process sometimes may be avoided when plastics are obtained from well-known sources. For instance, a beverage bottle manufacturer produced defective HDPE bottles. Such defective bottles were later transferred to other manufacturers producing HDPE bottles for engine oils. Since the defective bottles were rejected during the bottle quality sorting period and did not contain any beverage beforehand, such bottles could be directly used after undergoing a crushing process, to be reproduced into the engine oil bottle shapes. Another example is leftover skeleton plastic sheets of

Fig. 1.13 (A) Crusher machine for rigid recycled plastics; (B) Shredder machine for recycled plastic films.

thermoforming that can actually be crushed and reproduced by a sheet extruder into plastic sheets again for thermoforming.

After the collected plastics have been cleaned, they undergo a crushing or shredding process (refer to Fig. 1.13) to transform them into pellet size particles for the compounding and moulding process. Commonly, the coloured recycled plastics are more suitable to produce dark coloured products or black coloured products, unless the recycled plastic materials are colourless initially. Nonetheless, as compared to the virgin plastics, most of the recycled or reprocessed plastics may have a visible yellowish colour, indicating degradation has occurred, which can affect the colour setting of the end products when mixed with virgin plastics during the moulding process.

The recycled plastics are usually used by blending together with virgin plastics, since the quality of the plastics can fluctuate depending on the source of collection. This can greatly affect the processability of the recycled plastics if virgin plastics are not added to stabilize the impacts. For instance, the recycled plastics usually have a high melt flow index (MFI) resulting from the degradation of long chain macromolecule chains. The recycled plastics are added into virgin plastics so that the melt viscosity can be maintained at an acceptable value without further efforts to reset the processing parameters.

Most of the time, the recycled plastics can be added into the stream of existing plastic processing facilities, such as blow moulding, injection moulding and extrusion processes. In addition, there are also plastic processors producing recycling plastic compound to add to virgin plastic for reduction of the cost of processing to cater to the demands of manufacturers who do not have a compounding facility in their factories. Such recycled plastics usually have better quality, with the custom-made ability to meet the requirements of the manufacturers. Physical recycling is the most common and most straightforward, which is well adopted in the industries involved in the valorization of plastic wastes.

1.6.2 General operation principles of chemical recycling

Chemical recycling of plastics is also known as pyrolysis, which concerns heating the waste plastics at high temperature in an inert condition so that the waste plastics are transformed into pyrolysis oil, for energy recovery by combustion or separation into valuable chemicals. The typical pyrolysis of plastics can be affected by (1) reactor type, (2) type of fluidizing gas and its rate, (3) residence time and feedstock, (4) temperature, (5) pressure and (6) catalyst. As stated by Dwivedi, Mishra, Mondal, and Srivastava (2019), the typical pyrolysis output corresponding to plastic types is summarized in Table 1.13 and the properties of pyrolysis oil versus conventional fuel are summarized in Table 1.14.

On the other hand, studies have also been done to identify the composition of pyrolysis products from polyethylene, polypropylene and polystyrene using a fluidized bed reactor with 30 kg/h, as summarized in Table 1.15. In order to recover the valuable chemicals from the pyrolysis oil, a developed separation process is needed. A typical pyrolysis pilot plant is illustrated in Fig. 1.14. Although pyrolysis is a mature technology, there are still many obstacles that need to be overcome for chemical recycling of plastic using the pyrolysis process in order to obtain desirable product yield and quality results from the collected solid plastic waste. In particular, the variety of additives added into plastic such as mineral filler, e.g. calcium carbonate, flame retardant, plasticizer, weathering stabilizer, colour lubricants, etc., requires different pyrolysis parameters. Moreover, the presence of additives can also poison the catalysts of the catalytic pyrolysis process. As a result,

Table 1.13 Plastic pyrolysis product yield corresponding to process parameter.

Polymer	Thermal decomposition mode	Reactor type	Pyrolysis process parameters				Product yield		
			Temp. (°C)	*Heating rate (°C/min)*	*Pressure*	*Duration (min)*	*Gas (wt.%)*	*Liquid (wt.%)*	*Solid (wt.%)*
PET		Fixed bed	500	10	–	–	76.9	23.1	0
		–	500	6	1 atm	–	52.13	38.89	8.98
HDPE	Random chain scissioning resulting in monomers and oligomers	Horizontal steel	350	20	–	30	17.24	80.88	1.88
		Batch	450	–	–	60	5.8	74.5	19.7
		Semibatch	400	7	1 atm	–	16	82	2
			450	25	1 atm	–	4.1	91.2	4.7
		Fluidized bed	500	–	–	60	10	85	5
			650	–	–	20–25	31.5	68.5	0
LDPE		Batch	430	3	–	–	8.2	75.6	7.5
			500	6	1 atm	–	19.43	80.41	0.16
			550	5	–	–	14.6	93.1	0
		Pressurized batch	425	10	0.8–4.3 MPa	60	10	89.5	0.5
		Fixed bed	500	10	–	20	5	95	0
		Fluidized bed	600	–	1 atm	–	24.2	51	0
PVC	Chain-stripping (side chain reactions of substituents eliminating reactive substitutes (HCl), dehydrogenation and cyclization)	Vacuum batch	520	10	2 kPa	–	0.34	12.79	28.13
		Fixed bed	500	10	–	–	87.7	12.3	0
PP	Random chain fragmentation	Horizontal steel	300	20	–	30	28.84	69.82	1.34

Continued

Table 1.13 Plastic pyrolysis product yield corresponding to process parameter—cont'd

Polymer	Thermal decomposition mode	Reactor type	Pyrolysis process parameters				Product yield		
			Temp. (°C)	*Heating rate (°C/min)*	*Pressure*	*Duration (min)*	*Gas (wt.%)*	*Liquid (wt.%)*	*Solid (wt.%)*
		Batch	380	3	1 atm	–	6.6	80.1	13.3
			740	–	–	–	49.6	48.8	1.6
		Semibatch	400	7	1 atm	–	13	85	2
			450	25	1 atm	–	4.1	92.3	3.6
			500	6	1 atm	–	17.76	82.12	0.12
PS	Combination of chain rupture and unzipping, formation of oligomers	Batch	500	–	–	150	3.27	96.73	0
			581	–	–	–	9.9	89.5	0.6
		Semibatch	400	7	1 atm	–	6	90	4
		Pressurized batch	425	10	0.31–1.6 MPa	60	2.50		

Based on Dwivedi et al. (2019). Reproduced with permission of Elsevier.

Table 1.14 General properties of pyrolysis oil derived from plastics and comparison with conventional fuel.

Source	High heating value (MJ/kg)	Density (g/cm^3)	Viscosity (mm^2/s)
PET	28.2	0.90	–
HDPE	45.86	0.79	2.1
LDPE	38–39	0.78	1.89
PVC	43.22	0.84	6.36
PP	40.8	0.86	4.09
PS	43.0	0.85	1.4
Diesel	46.67	0.81–0.87	2.0–5.0
Kerosene	43.0–46.2	0.81–0.87	–
Gasoline	43.4–46.5	0.71–0.77	1.17

Based on Dwivedi et al. (2019). Reproduced with permission of Elsevier.

Table 1.15 Composition of hydrocarbon and derived compounds produced by pyrolysis of neat plastics in the fluidized bed reactor (Kaminsky et al., 2004).

Feedstock	PE	PP	PS
Temperature (°C)	740	760	520
Hydrogen	0.8	0.7	–
Methane	23.8	28.2	0.06
Ethane	6.7	4.0	–
Ethene	20.0	13.9	0.04
Propane	0.08	0.09	–
Propene	5.6	3.7	–
Butene	0.6	0.4	–
Butadiene	1.6	0.4	–
Isoprene	0.2	0.2	–
Cyclopentadiene	1.9	0.8	–
Other aliphatic compounds	1.3	1.6	0.1
Benzene	19.2	18.2	0.07
Toluene	3.9	6.6	1.7
Xylenes, ethyl benzene	0.08	0.4	0.4
Styrene	0.5	1.0	76.8
α-Methyl styrene	0.1	0.2	2.2

Continued

Table 1.15 Composition of hydrocarbon and derived compounds produced by pyrolysis of neat plastics in the fluidized bed reactor (Kaminsky et al., 2004)—cont'd

Feedstock	PE	PP	PS
Indane, indene	0.5	1.0	0.6
Naphthalene	2.8	3.5	0.04
Methyl naphthalene	0.6	0.9	–
Diphenyl	0.3	0.4	–
Fluorene	0.2	0.3	–
Phenanthracene/anthracene	0.5	0.7	–
Pyrene	0.3	0.2	–
Other aromatic compounds	6.9	11.1	18.6
Soot, fillers	1.8	1.7	–
Total H_2, C_1–C_4 (gasses)	59.1	51.4	0.1
Total oil	39.1	46.9	99.8

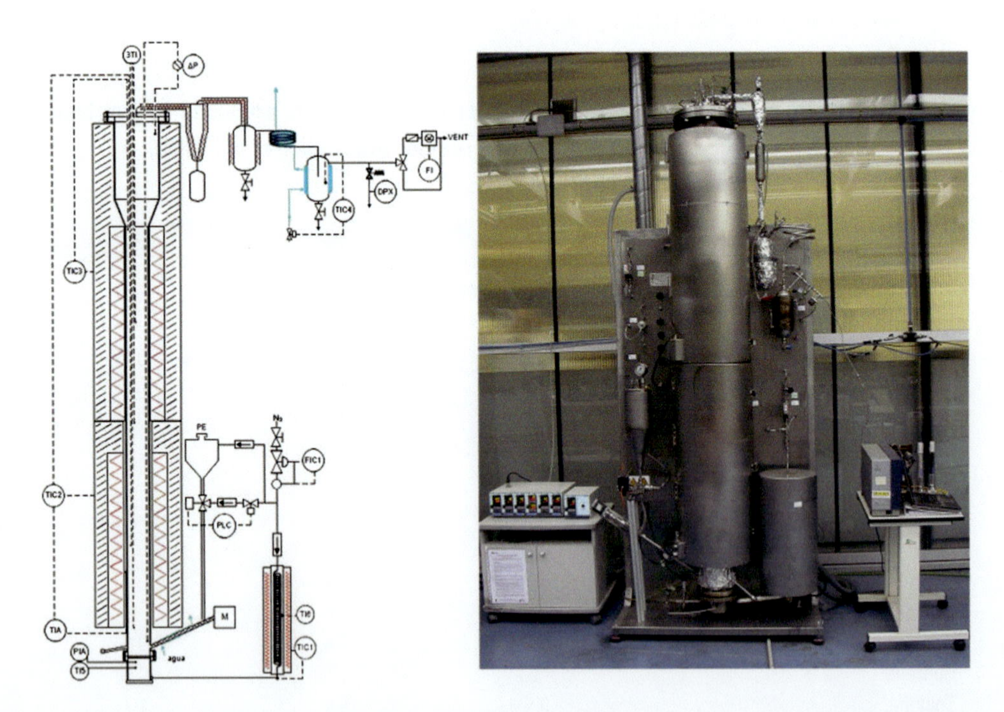

Fig. 1.14 Process instrumentation diagram and actual picture of pyrolysis pilot plant. Reproduced from Maartinez, L., Aguado, A., Moral. A., Irusta, R., (2011). Fluidized bed pyrolysis of HDPE: A study of the operating variables and the main fluidynamic parameters on the composition of production of gases. *Fuel Processing Technology, 92*, 221–228. Reproduced with permission of Elsevier.

most of the feedstock of solid plastic waste to the pyrolysis process must be properly segregated to safeguard the process conditions, while minimizing undesirable products, particularly toxic emissions, which can endanger the operators. Neutralization and treatment of carcinogenic compounds generated from the pyrolysis process can be very costly.

1.7 Conclusion

Plastics have brought innumerable advantages to humans; however, handling of plastic wastes has remained a significant challenge for decades. The commodity plastics such as polyethylene, polypropylene, polyvinyl chloride, polystyrene, polyethylene terephthalate, etc., still have very low recycling rates. Most of the time, the landfilling method remains the first choice for disposal of plastic wastes, although landfills are known to have long-term environmental impacts. Moreover, irresponsible human disposal of plastic wastes can further lead to accumulation of microplastics in the natural environment. Both researchers and industrial players have proposed a variety of approaches to minimize the impact of plastics on the environment; however, the most practical approach is through the efforts of consumers to reduce, reuse, and recycle plastics. Finally, at the end of life of plastics, people should select the most beneficial approach to treat plastic waste, including pyrolysis and incineration for energy recovery, before landfilling. Proper usage and handling of plastics are the responsibility of everyone, to safeguard the living environment for future generations, as well as improve our lives now.

References

Abed, R. M. M., Muthukrishnn, T., Al Khaburi, M., Al-Senafi, F., Munam, A., & Mahmoud, H. (2020). Degradability and biofouling of oxo-biodegradable polyethylene in planktonic and benthic zones of the Arabian Gulf. *Marine Pollution Bulletin, 150,* 110639.

Alfarraj, A., & Nauman, E. B. (2004). Super HIPS: Improved high impact polystyrene with two sources of rubber particles. *Polymer, 45,* 8435–8442.

Bandopadhyay, S., Martin-Closas, L., Pelacho, A. M., & DeBruyn, J. M. (2019). Biodegradable plastic mulch films: Impacts on soil microbial communities and ecosystem functions. *Frontiers in Microbiology, 9,* 819.

Benson, N. U., Bassey, D. E., & Palanisami, T. (2021). COVID pollution: Impact of COVID-19 pandemic on global plastic waste footprint. *Heliyon, 7,* e06343.

Brennecke, D., Buarte, B., Paiva, F., Cacador, I., & Canning-Clode, J. (2016). Microplastics as vector for heavey metal contamination from the marine environment. *Estuarine, Coastal and Shelf Science, 178,* 189–195.

Cao, D., Wang, X., Luo, X., Liu, G., & Zheng, H. (2017). Effects of polystyrene microplastics on the fitness of earthworms in an agricultural soil. *IOP Conference Series Earth and Environmental Science, 61*, 012148.
CNN Report (2019). Plastic waste dumped in Malaysia will be returned to UK, US and other.
Dan, Z., Yang, D., Zheng, Y., Yang, Y., He, Y., Luo, L., & Zhou, Y. (2019). Current progress in the adsorption, transport and biodegradation of antibiotics in soil. *Journal of Environmental Management, 251*, 109598.
Du, C., Liang, H., Li, Z., & Gong, J. (2020). Pollution characteristics of microplatics in soils in southeastern suburbs of Baoding City, China. *International Journal of Environmental Research and Public Health, 17*, 845.
Dwivedi, P., Mishra, P. K., Mondal, M. K., & Srivastava, N. (2019). Non-biodegradable polymeric waste pyrolysis for energy recovery. *Heliyon, 5*, e02198.
Financial Times. (2020a). *Chinese companies rush to exploit global medical equipment shortages.*
Financial Times. (2020b). *EU warns of global bidding war for medical equipment.*
Fried, J. R. (2003). *Polymer science and technology* (2nd ed.). Prentice Hall.
Kaminsky, W., Predel, M., & Sakiki, A. (2004). Feedstock recycling of polymers by pyrolysis in a fluidised bed. *Polymer Degradation and Stability, 85*, 1045–1050.
Krevelen, & Nijenhuis. (2009). Processing properties. In *Properties of polymers: Their correlation with chemical structure; their numerical estimation and prediction from additive group contributions* (4th ed., pp. 799–818). Elsevier (chapter 24).
Kruger, M., Kauertiz, B., & Detzel, A. (2009). *Life cycle assessment of food packaging made of Ingeo biopolymer and (r)PET. Final report.* Heidelberg, Germany: IFEU GmbH.
Lee, T. S., & Bee, S. T. (2019). *Polylactic acid* (2nd ed.). Elsevier.
Li, J., Zhang, K., & Zhang, H. (2018). Adsorption of antibiotics on microplastics. *Environmental Pollution, 237*, 460–467.
Liu, M., Lu, S., Song, Y., Lei, L., Hu, J., Lv, J., Zhou, W., Cao, C., Shi, H., Yang, X., & He, D. (2018). Microplastic and mesoplastic pollution in farmland soils in suburbs of Shanghai, China. *Environmental Pollution, 424*, 855–862.
Lwanga, E. H., Gertsen, H., Gooren, H., Peters, P., Salaki, T., Ploeg, K., Basseling, E., Koelmans, A. A., & Geissen, V. (2016). Microplastics in the terrestrial ecosystem: Implications for lumbricus terrestris (oligochaeta, lumbricidae). *Environmental Science and Technology, 50*, 2685–2691.
Markets and Markets. (2020). *Commodity plastics market by type (PE, PP, PVC, PS, ABS, PET, PMMA), end-use industry (packaging, construction, consumer goods, automotive, electronics, textiles, medical & pharmaceutical), and region—global forecast to 2025.* India: MarketsandMarkets Research Private Ltd.
Matsuo, M. (1969). Fine structures and facture processes in plastic/rubber two-phase polymer systems. II. Observation of crazing behaviours under the electron microscope. *Polymer Engineering and Science, 9*, 206–212.
Nikkei Asia. (2020). *Booming demand for PCs and phones squeezes teach supply chain.*
Ninwiwek, N., Hongsawat, P., Punyapalakul, P., & Prarat, P. (2019). Removal of the antibiotic sulfamethoxazole from environmental water by mesoporous silica-magnetic graphene oxide nanocomposite technology: adsorption characteristics, coadsorption and uptake mechanism. *Colloids and Surfaces A: Physicohemical and Engineering Aspects, 123*, 716.

Panno, S. V., Kelly, W. R., Scott, J., Zheng, W., McNeish, R. E., Holm, N., Hoellein, T. J., & Baranski, E. L. (2019). Microplastic contamination in karst groundwater system. *Groundwater, 57,* 189–196.

Parasha, N., & Hait, S. (2021). Plastics in the time of COVID-19 pandemic: Protector or polluter. *Science of The Total Environment, 759,* 144274.

Picheta, R. (2020). *Coronavirus is causing a flurry of plastic waste. Campaigners fear it may be permanent.* Source: https://edition.cnn.com/2020/05/04/world/coronavirus-plastic-waste-pollution-intl/index.html.

Pietrelli, L., Poeta, G., Bttisti, C., & Sighicelli, M. (2019). Characterization of plastic beach debris finalized to its removal: A proposal for a recycling scheme. *Environmental Science and Pollution Research, 24,* 16536–16542.

Polimeri Europa. (2009). *HIPS high impact polystyrene proprietary process technology.* Trade brochure.

Qi, H., Fu, D., Wang, Z., Gao, M., & Peng, L. (2020). Microplastics occurrence and spatial distribution oin seawater and sediment of Haikou Bay in the northern South China Sea. *Estuarine, Coastal and Shelf Science, 239,* 106757.

Rebelein, A., Int-Veen, I., Kammann, U., & Scharsack, J. P. (2021). Microplastic fibers—Underestimated threat to aquatic organisms. *Science of the Total Environment, 777,* 146045.

Shi, M., Sun, Y., Wang, Z., He, G., Quan, H., & He, H. (2019). Plastic film mulching increased the accumulation and human health risks of phthalate esters in wheat grains. *Environmental Pollution, 250,* 1–7.

Shruti, V. C., Perez-Guevara, F., Roy, P. D., Elizalde-Martinez, I., & Kutralam-Muniasamy, G. (2020). Identification and characterization of single use oxo/biodegradable plastics from Mexico City, Mexico: Is the advertise labelling useful. *Science of the Total Environment, 739,* 140358.

Tang, Y., Liu, Y., Chen, Y., Zhang, W., Zhao, J., He, S., ... Yang, Z. (2021). A review: Research progress on microplastic pollutants in aquatic environments. *Science and the Total Environment, 766,* 142572.

Tiseo, I. (2020). *Plastic waste worldwide-statistics & facts.* https://www.statista.com/topics/5401/global-plastic-waste/.

Tiseo, I. Available at https://www.statista.com/statistics/720231/global-polyethylene-terephthalate-production-capacity-distribution-by-region/. 2021. (Accessed 31 December 2021).

2

Eco-profile of plastics

2.1 Introduction

In the last decades, the demand for polymer materials in various industries and applications has led to severe detrimental effects on the environment. The basic environmental problem is due to excessive usage of resources and pollution in production industries. The production processes, in fact, contribute to negative environmental impacts such as global warming, from the raw material extraction process, processing methods, energy conversion processes and transportation of the products to the desired location, as well as waste disposal during the production process, or as waste after usage.

Global warming is a phenomenon of increasing global temperature, which is attributed to greenhouse gas emissions like carbon dioxide gas. Greenhouse gases possess a greater ability of trapping heat from the sun's radiation compared to normal gases, and thus heat up the atmosphere. The production and processing of raw polymers and their products cause carbon emissions throughout their life cycle. The production of plastic materials from crude oil can significantly contribute to global warming. In order to select suitable products or methods to produce polymer materials with the least environmental impact, an eco-profile study of commodity plastics is conducted to assist in the decision-making for more eco-friendly products or methods. The combination of information, including some additional factors such as economic and efficiency statistics, can be used to assess the suitability of a product or method.

The life cycle assessment (LCA) method is used to determine and assess the environmental impacts attributed to the activities in the life cycle production of a product from available raw materials that are more ecologically beneficial to the environment. The data obtained from the LCA method can help to identify the pathways that provide the environmental effects transmitted between media and life cycle stages. The application of a life cycle assessment can

Plastics and Sustainability. https://doi.org/10.1016/B978-0-12-824489-0.00010-6
Copyright © 2023 Elsevier Inc. All rights reserved.

track and record the impact of change on the environment for the entire production process and enable individuals to clearly describe the environmental trade-offs associated with alternative products or processes. The LCA method is helpful in identifying a production route for a product with less environmental impact than a currently available production method.

The environmental impact analysis evaluated in the LCA includes the potential factors related to the production life cycle of products, such as the consumption of natural resources, total waste generation in the life cycle of a product, energy consumption, release of greenhouse gas emissions, discharges of wastewater, etc. In short, the LCA is a useful methodological tool to determine the environmental sustainability of the production of a product by identifying the reaction or processing method with better environmental performance. In plastics-manufacturing polymers, the LCA enables polymer material manufacturers to compare the environmental impact of different products and production systems, or of the same product with different production systems or different recycling methods. From these comparisons, the manufacturers are able to select the more efficient methods and concepts in designing their production line and producing their products. Moreover, the LCA can highlight those companies involved, to comply with constantly evolving regulatory requirements for solid waste, lingering hazardous chemicals, released emissions and discharged wastewater. Furthermore, life cycle solutions for environmental protection and energy cost reduction are starting to yield economic benefits in the form of higher manufacturing efficiency, enhanced quality of products and fewer environmental consequences.

The LCA method consists of the following phases: goal and scope definition, life cycle inventory analysis, life cycle impact assessment analysis and life cycle interpretation, as summarized in Table 2.1. The stage of goal definition basically describes the objectives, scope and the functional unit of a production system to provide better understanding before the selection process. Inventory analysis is important in providing detailed information about the effects on all the environmental inputs and outputs.

2.2 Commodity polymers

2.2.1 Polyethylene

The use of crude oil-based polymer end-products is rapidly growing due to high market demands. Among the various types of polymers available, polyethylene is one of the common

Table 2.1 Stages of life cycle assessment (LCA).

Phases of LCA method	Description
Goal and scope definition	• Describe the system of production in terms of the boundaries of the system and the functional unit • Define and identify the purpose, objective, scope and boundaries of the study and the functional unit • Identify the alternative products or services in a functional unit that can be compared and analysed by the practitioners for better selection
Life cycle inventory/inventory analysis	• Estimate the consumption of the natural resources in the life cycle of the product • Predict and estimate the amounts of solid waste generated during the life cycle of the product, the release of gas emissions including harmful gases and greenhouse gas emissions during the life cycle of the product • Inventory analysis covers detailed data information on all the environmental inputs such as raw material and total consumption of energy • The environmental output for different stages in the life cycle, such as the release of emissions to water and air and the generation of solid waste
Life cycle impact assessment/ impact assessment	• Assessment process of environmental impacts from inventory and the determination of overall environmental performance of the product • Provide evaluation of the potential environmental impacts (such as climate change, toxicological stress, etc.) of the resource extractions, solid waste generation and emission release that contribute to inventory
Life cycle interpretation	• Present in every stage of an LCA • This stage is to improve or redesign the production processes in term of reduced cost, reduced waste, reduced raw material usages, improvement in the process's safety, etc.

nondegradable polymers used in various applications such as packaging, toys, etc. The excellent properties of polyethylene, especially low-density polyethylene (LDPE), in terms of mechanical properties, ease of processability and others, have increased its commercial and industrial value. However, the high consumption of polyethylene in various applications and industries has strongly contributed to environmental impacts due to overconsumption of fossil resources, excessive release of greenhouse gases and other harmful emissions and the littering of landscapes from the disposal of polyethylene products. A study conducted by Franklin Associates (2020) gives the total energy demand analysis of LDPE resin, including all the renewable energy and nonrenewable energy used in the process, transportation and material feedstock stages, as

Table 2.2 The total energy used in the production process of LDPE resin.

Stages	Nonrenewable energy (GJ)	Renewable energy (GJ)	Total energy demand (GJ)
Based on one functional unit: 1 functional unit = 1000 kg LDPE			
Cradle-to-olefins	69.3	0.076	69.4
Production of pristine LDPE resins	10.5	0.39	10.9
Total	79.8	0.47	80.3
Total percentage (%)			
Cradle-to-olefins	86.3	0.1	86.4
Production of pristine LDPE resins	13.1	0.5	13.6
Total	99.4	0.6	100

summarized in Table 2.2. The energy demands of the process stage are generated from various energy sources, such as fossil fuel, hydropower, nuclear, wind, etc. For the transportation stage, fuel energy includes the energy used to transport the fuels to the process. Finally, feedstock energy mainly refers to the energy sources such as oil, natural gas, etc. used as material feedstock in the production of LDPE resin. The total energy needed to produce a functional unit of 1000 kg of LDPE resin is 80.3 GJ. As seen in Table 2.2, 99.4% of the total energy used in the life cycle of LDPE resin production is nonrenewable energy generated from fossil fuel (petroleum, natural gas, coal, etc.), hydropower, nuclear and other energy sources. In addition, nonrenewable fuel energy is also used in the transportation of fuel to the LDPE production process. The total energy consumption in LDPE production is summarized in Table 2.3. The material feedstock fuel used to produce LDPE is mostly natural gas and petroleum, with the percentage of natural gas/petroleum in the total energy consumption at almost 94%.

The environmental impact categories of life cycle analysis, such as global warming potential, acidification potential, etc., related to LDPE resin are summarized in Table 2.4. The CO_2 emissions are mainly due to the combustion of petroleum, natural gas, etc. in the production of olefins, with a contribution of 64% to global warming potential. The remaining 36% of the release of CO_2 is mainly attributed to the production of virgin LDPE resin. The global warming potential of most production processes is mainly

Table 2.3 The energy consumption of LDPE resin production by type of fuel energy.

Stages	Natural gas (GJ)	Petroleum (GJ)	Coal (GJ)	Other energy (GJ)	Total energy (GJ)
Based on one functional unit: 1 functional unit = 1000 kg LDPE					
Cradle-to-olefins	63.40	5.43	0.29	0.267	69.4
Production of virgin LDPE resin	6.42	0.13	2.37	2.000	10.9
Total	69.80	5.56	2.65	2.270	80.3
Total percentage (%)					
Cradle-to-olefins	79.0	6.8	0.4	0.3	86.4
Production of virgin LDPE resin	8.0	0.2	2.9	2.5	13.6
Total	87.0	6.9	3.3	2.8	100.0

Table 2.4 Various environmental impact categories (global warming potential, acidification potential, etc.) of LDPE resins.

Impact categories	Cradle-to-olefins	Production of virgin LDPE resin
Global warming potential (GWP)	1243 kgCO_2 eq (64%)	684 kg CO_2 eq (36%)
Acidification potential (AP)	3.89 kg SO_2 eq (59.4%)	2.65 kg SO_2 eq (40.6%)
Eutrophication potential (EP)	0.23 kg N eq (76.4%)	0.071 kg N eq (23.6%)
Ozone depletion	1.2×10^{-6} (97.3%)	3.4×10^{-8} (2.7%)

contributed by the release of CO derived from fossil fuel such as petroleum and natural gas. Related to acidification potential, in this study the combustion of petroleum and natural gas and the production of ethylene olefins released 3.89 kg SO_2 eq based on 1000 kg of LDPE resins, with a total of 59.4% of the contribution, as shown in Table 2.4. The assessment of the acidification potential is to evaluate the potential of the release of acidic emissions harmful to the environment and this assessment focuses on the release of SO_2. As seen in Table 2.4, the emissions from the production of ethylene olefins and LDPE resins were found to have an insignificant impact on ozone depletion.

High-density polyethylene (HDPE) and low-density polyethylene (LDPE) are two common types of polyethylene used in various applications due to their excellent properties in terms of water and gas resistance. Ahuja and Sharma (2017) conducted a life cycle assessment analysis on the environmental impact of HDPE and LDPE plastic bags. The total heat energy used to produce a single functional unit (one functional unit is equal to 500 plastic bags) is summarized in Table 2.5. Referring to the table, the heat energy used to produce LDPE plastic bags was observed to be much higher than the total heat energy used by HDPE and HDPE with degradable additives. They also assessed the environmental impact of production of HDPE and LDPE plastic bags, summarized in Table 2.6. According to this table, the production of HDPE and LDPE plastic bags was observed to have a similar effect on the environmental impact of releasing harmful emissions into the environment. However, HDPE plastic bags are still found to have more negative impacts than LDPE plastic bags in all impact categories. Acidification potential refers to the release of emissions that can pollute rivers and oceans, while ecotoxicity implies emission release during the production of plastic bags that can pollute the environment.

Liptow and Tillman (2012) conducted a life cycle assessment analysis on sugarcane-based polyethylene and crude oil-based polyethylene. The production of plastics products has grown rapidly due to the high demand for polymer materials, and polyethylene is one of the largest consumption polymers in the world.

Table 2.5 Total heat energy used to produce a functional unit of various types of plastic bags (one functional unit equals 500 plastic bags).

Types of plastic bag	Total heat energy
High-density polyethylene (HDPE)	12
High-density polyethylene with degradable additive	13
Low-density polyethylene (LDPE)	80
Polypropylene (PP)	150
Polypropylene with additive	155

Table 2.6 The comparison of environmental impact between HDPE and LDPE plastic bags.

Impact categories	Comparison between the HDPE plastic bags and LDPE plastic bags, 100%	
	HDPE	*LDPE*
Acidification	100%	80%
Ecotoxicity	100%	85%
Carcinogens	70%	50%
Respiratory organics	98%	90%

Liptow and Tillman (2012) assessed and compared the environmental impacts, including global warming potential (GWP) and primary energy consumption, of the production process of sugarcane-based LDPE and crude oil-based LDPE. Referring to the analysis in Fig. 2.1, the total energy consumption of sugarcane-based LDPE is observed to be higher than crude oil-based LDPE. Basically, the sugarcane-based LDPE consumes significantly less fossil-fuel energy than crude oil-based LDPE. However, the energy consumption of renewable energy for sugarcane-based LDPE is significantly higher than crude oil-based LDPE and this results in the higher total energy consumption of sugarcane-based LDPE. On the other hand, they also showed that global warming potential (GWP) of sugarcane-based LDPE is more positive than crude oil-based LDPE. The global warming potential (GWP) is also strongly influenced by the land use change (LUC) emissions. However, their assessment of GWP excluded the LUC emissions and was unable to provide clearer information on GWP. For the impact categories of acidification potential (ACP), photo ozone creation, and eutrophication impacts, the activity that had the greatest environmental impacts in the three impact categories was the long-distance sea transport for both sugarcane-based LDPE and crude oil-based LDPE. This is mainly attributed to the release of high amounts of nitrogen oxide (NO_x), carbon dioxide (SO_2) and carbon monoxide (CO) from this activity. The production of ethanol in the sugarcane-based LDPE contributed a significant release of NO_x due to the combustion of bagasse during the production of ethanol to power the process. On the other hand, the high electricity consumption of the polymerization process for sugarcane-based LDPE greatly contributed to the release of SO_2 and CO emissions.

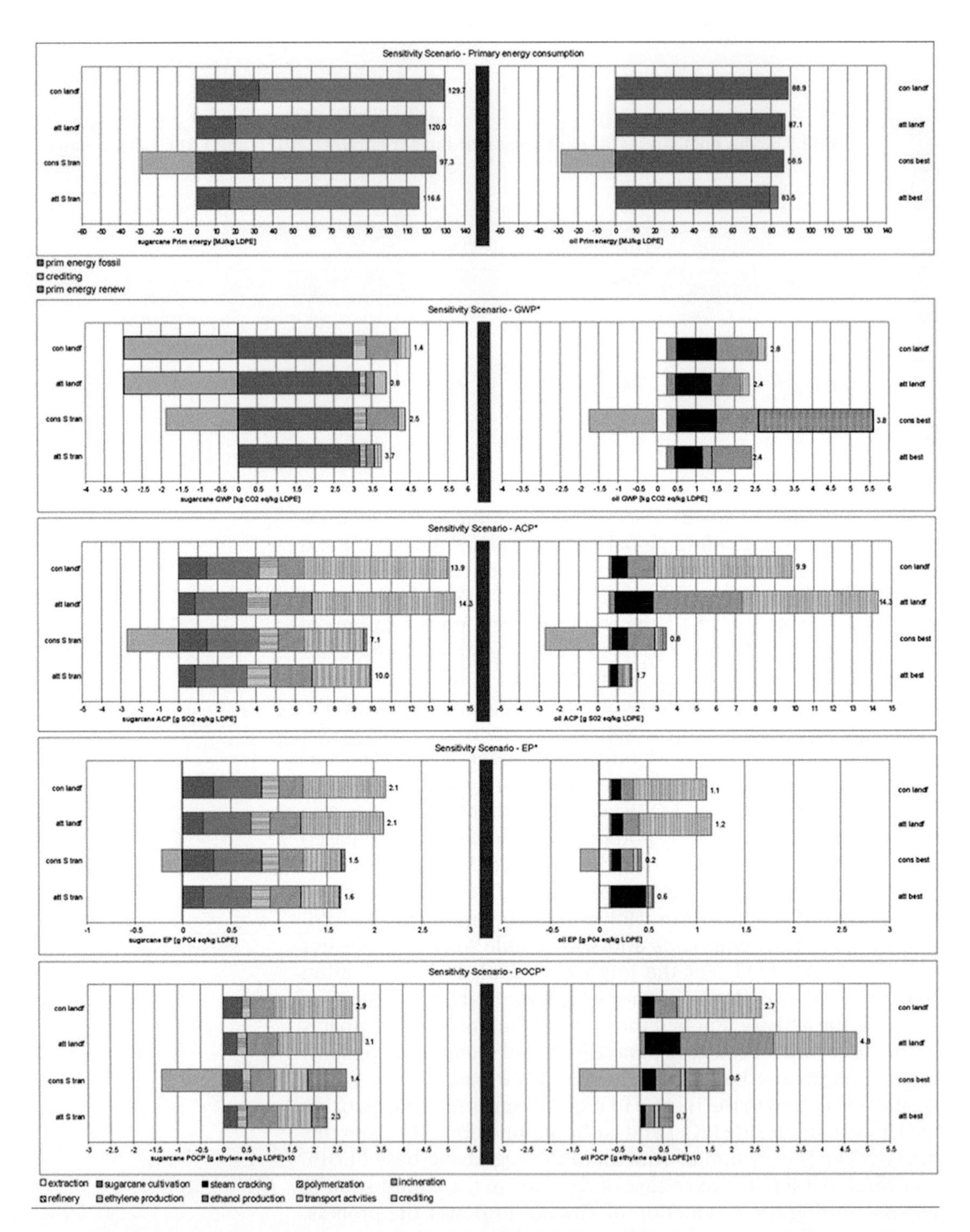

Fig. 2.1 The primary energy consumption, global warming potential (GWP), acidification potential (ACP), eutrophication potential (EP) and photochemical ozone creation potential (POCP) of the sugarcane-based and crude oil-based LDPE. From *Liptow, C., Tillman, A. M. (2012). A comparative life cycle assessment study of polyethylene based on sugarcane and crude oil. Journal of Industrial Ecology, 16(3): 420–435, with permission of Wiley.*

2.2.2 Polypropylene

Polypropylene (PP) is a common thermoplastic polymer with excellent properties, including high gas and water permeability resistance, mechanical properties, flame resistance, high heat distortion temperature and others. Polypropylene is widely used in polymer materials in the plastic manufacturing industry to produce various end products, especially plastic packaging. The use of polypropylene in the packaging industry is found to be 16% of the worldwide plastic materials (Alsabri et al., 2021). The incorporation of natural and artificial fibres in polypropylene has increasingly gained attention because of the excellent thermal and mechanical properties. In general, polypropylene is produced from crude oil or petroleum by derivation from propylene, which is an olefin monomer, as shown in Fig. 2.2. The life cycle assessment of polypropylene end-products has been applied to evaluate and investigate the total energy consumption, the use of various natural resources, the release of harmful pollutants into water and the generation of industrial solid waste that would occupy landfills during disposal. The life cycle analysis of polypropylene final products has been conducted starting from the extraction of raw materials (crude oils and natural gases) and the production phase of polypropylene resin through the end-of-life stage, as shown in Fig. 2.2.

Various life cycle assessments of polypropylene end-products have been conducted to investigate and analyse the environmental impact of polypropylene-based end-products. The life cycle assessment analysis of polypropylene from petroleum and natural gas is summarized in Fig. 2.3. Polypropylene is commonly used as a polymer base in various industries or applications, due to its excellent properties in terms of mechanical, physical, and thermal stability (able to withstand heating, cooling and reheating processes without significant degradation effects). There are a few grades of polypropylene and each grade is basically used for different applications due to the differences in chemical composition. Ingarao et al. (2016) has conducted an analysis of the life cycle energy and CO_2 emissions of three different materials, which are tin steel, polypropylene, and glass during the usage of a single functional unit phase. They observed that the production of polypropylene posed a lesser environmental impact than the production of tin steel and glass beverage packaging in terms of energy consumption and the release of CO_2 emissions. This is mainly attributable to the higher weight characteristic of glass and tin steel in comparison to polypropylene packaging. This has further provided less energy consumption and CO_2 emissions for polypropylene

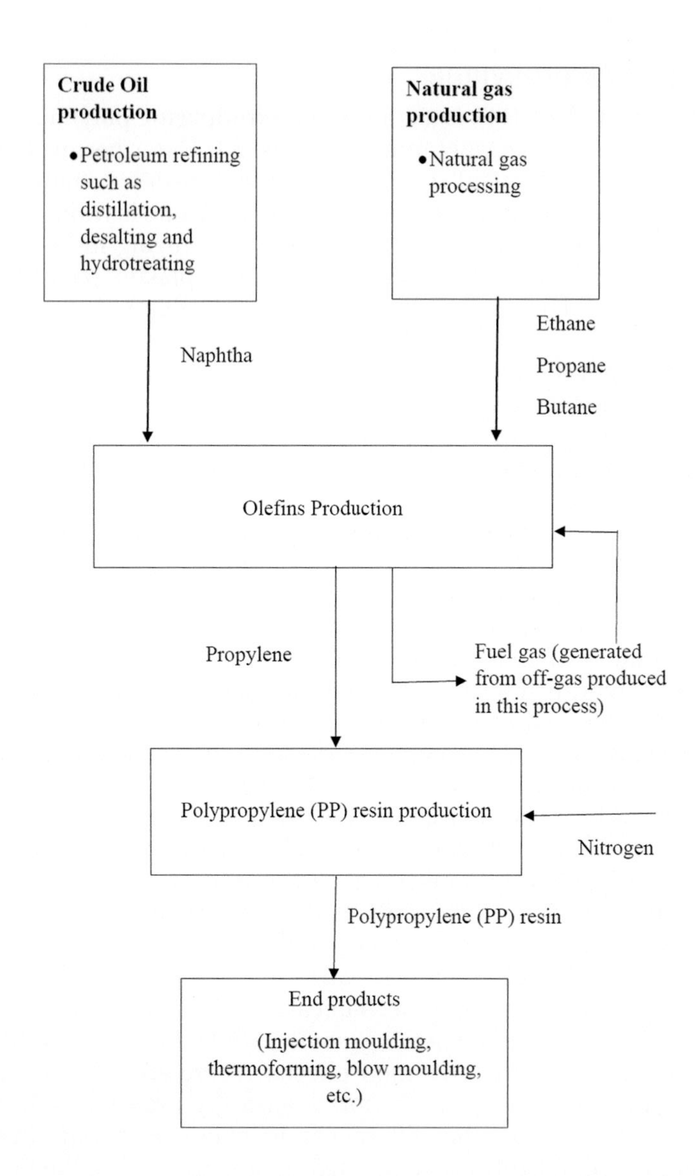

Fig. 2.2 Life cycle assessment of the production of polypropylene from petroleum.

than for other packaging materials in all three cycle stages, which are materials, manufacturing and transportation, as summarized in Fig. 2.3.

In research conducted by Mannheim and Simenfalvi (2020), they focused on the life cycle analysis in reducing the environmental impacts of the manufacturing phase. Mannheim and Simenfalvi (2020) assessed the environmental burden of polypropylene products throughout a life cycle analysis study on three different production scenarios: production without looping

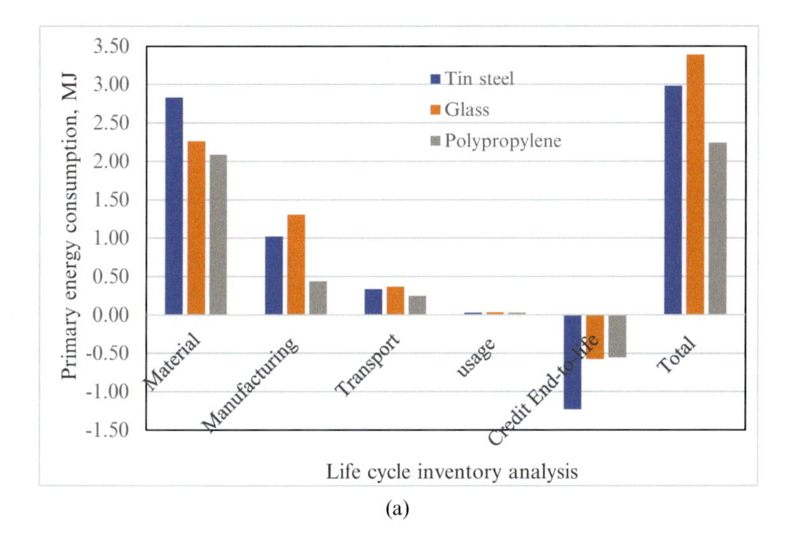

(a)

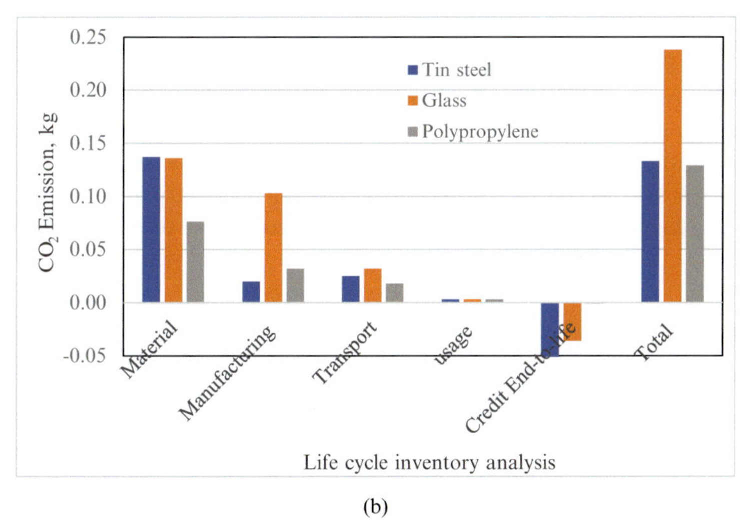

(b)

Fig. 2.3 The results of life cycle inventory analysis. Data from Ingarao, G., Licata, S., Sciortino, M., Planeta, D., Lorenzo, R.P., Fratini, L. (2016) Life cycle energy and Co_2 emissions analysis of food packaging: An insight into the methodology from an Italian perspective. *International Journal of Sustainable Engineering, 10*(1), 31–43.

method, production with water looping only and production with recirculated plastic scrap looping and process water looping. The looping method can use a dummy process in which the output flow can be connected into the same process as the input. In the scenario of water looping only, 80% of the water flow is recirculated in a closed loop and the remaining 20% is discharged and treated as municipal wastewater. In this scenario, the injection moulding with lower environmental load is minimized and the life cycle analysis is completed with municipal wastewater treatment. In the third scenario, the recovered materials flow includes the process water and product loss with a looping method. The product loss in the third scenario is recirculated and recycled as polymer scraps in the injection moulding process. The recycling of PP scraps was observed to reduce the environmental impact, as shown in Fig. 2.4.

Alsabri et al. (2021) conducted a life cycle assessment analysis on the environment impact of the polypropylene manufacturing process. Referring to the results in Table 2.7, the CO_2 emission equivalent for a functional unit of 1 ton PP is estimated to be 1586 kg CO_2 eq. This observation indicates that the weight of CO_2 emission was significantly higher than that of the product. The poor environmental impact of the PP manufacturing process also reflected the necessity of exploring the application of new technologies or performing some modifications to reduce the release of CO_2, to help in reducing the GWP impact. The heat and energy used in the PP manufacturing process can be refined by replacing with heat and energy from renewable sources. The application of cleaner energy could significantly reduce the overall GWP impact and the production cost of PP. This is because the cost of renewable energy is now lower than that of fossil-based energy (Alsabri et al., 2021; Qadir et al., 2021). Terrestrial acidification indicates the resistance in soil fertility due to the accumulation of nitrogen and sulphur-based acidic discharged emissions, such as NO_x, SO_2 and NH_3. In this study, the total impact of

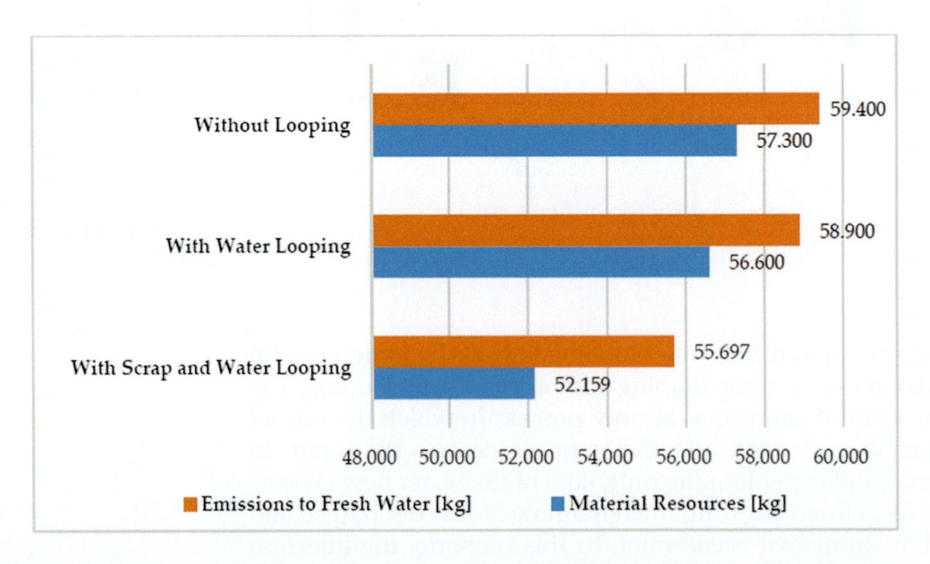

Fig. 2.4 Material resources and emissions to fresh water (in unit kg) of the production stage with transport under functional unit of 25 kg polypropylene product. From Mannheim, V., Simenfalvi, Z. (2020). The life cycle of polypropylene products: Reducing environmental impacts in the manufacturing phase. *Polymers, 12*, 1–18. Open access source.

Table 2.7 The emissions released (for various impact categories) to atmosphere during the manufacturing process of PP.

Impact categories (unit)	Emission per 1 ton of PP pellets
Terrestrial Acidification (TA), kg SO_4 eq.	4.99
Global Warming Potential (GWP), kg CO_2 eq	1586.35
Petrochemical Oxidant Formation (POF), kg NMVOC	4.24
Fossil Resource Depletion (FD), kg oil eq	1722.90
Human Toxicity (HT), kg 1.4-DB eq	77.00

terrestrial acidification was focused on the release of SO_4. The emissions from terrestrial acidification and petrochemical oxidant formation are reasonable. However, the release from fossil resource depletion is 1722.90 kg oil eq, which is quite high and has a negative environmental impact. This is mainly due to the high usage of fossil-based resources in the PP manufacturing process, and it is recommended to replace the fossil-based resources with other alternatives, such as biofuel.

2.2.3 Polystyrene

Polystyrene is a common packaging material used in plastic cups or containers for dairy products (such as yogurt, milk, ice cream), frozen food, dried food, etc. Polystyrene is also widely used in an expanded form to produce packaging materials such as trays and eggcups for fish, meat and vegetables (Zabaniotou & Kassidi, 2003). The manufacturing of polystyrene resin involves four main production processes: the production of ethylene, the production of benzene, the production of styrene by reacting the benzene and ethylene together and finally the polymerization of styrene to form polystyrene (Suwanmanee et al., 2012). The total energy consumption of the manufacturing process of polystyrene resin and the production of a polystyrene box has been summarized in Table 2.8. Referring to this table, the energy required for the extraction process using natural gas is 11.34 MJ/kg of NG, while the required energy for the extraction process using crude oil is 24.30 MJ/kg of crude oil. The total energy consumption for the extraction process is 46.38 MJ/kg PS resins.

Table 2.8 The energy consumption at different stages of the manufacturing of polystyrene resin to the delivery of the products.

Stages	Description	Energy consumption (MJ)
1.	Extraction of fossil fuel (Sheehan, Cmobreco, Duffield, Graboski, & Shapouri, 1998):	11.34 MJ/kg NG
	• Natural gas (NG)	24.30 MJ/kg Crude oil
	• Crude oil	46.38 MJ/kg PS resins
	• Total energy fuel used in extraction (0.33 kg NG and 0.63 kg crude oil)	
2.	The production of polystyrene resins (Razza, Fieschi, Innocenti, & Bastioli, 2009)	39.75 MJ/kg PS resins
	• Derived from 0.37 kg NG, 0.34 kg Crude oil and 0.37 kg Coal	
3.	Thermoforming of polystyrene box	
	• Electricity	4.75 MJ/kg PS resin
4.	Transportation (energy fuel consumption)	
	• Delivery of polystyrene box to customer	284.19 MJ/FU

The total energy consumption of the production of polystyrene resin is approximately 39.75 MJ/kg of polystyrene resins. Natural gas, coal and crude oil are the types of fuel used in the production of polystyrene resin. The total energy consumption of the polystyrene resin manufacturing process from the extraction process is 86.13 MJ/kg polystyrene resins (as calculated in Table 2.8). The total energy consumption analysed for the manufacturing of polystyrene resin by Vink et al. (2010) was 87 MJ/kg polystyrene resins (as tabulated in Table 2.9), which is close to 86.13 MJ/kg polystyrene resins, as discussed earlier. In addition, the thermoforming of polystyrene resin into polystyrene boxes requires 4.75 MJ/kg PS; the energy consumption in the thermoforming stage is mainly attributed to the melting process, which requires the use of electricity for heating purposes. The transportation in this cradle-to-grave cycle requires a considerably high energy consumption derived from natural gas, petroleum and coal.

Table 2.10 lists the stages in the manufacturing process of a polystyrene container that contribute to the environmental impacts of global warming potential, acidification potential, eutrophication potential, respiratory organics and aquatic

Table 2.9 The total energy consumption and global warming potential of the manufacturing process of polystyrene resin (cradle to polymer).

Description	Value
Total energy consumption	87 MJ/kg polystyrene resins
Global warming potential	3.4 kg CO_2 eq/kg polystyrene resins

Data from Vink, E. T., Davies, S., Kolstad, J. J. (2010). The eco-profile for current Ingeo polylactide production. Industrial Biotechnology, *167–180.*

Table 2.10 The impact categories of the manufacturing of polystyrene resin for the production of containers.

Stages	GWP, kg CO_2 eq	AP, kg SO_2 eq	EP, kg PO_4 eq	Respiratory organics, kg ethylene	Aquatic ecotoxicity, kg TEG
Production of PS resin	70	0.47	1.97×10^{-4}	5.60×10^{-2}	9240
Extrusion	12	0.05	2.89×10^{-4}	2.60×10^{-3}	700
Thermoforming	18	0.09	9.18×10^{-4}	6.75×10^{-3}	1150
Electricity	3	0.02	3.79×10^{-5}	1.61×10^{-4}	96
Transportation	31.7	0.22	4.07×10^{-3}	5.94×10^{-2}	12,600

Remarks: The environmental impact of the production of 1000 units of containers.
GWP, global warming potential; *AP*, acidification potential; *EP*, eutrophication potential; *TEG*, triethylene glycol.

ecotoxicity through a cradle-to-grave life cycle analysis. According to Table 2.10, the production stage of polystyrene resin (from raw materials to polystyrene resin) contributed to the highest global warming potential with the release of the highest amount of CO_2 (70 kg CO_2 eq per 1000 units container or 24.96 kg PS). This is mainly attributed to the energy consumption in the extraction and polymerization processes (as shown in Table 2.8), largely derived from fossil fuels such as natural gas, coal and crude oil. The consumption of natural gas, coal and crude oil in the production of polystyrene resin is the main contributor to release of carbon dioxide (CO_2) to the environment, thus inducing global warming potential. Referring to Table 2.10, transportation contributed a considerably high emission of CO_2 in the manufacturing process of polystyrene containers. This is mainly attributed to the transportation need for the consumption of fossil fuel such as raw crude oil, natural gas, coal, etc. (Suwanmanee et al., 2012), which has greatly contributed to the release of carbon dioxide (CO_2), carbon monoxide (CO), nitrogen oxides (NO_x), SO_2, air acidification, etc. (Marten & Hicks, 2018). The high CO_2 emissions in transportation have contributed to the second highest global warming potential value.

Polystyrene resin production has contributed to the highest acidification potential, with high SO_2 emissions. This is due to the production of ethylene, benzene and ethylene styrene and the polymerization of ethylene styrene has high consumption of fossil fuel (as shown in Table 2.9) and SO_2 emissions, which contributed to the highest acidification potential (Suwanmanee et al., 2012). The transportation in the manufacturing process of polystyrene containers provided the second highest value in the acidification impact, from high SO_2 emissions. As mentioned earlier, the consumption of fossil fuels such as natural gas, coal, etc. in the transportation stage leads to hazardous SO_2 emissions and thus induces the acidification potential. The extrusion and thermoforming stages caused the least contribution to acidification potential, due to these stages' use of electricity, which has low impact on acidification potential. Referring to Table 2.10, transportation in the manufacturing process of polystyrene containers had the highest impact on eutrophication potential with 4.07×10^{-3} kg PO_4 eq. The energy derived from the combustion of natural gas, coal, etc. in the transportation stage had the highest impact on eutrophication potential due to PO_4 emissions. The production process of polystyrene resin was observed to have the least impact on eutrophication potential. The transportation stage in the manufacturing process of polystyrene resins also had the highest impact value in the categories of respiratory organics and aquatic ecotoxicity. This was followed by the production process of polystyrene resin as the second highest contributor to respiratory organics and aquatic ecotoxicity.

2.2.4 Polyvinyl chloride

Polyvinyl chloride (PVC) is one of the most widely applied and used plastics in the world. PVC is a polymer with an amorphous structure, produced through the polymerization of vinyl chloride monomers. The ease of processability, low price and recyclable behaviour of PVC have increased its demand and consumption in many different applications. The high consumption of PVC materials can have a significant impact on the environment due to the release of harmful chemical substances. Generally, the consumption of energy in the manufacturing of PVC resins can be divided into two main stages: the extraction and refining of raw materials to produce the ethylene dichloride and vinyl chloride monomer and the production of virgin PVC resin. In the first stage, energy consumption includes the production of chloride through salt extraction and electrolysis, the production of ethylene (the processes of extraction, cracking, distillation and refining of natural gas and petroleum) and the chlorination and cracking process to produce vinyl chloride (Howell, 1991). The energy consumption in the second stage is mainly due to the polymerization of virgin PVC resins. The life cycle inventory analysis of Franklin Associates (2021) includes the energy demand, nonrenewable energy demand, renewable energy demand, total energy demand and total water consumption.

Referring to Table 2.11, the total energy demand for the manufacture of PVC resins is 57.8 GJ/1000 kg of PVC resins; 91.1% (52.6 GJ) of total energy is used in the stage of cradle-to-incoming materials, while the total energy used in the stage of virgin PVC resin production is only 8.9%. The production of ethylene dichloride and vinyl chloride monomers from raw materials of natural gas consumes a huge amount of energy in comparison to the polymerization of PVC resins. In addition, the total nonrenewable energy used in the manufacture of PVC resins is more than 99%. The nonrenewable energy includes the use of fossil fuels

Table 2.11 The total energy demand for the manufacturing of PVC resins from the raw materials for one functional unit of 1000 kg PVC resins.

Stages	Nonrenewable energy (GJ)	Renewable energy (GJ)	Total energy demand (GJ)
Cradle-to-incoming materials	52.20	0.40	52.60 (91.1%)
Production of virgin PVC resins	4.99	0.12	5.11 (8.9%)
Total	57.20 (99.1%)	0.52 (0.9%)	57.80 (100%)

such as petroleum, coal, etc. to generate the process energy, transportation energy and as feedstock to the production of olefins, etc. As an example, the production of ethylene is basically from natural gas, which is commonly used as a hydrocarbon feedstock for most polymer production, including PVC (Howell, 1991). The total water consumption of the PVC resins manufacturing process is summarized in Table 2.12. According to the assessment conducted by Franklin Associates (2021), most of the water is consumed in the stage of cradle-to-incoming materials, with 71% of total water consumption. Most water consumption in this stage is attributed to individual unit processes; however, the water consumption from fuel production in this stage is less significant. The water consumption of polymerization of virgin PVC resin is 4.16 m^3/1000 kg PVC resin; these results are observed to be slightly higher than the assessment results of Comanita et al. (2015), approximately 3.1 m^3/1000 kg PVC resins.

In general, most of the global warming potential for any system is contributed by the release of carbon dioxide from the consumption of the fossil fuels. Referring to Table 2.7, 85% of the carbon dioxide release (which contributes to high global warming potential) is attributed to the stage of cradle-to-incoming materials. This observation is also found to be consistent with the assessment of Comanita et al. (2015). This is mainly because this stage includes the extraction process of raw materials, the synthesis of vinyl chloride monomers and the production of ethylene. The extraction of raw materials such as crude oil contributed to high carbon dioxide emissions and thus induced the GWP. Furthermore, the production of vinyl chloride monomer and ethylene requires high consumption of fossil fuel to generate energy, and this further contributed to high carbon dioxide emissions

Table 2.12 Total water consumption of the manufacturing of PVC resins from raw materials per one functional unit of 1000 kg PVC resins.

Stages	Total water consumption, m^3 per 1000 kg PVC resins	Percentage
Cradle-to-incoming materials	10.131	71%
Production of virgin PVC resins	4.16	29%
Total	14.29	100%

(Comanita et al., 2015). The polymerization of virgin PVC resins contributed 15% of global warming potential, as shown in Table 2.9, mainly attributed to the combustion of natural gas and coal in the boilers to generate electricity for the production of virgin PVC resins. The combustion of natural gas and coal has high carbon dioxide emissions and induced the high global warming potential.

Acidification impact in the manufacturing of PVC resin is mainly assessed based on the combustion of fossil fuel to generate electricity in the production of ethylene and vinyl chloride monomer and the process energy and transportation energy used in the production process. From Table 2.13, 84% of the acidification potential is contributed in the stage of cradle-to-incoming materials. This is mainly attributed to the production of ethylene and vinyl chloride monomer requiring high energy consumption, with the required energy generated from the combustion of fossil fuels and thus contributing to acidification potential. In addition, the emissions contributed by the production of virgin PVC resins are observed to be 16%. These emissions are basically released due to the combustion of fossil fuels in the boilers in the plant and contribute to the acidification potential.

The occurrence of eutrophication is mainly attributed to the release of excess nutrients to the environment, causing a rapid growth phenomenon of the aquatic plants. Referring to Table 2.13, 86% of eutrophication potential is contributed in the

Table 2.13 The impact categories (global warming potential, acidification potential, eutrophication potential and ozone depletion potential) of the manufacturing of PVC resins.

Stages	Global warming potential (GWP), kg CO_2 eq	Acidification potential (AP), kg SO_2 eq	Eutrophication potential (EP), kg N eq	Ozone depletion potential, kg CFC-11 eq
Cradle-to-increasing materials	1781 (85%)	5.14 (84%)	0.22 (86%)	0.00220 (89.5%)
Production of virgin PVC resins	341 (15%)	1.00 (16%)	0.04 (14%)	0.00026 (10.5%)
Total	2095 (100%)	6.14 (100%)	0.26 (100%)	0.00246 (100%)

Data from Franklin Associates. (2021). Cradle-to-gate life cycle analysis polyvinyl chloride (PVC) resin. American Chemistry Council (ACC) Plastics Division.

stage of cradle-to-incoming materials. The eutrophication in this stage is mainly due to the combustion of natural gas and coal to generate the electricity. According to Franklin Associates (2021), the emissions from the combustion of natural gas and coal are observed to be more than 50% of the eutrophication potential. The production of virgin PVC resins also released 14% of emissions that contributed to eutrophication potential. The eutrophication impact is mainly due to the combustion of natural gas and coal to generate electricity. On the other hand, the ozone depletion potential implies a weakening effect of protective ozone caused by the release of ozone depletion emissions such as chlorofluorocarbons (CFCs) and halons. The potential of the ozone depletion substances is categorized in accordance with the chemical reactivity and lifetime. According to Franklin Associates (2021), the stage of cradle-to-incoming materials contributed 89.5% of ozone depletion potential for the manufacturing of PVC resins. The release of carbon tetrachloride in the production of ethylene and vinyl chloride monomer contributed to a major percentage of the total ozone depletion potential.

2.2.5 Polyethylene terephthalate

Polyethylene terephthalate (PET) is a polymer produced through the polymerization of ethylene glycol and terephthalic acid. PET is mainly used in applications of packaging and textile production. Initially, the excellent properties of durability, mechanical strength and transparency of PET caused it to be used in disposable carbonated beverage bottles. However, the short useful life, large production volume and nonbiodegradability of PET have received attention from various researchers, who have conducted life cycle assessment analysis to investigate and assess the environmental impact of PET packaging. The first LCA was conducted in 1969 to investigate the environmental impact of the usage of PET as packaging bottles for carbonated drinks. Over the years, various LCA analyses and research studies on PET have been conducted to assess and find a better packaging material to replace it in carbonated drink packaging bottles. Alfarisi and Primadasa (2018) conducted a life cycle assessment on the environmental impact of the manufacturing process for PET packaging bottles. They found that the manufacturing process of PET contributes to 49 environmental impact categories. Among these categories, global warming potential, human toxicity and marine aquatic ecotoxicity are the three highest impact categories, as summarized in Table 2.14. The global warming potential is related to the release of greenhouse gases such as methane, carbon

Table 2.14 The top three impact assessment categories of the manufacturing process.

Impact categories	Value
Global warming potential	650.1 kg CO_2 eq
Human toxicity	267.0 kg 1,4-DB eq
Marine aquatic ecotoxicity	16.0 kg 1,4-DB eq

Data from Alfarisi, S., Primadasa, R. (2018). Carbon footprint and life cycle assessment of PET bottle manufacturing process. The 1st international conference on computer science and engineering technology, Universitas Muria Kudus, EAL.

dioxide, etc. to the environment. Referring to Fig. 2.5, the polymerization of PET resin was found to release the highest amount of CO_2, which is the main contributor to global warming potential. This is due to the high electricity usage during the production of PET via polymerization at the plant. On the other hand, the polymerization of PET was also found to release the highest emissions harmful to human health, as summarized in Fig. 2.5. This is attributed to the input to the polymerization process containing harmful materials such as chromium and mercury. The same observation was also found on the impact on marine aquatic ecotoxicity, which refers to the emissions released to the environment that are harmful to aquatic life.

Papong et al. (2014) conducted a life cycle assessment on the environmental impact of PET and PLA from water bottles. The assessment results of the environmental impact categories of the PET water bottles are summarized in Table 2.15. Production of 1000 units of PET bottles released overall 68 kg CO_2 eq and PET bottle production released approximately 42 kg CO_2 eq. This is attributed to the combustion of fossil fuels for electricity, which is mainly used in the PET bottle production stage. The researchers also found that the total required fossil energy demand during the production of PET bottles was 2120 MJ. In comparison to the manufacturing of PLA with a total required fossil energy demand of 700–800 MJ, the fossil energy used in the manufacturing of PET bottles is much higher than that in the PLA manufacturing process. On the other hand, the impact of PET bottle manufacturing on human toxicity was mainly contributed by the production of the raw materials of terephthalic acid and ethylene glycol. The total emissions attributed to human toxicity were 36 kg 1,4-DB eq, and 94% of this total harmful emission was attributed to the stage of production of PET from terephthalic acid and ethylene glycol. In a study conducted by Papong et al. (2014), they revealed that the manufacturing of PET

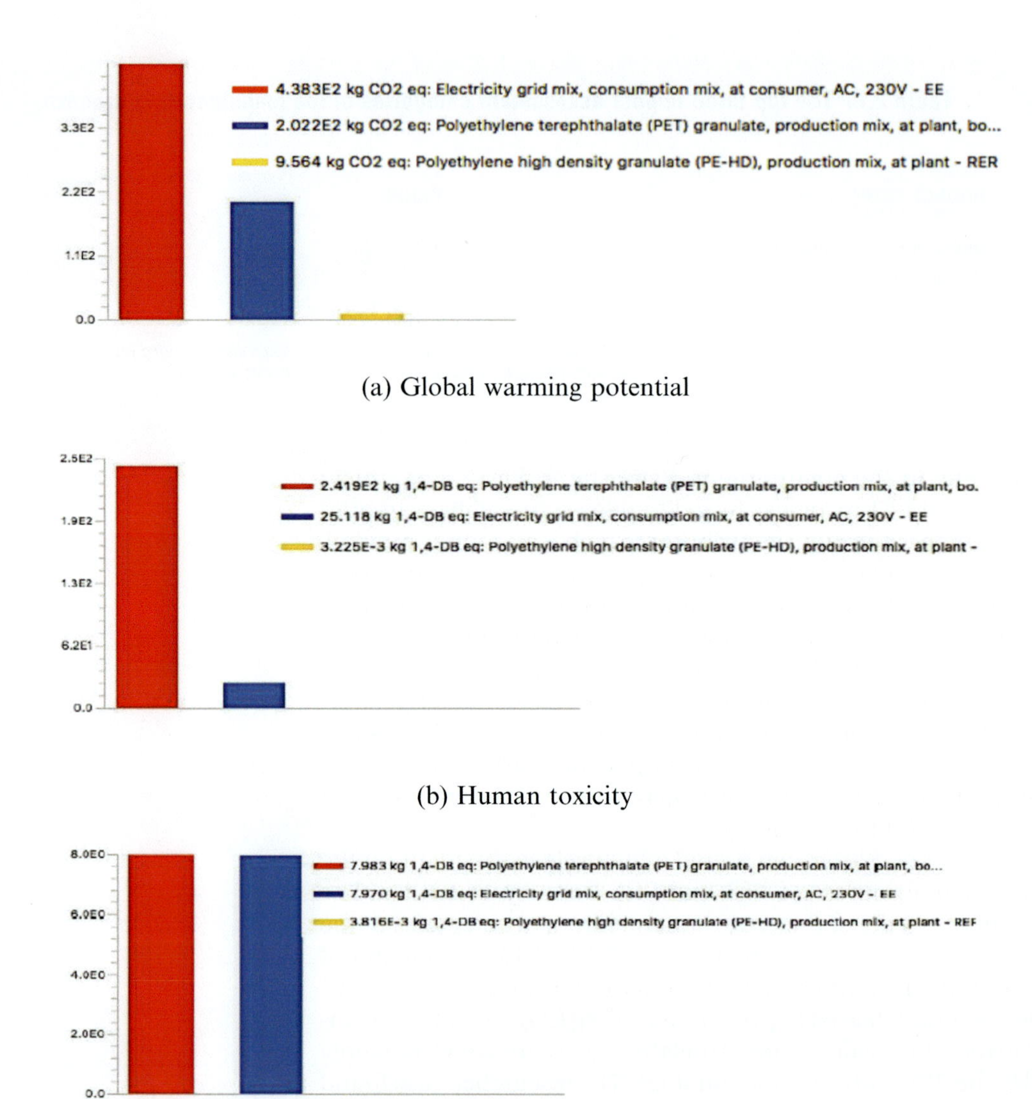

(a) Global warming potential

(b) Human toxicity

(c) Aquatic ecotoxicity

Fig. 2.5 The input materials contribute to the impact categories of global warming potential, human health and marine aquatic ecotoxicity. From Alfarisi, S., Primadasa, R. (2018). Carbon footprint and life cycle assessment of PET bottle manufacturing process. The 1st international conference on computer science and engineering technology, Universitas Muria Kudus, EAL. Open access source.

bottles has a significantly lower impact on eutrophication potential (EP) when compared to the manufacturing of PLA bottles. This is mainly attributed to the production of PET resin, which is a petrochemical catalysis process that poses a low chemical oxygen demand (COD). The manufacturing of PET bottles was also

Table 2.15 The impact categories of the manufacturing of PET bottles.

Impact categories	Overall value per 1000 units of bottles[a]	Value of the production stage of PET bottles
Global warming potential	68 kg CO_2 eq	42 kg CO_2 eq
Human toxicity	36 kg 1,4-DB eq	34 kg 1,4-DB eq
Eutrophication potential (EP)	0.027 kg PO_4 eq	0.014 kg PO_4 eq
Acidification potential	0.26 kg SO_4 eq	0.017 kg SO_4 eq

Stages: Production of raw material, raw material transport, PET production, PET resin transport and production of bottles.
[a]Based on one functional unit of 1000 units of water bottles.

observed to have a lower impact on acidification than the manufacturing of PAL bottles. This observation is consistent with the findings of Gomes et al. (2019). Referring to Fig. 2.6, the impact on eutrophication potential and acidification potential of the manufacturing of PET bottles is lower than the impact caused by the PLA manufacturing process. Various research studies have been conducted to improve the environmental impact of PET bottles. Kang, Auras, and Singh (2017) investigated the effect of increasing the reclamation process of used PET bottles on the life cycle assessment of PET bottles. An increase in the efficiency of the reclamation process could effectively provide an improvement in the environmental impact of PET bottle production.

2.2.6 Polycarbonate and ABS

Polycarbonate is a durable polymer that is commonly used in eyewear lenses and exterior automotive components. The production of polycarbonate is always a concern due to the emissions released to the environment being highly dependent on the different ratios of the consumed input materials. If the input materials are closely monitored, the released emissions would be greatly reduced and better efficiency would be achieved. Thus the life cycle assessment is important in analysing the manufacturing process of polycarbonate products. In general, polycarbonate can be produced through two different routes: through the reaction of phosgene with bisphenol A and with a nonphosgene method, as shown in Fig. 2.7 (Altuwair, 2017; Fukuoka et al.,

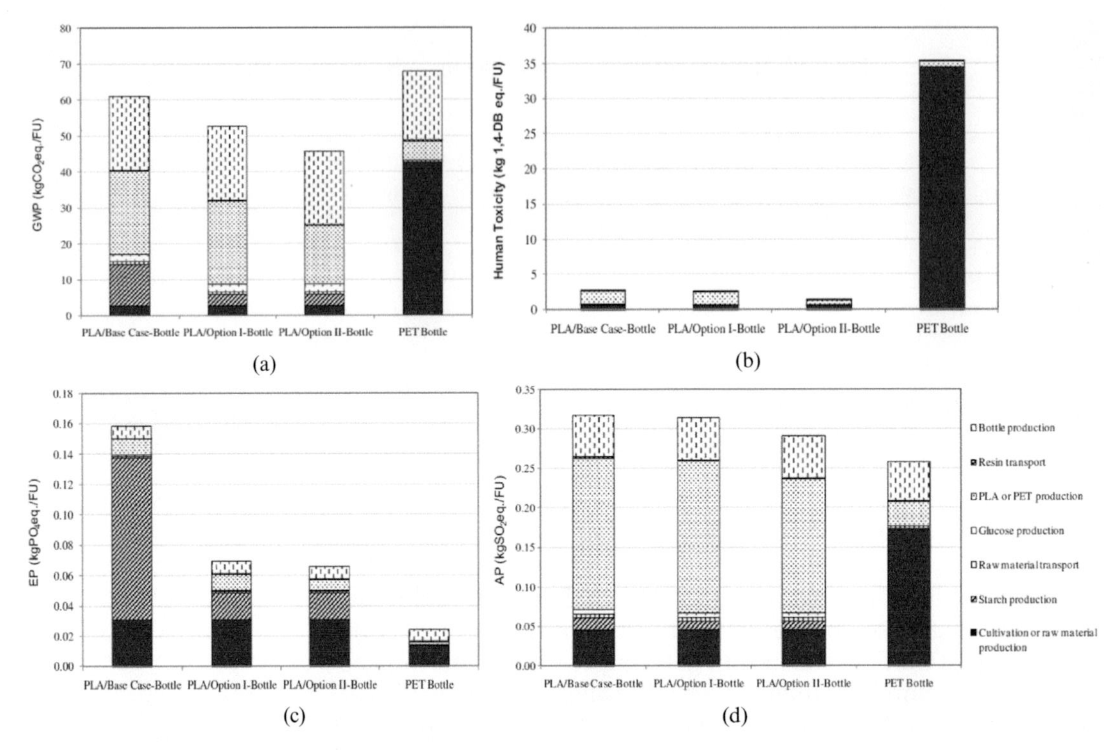

Fig. 2.6 The impact catergories of the manufacturing of PET and PLA bottles. Modified from *Papong, S., Malakul, P., Trungkavashirakun, R., Wenunun, P., Chom-in, T., Nithitanakul, M., & Sarobol, E. (2014). Comparative assessment of the environmental profile of PLA and PET drinking water bottles from a life cycle perspective. Journal of Cleaner Production, 65, 539–550. With permission of Elsevier.*

2010). The carbon monoxide used in the production process is mainly obtained from the pyrolysis of crude oil, coal and natural gas. Basically, the production of polycarbonate involves several stages, which are the chlorination process, oxychlorination process and vinylation process. The chlorination process in general is expected to release the most undesirable acidic products. A comparison between the production of polycarbonate using the phosgene method and the nonphosgene method, in terms of their impact on human health and environmental impact, is shown in Table 2.16. The phosgene process method is an indirect reaction method that produces polycarbonate by reacting the phosgene ($COCl_2$) with bisphenol A, and it has more process stages than the nonphosgene process method. In this process method, the bisphenol A is indirectly produced by the reaction between phenol and acetone, as one of the additional stages in the production of polycarbonate.

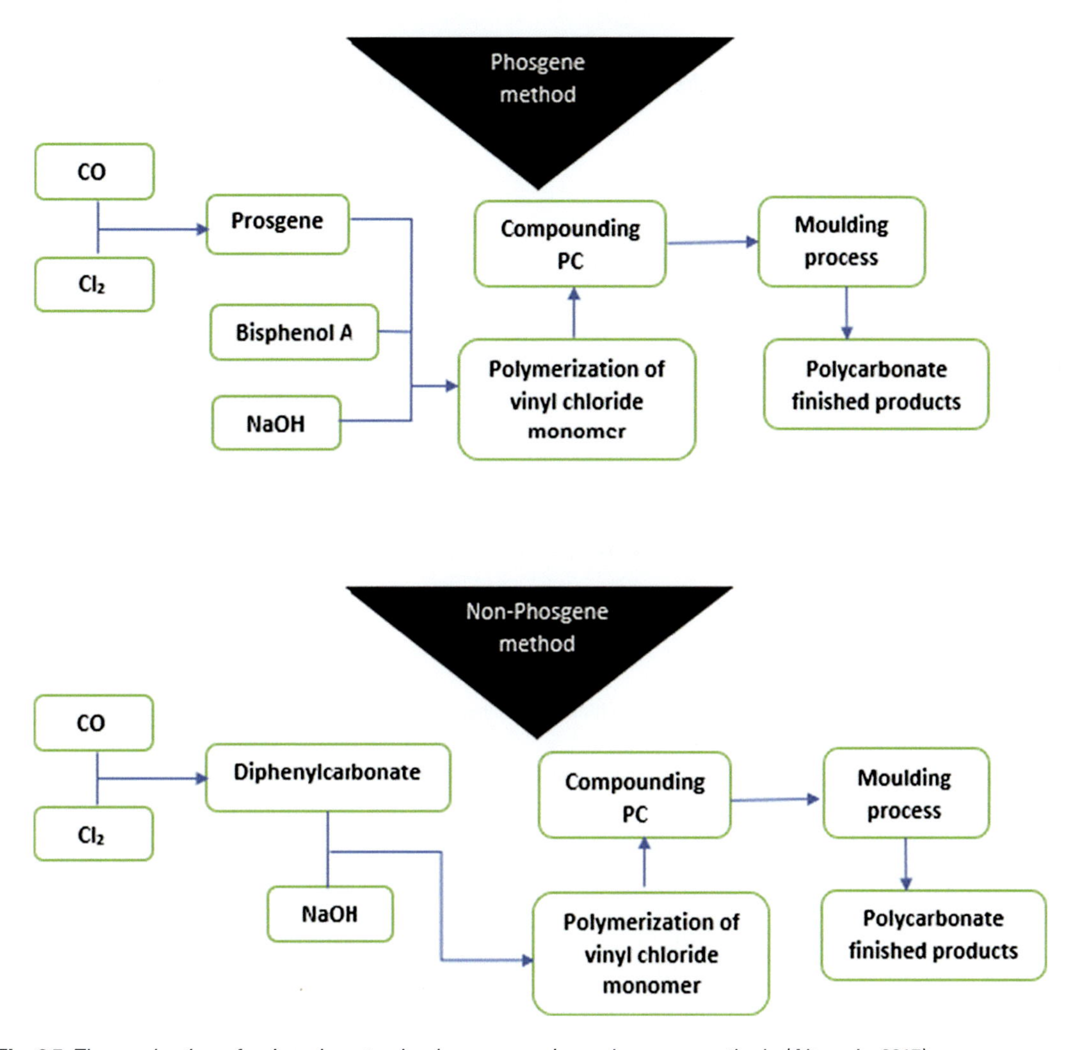

Fig. 2.7 The production of polycarbonate via phosgene and nonphosgene methods (Altuwair, 2017).

Polycarbonate in the nonphosgene process method is directly produced by using phenol with carbon monoxide. The production of polycarbonate with the phosgene method also produces a high yield level of undesirable by-products, such as phenol, up to 150–200 kg/t. In addition, the phosgene process method also releases high corrosive and acidic by-products that are harmful to human health and the environment. The nonphosgene process, on the other hand, is found to release fewer emissions with lower acidification potential and lower impact on human health and environment.

Table 2.16 Comparison of the production of polycarbonate via phosgene and nonphosgene methods.

Process description	Phosgene method	Nonphosgene method
Process design	Indirect reaction process in which the bisphenol A is produced indirectly by reacting phenol with acetone (Altuwair, 2017)	Direct reaction process with reaction of phenol with carbon monoxide (CO) (Altuwair, 2017). Also, can be produced directly from diphenyl carbonate monomer and carbon dioxide (CO_2) (Fukuoka et al., 2010)
Number of process stages	More stages of process involved	Fewer stages of process involved
Acidification impact	Produce high corrosive and acidic materials	Low acidic materials
Process outputs	Sodium chloride NaOH low conversion	Water and carbon monoxide (CO) high conversion
Catalyst options	Base catalysts	Pd catalysts
Environment impact	High risk of human health and environmental impact	Low risk of human health and environmental impact
Characterization of risk	High risk of human health impact, high cost due to high energy consumption	Lower risk of human health impact, lower cost due to low consumption of energy

Vink et al. (2010) summarized the total energy consumption and global warming potential in the production of polypropylene, ABS, polycarbonate and nylon 6 in terms of CO_2 eq. They observed that the production of polycarbonate had lower energy demand as compared to ABS and nylon 6. In addition, polycarbonate was observed to have lower carbon dioxide emissions with a lower global warming potential effect as compared to ABS and nylon 6. In a study conducted by Schwarz et al. (2021), the researchers found that the production of acrylonitrile butadiene styrene (ABS) resin has a higher environmental impact, which is associated with the dissolution treatment. However, when compared to polypropylene, the polycarbonate was found to have higher energy demand because the production of polycarbonate, whether using the phosgene process method or the nonphosgene process method, tends to consume more energy and generate more harmful emissions such as carbon dioxide. On the other

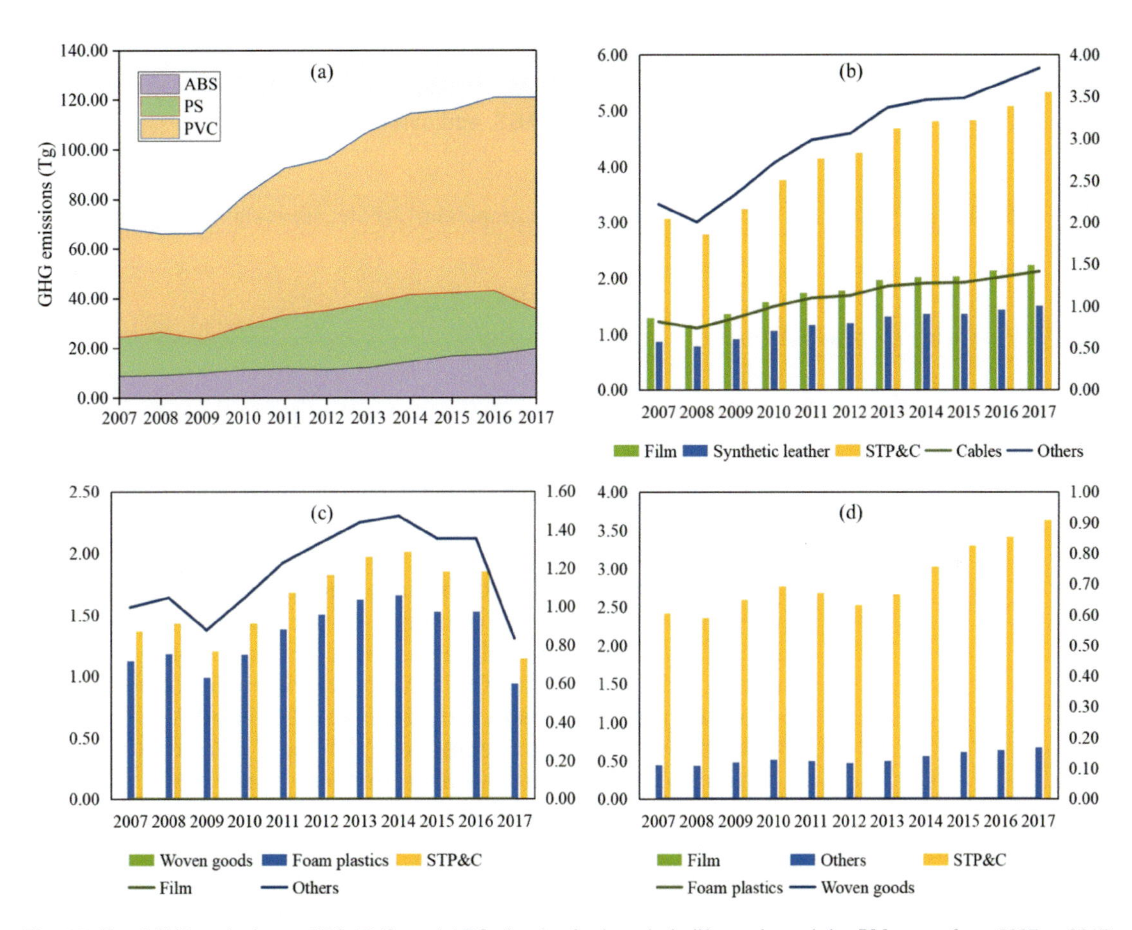

Fig. 2.8 Total GHG emissions of PS, PVC, and ABS plastics in the whole life cycle and the PM stage from 2007 to 2017 ((a) total GHG emissions, (b) PS, (c) PVC, (d) ABS). From Chu, J., Zhou, Y., Cai, Y., Wang, X., Li, C., and Liu, Q. (2022) Life-cycle greenhouse gas emissions and the associated carbon-peak strategies for PS, PVC, and ABS plastics in China. Resources, Construction & Recycling, 182: 106295–106306, with permission of Elsevier.

hand, a life cycle assessment of the greenhouse gas (GHG) emission of acrylonitrile butadiene styrene (ABS) released from 2007 to 2017 was conducted by Chu et al. (2022) and compared with PVC and polystyrene. Referring to Fig. 2.8, the GHG emissions from 2007 to 2017 were observed to steadily increase from 8.71 Tg CO_{2e} (2007) to 19.72 Tg CO_{2e} (2017). The annual growth rate of GHG emissions in the period of 2007–17 was 8.7%. In the waste management and recycling process of ABS resin, the GHG emissions in the stages of landfill, incineration, recycling and untreated reached a low point in 2012, attributed to the policy intervention of China (Table 2.17).

Table 2.17 The global warming potential and energy consumption of polypropylene, polycarbonate, ABS, and nylon 6.

Parameter	Polypropylene	Polycarbonate	ABS	Nylon 6
Global warming potential, kg CO_2 eq/kg polymer	1.9	3.8	7.6	9.1
Energy consumption, MJ/kg polymer	73	95	113	120

Data from Vink, E. T., Davies, S. and Kolstad, J. J. (2010). The eco-profile for current Ingeo polylactide production. Industrial Biotechnology, *167–180.*

2.3 Biodegradable polymers and biobased polymers

In recent years, the disposal of nondegradable commodity polymers and their environmental impacts have become major concerns. Single-use polymers have received global attention due to their severe harmful impacts on the environment, particularly marine and freshwater organisms Zaki and Aris (2022). In fact, the wastes of nondegradable polymers were found to take decades to degrade in landfills (Elmowafy et al., 2019). Therefore the recycling of conventional polymers and development of biodegradable polymers are practical methods to minimize impact on the environment. Nonetheless, the usage of recycled plastics is limited to certain applications such as furniture, household appliances or clothes that are not in contact with food, due to hygiene concerns. Similarly, limited knowledge of the environmental distribution and ecological risks of the manufacturing process of biodegradable polymers has also restricted the production and application of biodegradable polymers (Zhao et al., 2020).

Biodegradable polymers that can decompose in a shorter period of time are more preferable than the conventional polymers for single-use plastics. However, the carbon extraction during the manufacturing process of biodegradable polymers can have serious detrimental impacts on the environment. Furthermore, the economic and social development related to biodegradable polymers is another crucial factor that affects global acceptance. Thus the life cycle and sustainability studies of biodegradable polymers should also include waste disposal methods, material and utilities costs during the manufacturing process, and the awareness of relevant knowledge of society on the environment impacts. These factors are basically related and affect each other in direct and indirect ways, and subsequently affect the sustainability score of biodegradable polymers. According to Atiwesh, Mikhael, Parrish, Banoub, and Le (2021), the greater savings in energy and lower depletion of natural fossil resources in biobased plastics production could significantly reduce the impact on the global warming issue. The main environmental impact of the life cycle of biopolymers is related to carbon extraction from processing of the raw materials, the manufacturing process and transportation. Hence, a thorough comparison of the life cycles of biodegradable, biobased and conventional fossil plastics is important to determine the respective pros and cons of each type of plastic.

2.3.1 Polylactic acid[a]

First, Simon, Amor, and Földényi (2016) compared the impacts of aluminium, polyethylene terephthalate (PET), PLA, cartons and glass beverage bottles. They found that recycling of the post-consumer bottles tends to generate the least greenhouse gases (GHGs). This finding was an encouraging support of the global effort to promote recycling as a promising approach to mitigate plastic pollution. In this study PLA exhibited the lowest GHG of 66 kg CO_2-eq followed by the large 1.5 L PET bottle with 85 kg CO_2-eq, and cartons at 88 kg CO_2-eq. This aligns with the general perception that PLA produced from agricultural sources is the most environmentally friendly product. But when PLA bottles have undergone incineration and landfilling, the GHG can increase severalfold, to 498 CO_2-eq and 500 CO_2-eq, respectively. This indicates PLA still has negative impacts on the environment at the end-of-life stage. Similar impacts can be seen for glass, PET, aluminium, and paper cartons when incinerated or landfilled. Another study by Papong et al. (2014) reported that PLA overall has a lesser impact compared to PET bottle production. Ironically, the eutrophication and acidification potential of PLA is higher, due to the agricultural sources of input that require cultivation using fertilizers, herbicides, pesticides, diesel for the harvester, and long-distance transportation to the factories for the polymerization process. In addition, the washout of the fertilizers by rainwater or irrigation water can cause secondary pollution in lakes and rivers. On the other hand, in production of PET, the process only involves chemicals that are well contained in factories where the exposure to hydrocarbons, chemicals, catalysts, and residues is more manageable, under the control of factory operators. As

[a] Related reading also can be found in Chapter 7.

such, the possibility of eutrophication for PET is much lower. Nonetheless, one of the advantages of PLA production is that the electricity to operate it can be partially replaced by combustion of agricultural residues; also, a large farmland area can accommodate the installation of wind turbines to harvest wind energy to reduce dependency on nonrenewable fossil fuels in the factory process.

In addition, Cheroennet, Pongpinyopap, Leejarkpai, and Suwanmanee (2017) conducted an analysis to determine the sustainability of PLA from the perspective of water consumption. They made a comparison of biobased box production from polystyrene, PLA from sugarcane derivation (PLA-S), PLA from sugarcane derivation with blended starch (PLA-S/starch) and polybutylene succinate (PBS). In addition, the water footprint (WF) assessment was further divided into three types of water footprints: green WF, blue WF and grey WF. The green WF is the ratio of rainwater to crop yield from the field during the growing period, the blue WF is the ratio of irrigation water used to crop yield from the field during the growing period, and the grey WF is the volume of water required to dilute the pollutant concentration. This detailed analysis of water consumption revealed that PLA-S actually consumes more water to produce biobased boxes due to the fact that growing sugarcane requires a large volume of water from rainwater and irrigation water sources. Thus PLA-S has a water footprint of $1.11\,m^3$ with 36.14% from green water contribution, 49.82% from blue water contribution and a grey water contribution of 14.05%. PS as the nonrenewable polymer material has $0.70\,m^3$ with 100% contribution from blue water. This is followed by the PLA-S/starch containing starch blending in PLA-S, which requires $0.55\,m^3$ water to produce per box and the smallest WF is PBS at $0.38\,m^3$. On the other hand, the addition of starch into PLA-S showed a reduction of WF for PLA-S of 50.42%. This is mainly attributed to the higher quantities of water required to grow sugarcane as compared to cassava starch. It was further proven by Cheroennet et al. (2017) that irrigation water for planting of sugarcane was significantly higher than that for cassava, which results in the blue WF for PLA-S and PLA-S/starch being 0.55 and $0.27\,m^3$. The grey water for PLA-S is also higher than PLA-S/starch, because PLA-S requires 4.06 kg to produce each PLA-S box, whereas the sugarcane and cassava are 1.75 and 0.11 kg to produce each PLA-S/starch box. The reason is that blending of starch involves a simple process in which raw starch is blended into the PLA-S. However, production of PLA-S involves complicated chemical reactions and tends to mass losses along the process, and additional utilities are required to support the operation as well. Hence, PLA-S/starch exhibits favourable results in terms of less material consumption to produce a biobased box. In addition, Cheroennet et al. (2017) found that the carbon footprint (CF) of the PLAS (0.675 kg CO_2 equivalent) is the highest between PS (0.05 kg CO_2 equivalent) and PLAS-starch (0.303 kg CO_2 equivalent). This is actually related to the usage of plastic material to make the boxes. For instance, PS boxes can be produced with a very small amount of resin due to PS foam boxes being very low density (~0.053 kg PS/box) with outstanding physicomechanical properties. PLAS requires ~0.243 kg PLAS to produce a PLAS box, and subsequently 0.105 kg PLAS pellets and 0.032 kg of cassava starch to produce a PLAS-starch box. In other words, depending on the nature of the plastic materials, a stronger material requires smaller quantities (lower thickness) to produce an equivalent performance of products, or vice versa.

PLA producers NatureWorks have presented several publications related to the LCA profile of PLA. First, Vink, Rabago, Glassner, and Gruber (2003) with the first LCA compared the PLA production with major energy input from fossil fuels. As illustrated in Fig. 2.9, the consumption of fossil energy was 54.1 MJ/kg for PLA and the renewable energy was 28.4 MJ/kg. In comparison with other petroleum-based polymers, PLA is outstanding at utilizing less fossil fuel inputs, as shown in Fig. 2.10. (Vink et al.,

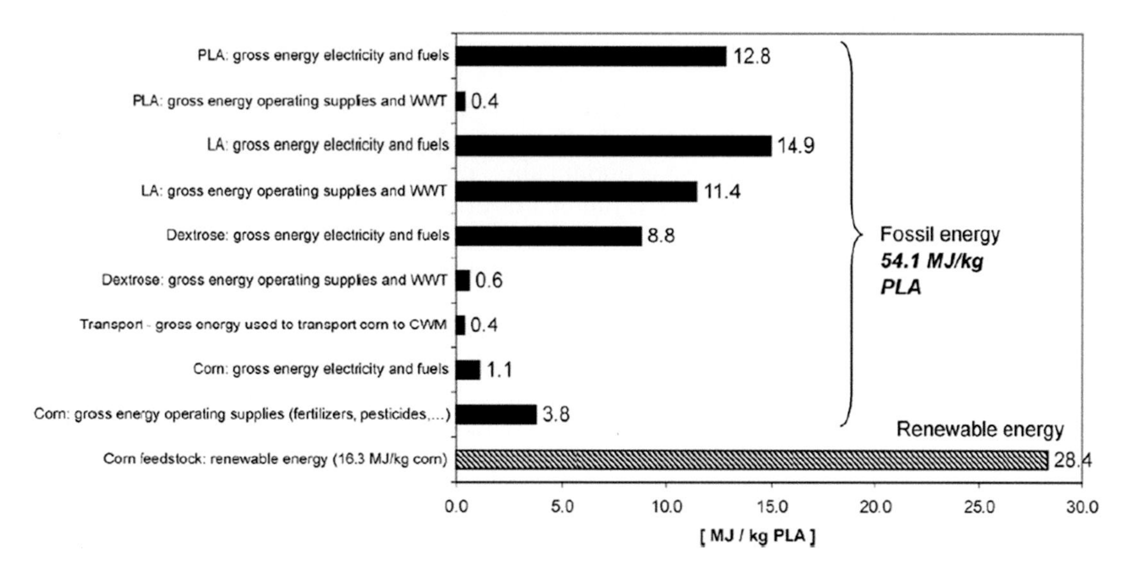

Fig. 2.9 Gross energy consumption for production of PLA. *LA*, lactide; *WWT*, wastewater treatment. From Vink, E. T. H., Rabago, K. R., Glassner, D. A., Gruber, P. R. (2003). Applications of lie cycle assessment to NatureWorks™ polylactide (PLA) production. *Polymer Degradation and Stability, 80*, 403–419, with permission of Elsevier.

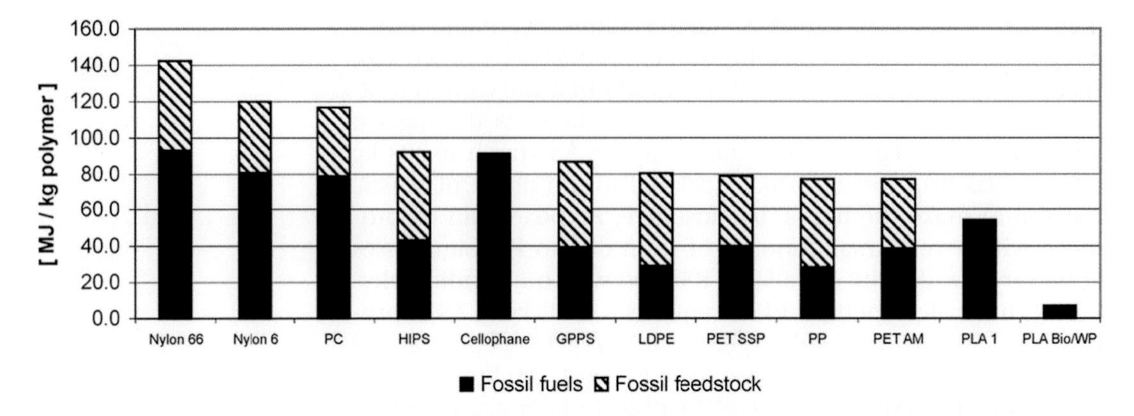

Fig. 2.10 Comparison of fossil fuel energy consumption for petroleum-based polymers and PLA. The cross-hashed part of the bars represents the fossil energy used as chemical feedstock (the fossil resource to build polymer chains). The solid part of each bar represents the gross fossil energy used for fuel and operation supplies used to drive the production processes. *PC*, polycarbonate; *HIPS*, high impact polystyrene; *GPPS*, general-purpose polystyrene; *LDPE*, low-density polyethylene; *PET SSP*, polyethylene terephthalate, solid state polymerization (bottle grade); *PP*, polypropylene; *PET AM*, polyethylene terephthalate, amorphous (fibres and film grade); *PLA1*, PLA without adoption of biomass and wind power; *PLA B/WP*, PLA with adoption of biomass and wind power. From Vink, E. T. H., Rabago, K. R., Glassner, D. A., Gruber, P. R. (2003). Applications of lie cycle assessment to NatureWorks™ polylactide (PLA) production. *Polymer Degradation and Stability*, 80, 403–419, with permission of Elsevier.

2003) proposed that optimization of the substitution for fossil energy with biomass/wind power could improve the sustainability of PLA. Although the implementation of a renewable energy initiative is attractive, the feasibility analysis of the workability was still unfavourable for NatureWorks, because they found that the PLA plant in Nebraska was not located on a site suitable to produce economically competitive wind resources. Also, Nebraska is a public power state or, in other words, NatureWorks is obliged to purchase electricity from its local utility. In order to resolve the obstacles, NatureWorks had to opt for a Renewable Energy Certificate (REC) to reduce (1) indirect emissions from electricity production at 1.561 CO_2 eq. (kg/kg PLA) and (2) fuel, material, corn production and reclamation at 1.244 CO_2 eq. (kg/kg PLA). In fact, the REC is a sort of carbon credit trade that encourages companies that produce renewable energy to trade off in the voluntary market to foster green energy development. The carbon credit can be traded to other businesses that, although unable to produce renewable energy efficiency, are eager to participate in the renewable energy industry to promote lower emissions in their economic activities. As a result, Vink et al. (2007) reported a 90% reduction of carbon emissions after the purchase of RECs, as shown in Table 2.18. Subsequent process improvement was reported by Vink et al. (2010) for NatureWorks.

Table 2.18 Scenario of PLA emission with the purchasing of renewable energy certificate (REC).

Process	CO_2 eq. (PLA) (kg/kg PLA)	
	Before purchase REC	*After purchase REC*
(1) NatureWorks/Cargill site, direct emissions	1.038	1.038
(2) Indirect emissions from electricity production	1.561	1.561
(3) Fuel, material, corn production, reclamation	1.244	1.244
Corn feedstock-CO_2 uptake	−1.820	−1.820
REC purchased to offset electricity emission from (1)	–	−1.553
REC purchase to offset electricity emissions from (2)	–	−0.197
Total	2.023	0.272

Based on Sin, L. T., Tueen, B. S. (2019). Polylactic acid, a practical guide for the processing, manufacturing, and applications of PLA, 2nd ed., Elsevier, with permission of Elsevier.

2.3.2 Starch-based polymeric material

In general, starch-based polymeric materials are a blend of starch with conventional fossil plastics or biodegradable plastics. The purpose of blending is to reduce the portion of fossil plastics or to induce biodegradability of the plastic composite materials. Meanwhile, starch is blended with biodegradable plastic with the intention to reduce the cost of the compound. This is because most of the biodegradable plastics resins are expensive due to the high cost of production. Regardless of the intention, the general public perception is that blending the starch into the fossil plastic can transform the plastic material to become more environmentally friendly. Thus it is important to find out whether this perception is true. A comprehensive review by Shen and Patel (2008) has made a comparison of starch-added plastic compound with conventional neat plastic resin, as shown in Table 2.19. Obviously, the starch-added plastic required less nonrenewable energy than the fossil-based plastics. Nonetheless, there is concern that the weaker or inferior properties of starch-added plastic require a larger mass to meet the functional requirements, such as similar strength, rigidity and flexibility (comparison of grocery bags for PP and starch-added plastic). This could eventually result in an insignificant difference for both nonrenewable energy use and GHG emissions for starch-added and nonstarch-added plastics.

In addition, Broeren et al. (2017) have done a comparison between virgin starch-added plastics and reclaimed starch-added plastics. Referring to Fig. 2.11, it can obviously be seen that the nonrenewable energy use for fossil type plastics of polypropylene

Table 2.19 Summary of energy and GHG emissions per functional unit plastic products; products listed are all commercial products manufactured by state-of-the-art technologies.

Type of plastic	Functional unit	Cradle-to-gate nonrenewable energy use[a] (MJ/FU)	Type of waste treatment assumed for calculation of emissions	GHG emissions (kg CO_2 eq./FU)
Loose fills				
Starch loose fills	1 m^3 (10 kg)	492	Wastewater treatment plant	21
Starch loose fills	1 m^3 (12 kg)	277	30% incineration, 70% landfilling	33.5
EPS loose fill	1 m^3 (4.5 kg)	680	Incineration	56
EPS loose fill	1 m^3 (4 kg)	453	30% incineration, 70% landfilling	22.5
EPS loose fill (by recycling of PS waste)	1 m^3 (4 kg)	361	30% incineration, 70% landfilling	18.6
Films				
TPS film	100 m^2	649	80% incineration, 20% landfilling	25.3
PE film	100 m^2	1340	80% incineration, 20% landfilling	66.7
Grocery bags[b]				
50% starch + PBS/A (single use)	3.12 kg	n/a	70.5% landfill.; 10% compost.; 0.5% litter; 19% reuse	2.5
50% starch + PBAT (single use)	3.12 kg	n/a	70.5% landfill.; 10% compost.; 0.5% litter; 19% reuse	2.88
50% starch + PCL (single use)	4.21 kg	n/a	70.5% landfill.; 10% compost.; 0.5% litter; 19% reuse	4.96
HDPE (single use)	3.12 kg	n/a	78.5% landfill.; 2% recycle; 0.5% comp.; 19% reuse	6.13
PP (multiple use)	0.48 kg	n/a	99.% landfill.; 0.5% litter	1.95
LDPE (multiple use)	1.04 kg	n/a	97.5% landfill.; 2% recycle; 0.5% litter	2.76

Abbreviations: *TPS*, thermoplastic starch; *PVOH*, polyvinyl alcohol; *PCL*, polycaprocactone; *PBAT*, polybutylene adipate terephthalate; *EPS*, expandable polystyrene; *HDPE*, high-density polyethylene; *LDPE*, low-density polyethylene; *PET*, polyethylene terephthalate.

[a]Total of process energy and feedstock energy. Nonrenewable energy only, i.e. total fossil and nuclear energy. In the 'cradle-to-factory gate' concept the downstream system boundary coincides with the output of the polymer or the end product. Hence, no credits are ascribed to valuable by-products from waste management (steam, electricity, secondary materials).

[b]The functional unit is defined as the grocery bags needed for "a household carrying approximately 70 grocery items home from a supermarket each week for 52 weeks"; the functional unit is determined by the weight, the capacity (volume), and the lifetime of the bag. The volume of TPS bags and HDPE singlet bag are the same (6–8 items); the volume of the PP bag is 1.2 times the volume of the HDPE singlet bag; and the volume of the LDPE bag is three times the volume of the HDPE singlet bag. All the TPS bags and the HDPE singlet bag are for single use; the PP bag is multiuse and has a lifetime of 2 years; and the LDPE bag is multiuse and has a lifetime of 1 year.

Data from Shen, L., Patel, M. K. (2008). Life cycle assessment of polysaccharide materials: A review. Journal of Polymer and the Environment, 16, *154–167, with permission of Springer Nature.*

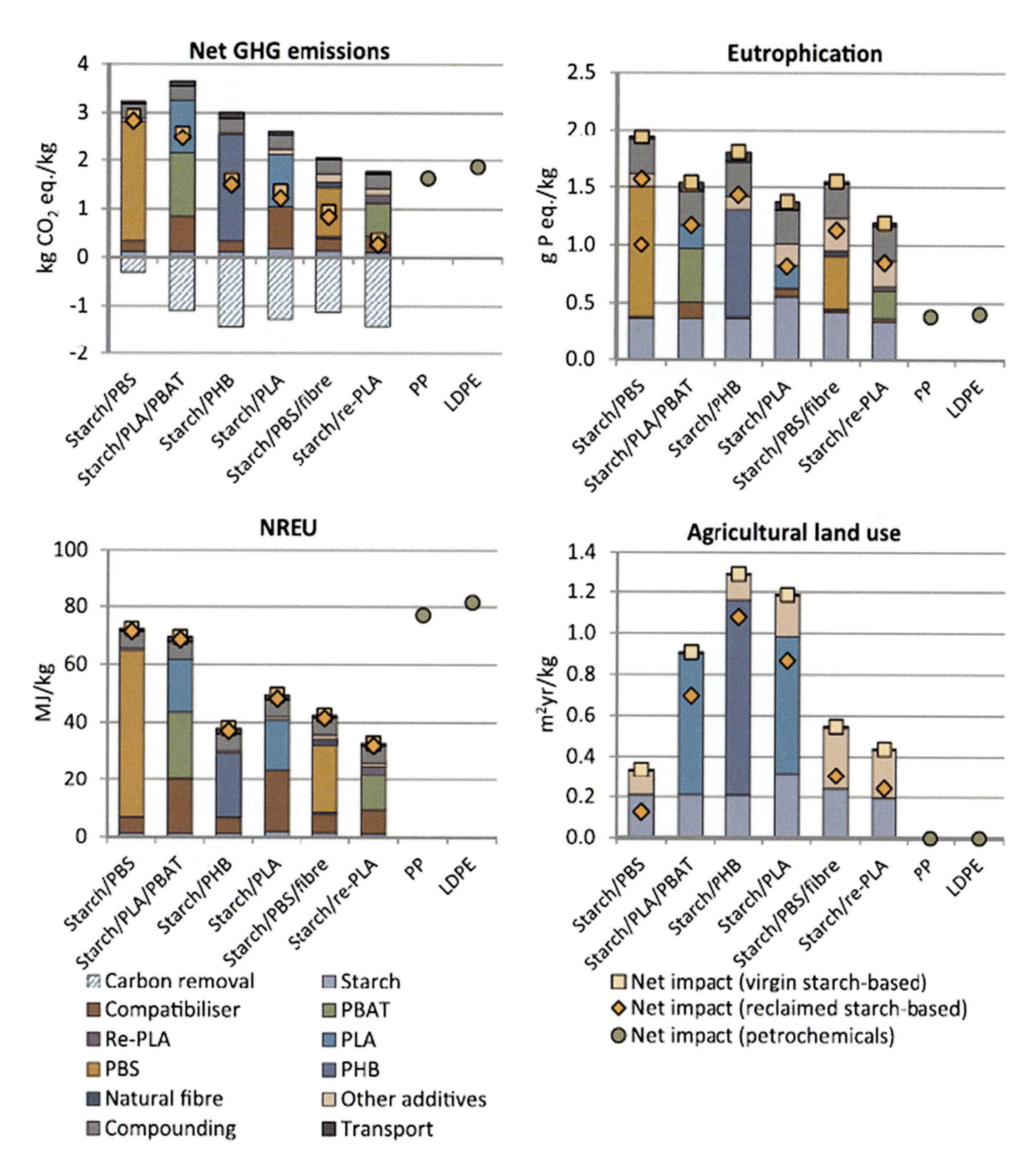

Fig. 2.11 GHG emission, nonrenewable energy usage (NREU), eutrophication, and agricultural land use for starch-based plastic. *Re-PLA*, recycled PLA; *PBS*, polybutylene succinate; *PBAT*, polybutylene adipate terephthalate; *PHB*, polyhydroxybutyrate. Based on Broeren, M. L. M., Kuling, L., Worrell, E., Shen, L. (2017). Environmental impact assessment of six starch plastics focusing on wastewater-derived and additives. *Resources, Conservation & Recycling*, *127*, 246, 255, with permission from Elsevier.

(PP) and low-density polyethylene (LDPE) are higher compared to starch-added plastics, whereas the GHG emissions of starch-added plastics are higher than LDPE and PP. This is mainly attributed to the planting and harvesting stages using heavy machinery, pesticides and herbicides and actually replacing the CO_2 intake from photosynthesis. Both starch/PBS and starch/PLA/PBAT account for higher net GHG emissions than PP and LDPE, because of the high percentage of plastic PBS (55%) and PLA:PBAT (45%:18%) content with a high percentage of petrochemical additives >20% added. This indicates that the ecofriendliness of the plastics is also dependent on the unavoidable additives package as well. On the other hand, the eutrophication and land use for starch-based plastics are undeniable higher than for PP and LDPE. When a comparison was made of virgin and reclaimed starch plastics, the analysis found that both GHG emissions and nonrenewable energy usage were similar, with reclaimed starch plastics being slightly lower. Meanwhile, the eutrophication and land use for reclaimed starch plastics were superior to virgin starch, mainly due to reclaimed starch being extracted from the wastewater stream, with the impacts of the planting stage being excluded. In conclusion, starch-added plastics are good alternatives to be considered to replace fossil-based plastics, but with compromises on eutrophication and land use.

2.3.3 Polyhydroxyalkanoates

The production of microbial polyhydroxyalkanoate (PHA) polymers is currently facing two main challenges before being commercialized. The first challenge is the fact that the production cost of PHA is very high, and it is unable to show any beneficial effect in cost and properties effectiveness over fossil-based plastics such as polypropylene, polyethylene, etc. The second challenge is to verify data to support the environmental friendliness of the production of PHA over the production of petroleum-based polymers. The available environmental analysis for PHA production is mainly based on simulations, since actual large-scale production plants for PHAs are not currently available. According to Gerngross and Slater (2000), various researchers at Monsanto designed a theoretical model facility by extending an existing corn-processing plant to extract the PHA from the corn. They found that the extraction process of PHA from the corn required the use of large amounts of solvent, which requires a huge recovery facility comparable to the existing petrochemical plastic factories. PHA is one of the aliphatic microbial polyesters with high stiffness behavior and high melting temperature. It is made from sugar extracted from corn plants and is a biodegradable polymer. The grown corn stover (corn residues such as leaves, cobs, stalks, etc.) is harvested using a conventional method and then delivered to the factory for complicated enzymatic, fermentation and extraction processes (Gerngross & Slater, 2000). Earlier, Gerngross (2001) tabulated the required energy for the fermentation production of PHA from glucose, as shown in Table 2.20, and

Table 2.20 Energy used to produce PHA from glucose using fermentation method.

Particular	Electricity (kWh)	Steam (kg)	Fossil fuel equivalent[a] (kg)
Fermentation			
Media sterilization[b]	None	0.45	0.02
Agitation[c]	0.32	None	0.09
Aeration[c]	1.27	None	0.35
Cooling[d]	0.76	None	0.21
Downstream			
Centrifugation and washing[e]	0.50	None	0.14
High pressure homogenizer[f]	1.97	None	0.54
Centrifugation and washing[e]	0.50	None	0.14
Evaporation[g]	None	0.33	0.02
Spray drying[h]	None	2.00	0.10
Total	5.32	2.78	1.59

[a]Amount of fossil fuel (kg) consumed to generate the electricity and steam listed in the same row. Conversion: Electrical energy (kWh) × 0.272 (kWh/kg) = FFE (kg). In 1997 the US average for producing 1 kWh of electrical power, from all power sources (including geothermal, hydroelectric, nuclear, and alternative power generation) required the direct combustion of 0.272 kg of fossil resources (10). These resources were 83% coal, 13% natural gas, and 3.5% petroleum. To convert steam (kg) to FFE (kg): Steam (kg) × Heat of evaporation of water (2400 kJ/kg)/47,219 kJ/kg (Heat of combustion of natural gas) = FFE (kg). Calculation assumes 100% efficiency, the use of natural gas for steam generation, and no heat loss.
[b]Medium is continuously sterilized to 143°C for 30 s; 68% of energy is recaptured through a heat exchanger and used to prewarm the incoming medium.
[c]Agitation and aeration for this type of aerobic fermentation with very high cell densities is estimated to require 5W/1 of power input. Power is delivered by mechanical agitation and compressed air and amounts to 1.59 kWh/kg of PHA.
[d]Cooling: Considering the use of a fully jacketed 114,000 L fermenter (3:1, H:D) that has additional cooling coils on the inside providing a total cooling area of 266 m^2 or 2.32 m^2/m^3. Under the assumed production schedule (48 h and 190 g/L of biomass formed) we have to remove approximately 17.6 W/L or a total of 2 million J/s (25). Coolant is provided through cooling towers and chillers which deliver about 48,700 and 18,000 kJ of cooling per kWh input, respectively.
[e]Centrifugation: Reduced the energy input by 66% to reflect the energy savings from substantially higher solids content.
[f]Cell disruption.
[g]Evaporation is performed with a triple effect evaporator. A preconcentrated slurry containing 30% solids is concentrated to 50% solids. This step requires the evaporation of 1.33 kg of water and requires 0.33 kg of steam per kg PHA. Further concentration is not feasible because of the high viscosity of the slurry.
[h]The final slurry is processed to a powder employing a spray dryer. Dry air is required to provide process heat and evaporate the remaining water. The energy required is generally about double the amount of the water that has to be evaporated from the slurry. Hence about 2 kg of steam is required to remove 1 kg of water, leaving us with 1 kg of dry powder.
Data from Gerngross, T.U. (2001). Polyhydroxyalkanoates: the answer to sustainable polymer production. ACS Symposium Series, 784, *10, with permission from Springer Nature.*

Table 2.21 Energy and fossil fuel equivalent required to produce PS and PHA.

	Polystyrene		PHA	
Particular	*Energy*	*Fossil fuel energy (kg)*[a]	*Energy*	*Fossil fuel equivalent (kg)*[a]
Production of raw materials	See detail as below[b]	1.78	31,218 kJ	0.80
Utilities				
Steam[c,d]	7.0 kg	0.40	2.78 kg	0.14
Electricity[c]	0.30 kWh	0.08	5.324 kWh	1.45
Total		2.26		2.39

[a]Fossil fuels required to produce the energy and raw material in the corresponding energy column taking into account the primary fuel usage patterns for each industry. Example: In the corn wet milling energy is generated from natural gas (37%), coal (48%), and petroleum (10%). Therefore, for each MJ of energy, currently consumed by the corn wet milling industry 370,000 kJ are generated by combusting 7.84 kg of natural gas, etc. Petroleum, coal, and gas have a heat of combustion of 42,000 kJ/kg, 25,788 kJ/kg, and 47,218 kJ/kg, respectively.
[b]Feedstock required to produce 1 kg of PS from crude oil. The feedstock for direct polymer synthesis is included as well as a fraction that is lost in the process and recycled for energy generation. Since this type of process does not allow for a clear allocation of feedstocks and energy, numbers include both feedstock and process energy.
[c]Energy (including steam and electrical power) to produce glucose, ammonia, and water. Calculations: 3.33 (Yield of PHA on glucose) $\times$ 8128 kJ + 4028 kJ (Energy to produce 0.109 kg of ammonia) + 120 kJ (Energy to provide 26 L of process water).
[d]The production of 1 kg of steam requires the combustion of 0.058 kg of residual fuel oil during PS production. PHA fermentation most likely will use natural gas for steam generation and therefore only requires 0.0508 kg of gas to produce the same amount of steam.

did a comparison of energy between PS and PHA, shown in Table 2.21. Gerngross (2001) showed that there are no fewer emissions for PHA over polystyrene (PS). Whenever products involve multiindustrial processes (agricultural to biotechnology to polymerization), the emissions tend to be more substantial.

Furthermore, during the PHA production process, the usage of fossil fuel is still required in the extraction process of PHA polymer from the plant and the total energy needed in the extraction process is still much higher than fossil fuel-based polymer, as summarized in Table 2.22. A study conducted by Akiyama et al. (2003) compared the fermentation of P(3HB-co-5mol%3HHx) copolymer from soybean oil as a carbon source and the fermentation of production of P(3HB) from glucose, with the total fossil fuel energy for both fermentation production processes being comparable. Meanwhile, according to the LCA analysis conducted by Akiyama et al. (2003), the production of PHA through two different fermentation methods was found to consume lower total fossil fuel energy than the petroleum-based polymers such as PP, LDPE, PS, etc., as shown in Table 2.22. They also found the

Table 2.22 Total fossil fuel energy used in the production of PHA via different methods.

Type of polymer	Total fossil fuel energy (cradle-to-factory gate), GJ/ton polymer
PHA grown in corn plants (Gerngross & Slater, 2000)	90
PHA (bacterial fermentation) (Gerngross & Slater, 2000)	81
PHA, poly(3-hydroxybutyrate-co-5mol% 3-hydroxyhexanoate), P(3HB-co-5mol%3HHx) fermentation of soybean oil with *Ralstonia eutropha* (Akiyama et al., 2003)	50
PHA, poly (3-hydroxybutyrate) from fermentation of glucose with *Aeromonas caviae* (Akiyama et al., 2003)	59
HDPE (Gerngross & Slater, 2000)	80
PET (bottle grade) (Gerngross & Slater, 2000)	77
PS (general-purpose usage) (Gerngross & Slater, 2000)	87
LDPE (Akiyama et al., 2003)	81
PP (Akiyama et al., 2003)	77

CO_2 emission of PHA copolymer from soybeans and corn was significantly lower than that of petroleum-based polymers. The reduction of CO_2 emissions during the soybean and corn plantation is mainly contributed by the absorption of CO_2 gas from air by the soybean and corn plants (Akiyama et al., 2003). In another study by Kim and Dale (2005), it was reported that the global warming associated with corn grain-based PHA was 1.6–4.1 kg-CO_2 eq/kg. In general, the most significant contribution to the environmental impact for the production of PHA is the PHA fermentation and recovery process, while the photochemical smog and eutrophication are due to the fertilizer and cultivation stages.

2.4 Supply chain management of plastics

The supply chain management of plastics, as shown in Fig. 2.12, is about the step flows of plastics, from the raw materials to polymers, followed by undergoing processing steps such as injection moulding, blown film, and thermoforming to make usable forms of plastic products, such as containers for food packaging, and finally reaching the hands of the consumers. After use, it is discarded or recycled. In the early days, the supply chain of plastics was in one direction, from production to landfill. This caused plastic wastes to pile up in the landfill or be discarded

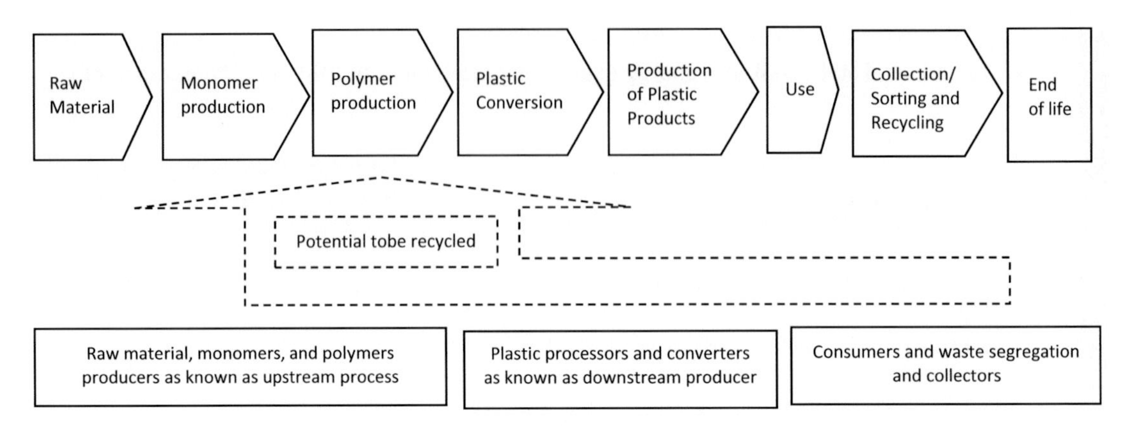

Fig. 2.12 Supply chain of plastic products.

everywhere, leading to catastrophic pollution, causing environmental and microplastics issues. The supply chain of plastics is now focusing on a circular approach, which involves recycling and waste-to-energy approaches to achieve a closed-supply supply chain in order to minimize the impacts on the environment.

As observed in Fig. 2.12, for each of the stages within the supply chain, there are certain improvements that can be implemented to minimize impacts on the environment, as summarized in Table 2.23. It is important to note that plastics technologies are always developing and scientists have made an effort to actively reduce the impacts of plastics on the environment.

From another perspective, Ryberg et al. (2019) examined global environmental losses of plastics across their value chains and found that plastic pollution can occur along the value chains. The researchers conducted the study because they found that there is a deficiency in the amount of plastic produced globally that is used in different sectors and eventually disposed of (end-of-life stage). As shown in Fig. 2.13, Ryberg et al. found that the mismanagement of municipal solid waste is the source of the major losses, with 4.1 million tones, which results from illegal dumping and landfilling at the end-of-life of plastic products. Meanwhile, the applications related to wastewater sludge accounted for the highest microplastics losses. Also a large quantity of microplastics are lost globally that are primarily made up of abrasions of polymer-containing products, such as tires, road markings, and city dust, and washing of synthetic textiles was found to contribute to global microplastics losses. Since wear and tear of plastics during their application is unavoidable throughout the application period, this means that the

Table 2.23 Improvements in plastics supply chain.

Stage	Potential and example of improvements
Raw material	The raw materials of plastics are usually derived from fossil sources with involvement of many types of chemicals as well, while certain inputs can still be substituted using renewable inputs from the agricultural sources. For instance, ethanol can be used to produce polyethylene via a dehydration process to turn ethanol into ethylene (Bedia et al., 2011)
Monomer production	One example of improvements that can be made is polyethylene terephthalate (PET) produced from 70% of terephthalic acid and 30% monoethylene glycol (MEG) by weight. For biobased PET, the MEG is produced from renewable raw materials using ethanol from sugarcane instead of fossil raw materials
Polymer production	The polymerization process involves a complicated catalytic reaction and is well developed. Nonetheless, there is opportunity to select the source of energy to be used for polymerization. For instance, the production of Ingeo polylactic acid by NatureWorks has partially substituted for fossil fuel by wind power, successfully reducing 80% fossil fuel dependency. For certain plants located in the petrochemical area, use of wind power can be impractical. One of the ways to reduce emissions is through the Renewable Energy Certificate (REC). The REC is a sort of carbon credit trade that encourages the companies that produce renewable energy to trade off in the voluntary market to foster green energy development. The carbon credit can be traded to other businesses who are unable to produce renewable energy efficiency but are still interested in participating in the renewable energy industry to promote lower emissions in their business activities
Plastic conversion	Plastic conversion is also known as a compounding process, which is intended to modify or improve the original plastic resin with outstanding mechanical and weathering resistance. A common improvement that is done is to reduce the plastic percentage by introducing fillers such as calcium carbonate, talc, kaolin clay, or wood floor. This can reduce up to 60% of the plastic portion with mineral filler or agricultural residues such as wood flour, and save costs as well. The drawback is that recyclability and segregation remain issues. This means that some plastic compounds require specific segregation and recycling processes to minimize incompatibility during the recycling
Production of plastic products	For product designers, it is important to develop plastic products that are durable, lightweight, and have the ability to keep the content, especially food, for longer times to avoid wastage. For instance, thermoforming is used to produce thin containers for food and cups for beverages, particularly for single-use products. For prolonged food storage, multilayer plastic films can be used to replace a thick single layer of plastic film, to reduce the penetration of oxygen from the air that causes the food to spoil. However, the drawback of multilayer plastic packaging is that it is difficult to recycle. Multilayer plastic packaging is most suitable to undergo a chemical recycling process or incineration to generate energy. On the other hand, for plastic products such as outdoor furniture, the plastic materials should be formulated with additional weathering resistance additives, so that the furniture

Continued

Table 2.23 Improvements in plastics supply chain—cont'd

Stage	Potential and example of improvements
	can withstand prolonged sunlight and moisture effects, against fading and chalking. Importantly, the plastic products producers should take responsibility for labeling the plastic types according to the International Resin Identification code, so that sorting of plastic wastes for recycling can be done easily. Also, with the current Extended Producer Responsibility policies implemented by many countries, the producers have the responsibility to treat their end-of-life products upon return by consumers
Use	As responsible consumers, selection of simple and recyclable packaging is of the utmost importance. Although multilayer plastic film packaging can prolong the shelf life of food containers, the drawback is that the multilayer packaging is difficult to be recycled. One of the methods is to bring one's own containers to pack takeaway food. On the other hand, there are many degradable plastic packaging styles on the market, but not all the degradable packaging is biodegradable. For instance, oxo-degradable packaging uses metallic salt as the pro-degradant, to promote breakdown of the plastic. This may cause microplastic pollution as well as accumulation of heavy metal particles resulting from the metallic salt traces. Inasmuch, consumers need to buy environmentally friendly plastic products to avoid unrealized environmental pollution
Collection/sorting and recycling	In order to minimize the effort in recycling plastic waste, consumers need to take responsibility to do the sorting according to the type of plastics. When carrying out the sorting process, consumers should ensure the plastic items are cleaned and parts are separated according to plastic types, including the stick-on labels. The recycling of plastic waste needs to be undertaken by competent recyclers to minimize secondary pollution in terms of water, air, and solid wastes discharged from the recycling process in the factories
End of life	After several recycling periods, the plastic materials are degraded and reach their end-of-life and must be fully disposed of. Medical wastes like facemasks, tubes, syringes, and gloves are also nonrecyclable due to hygiene concerns. At this stage, the plastic wastes are incinerated to generate thermal energy to drive turbines to transform into electrical energy, back to the circular economy chain

microplastics problem will remain as long as plastics are being widely used globally.

2.5 Conclusion

Plastics have been used by mankind for many decades, especially fossil types of plastics. While many research studies and great efforts have been made to improve environmental friendliness and economy feasibility of plastics, the exploitation of plastics use has undeniably caused serious environmental problems. One of the solutions to the plastics pollution problem is the plastic

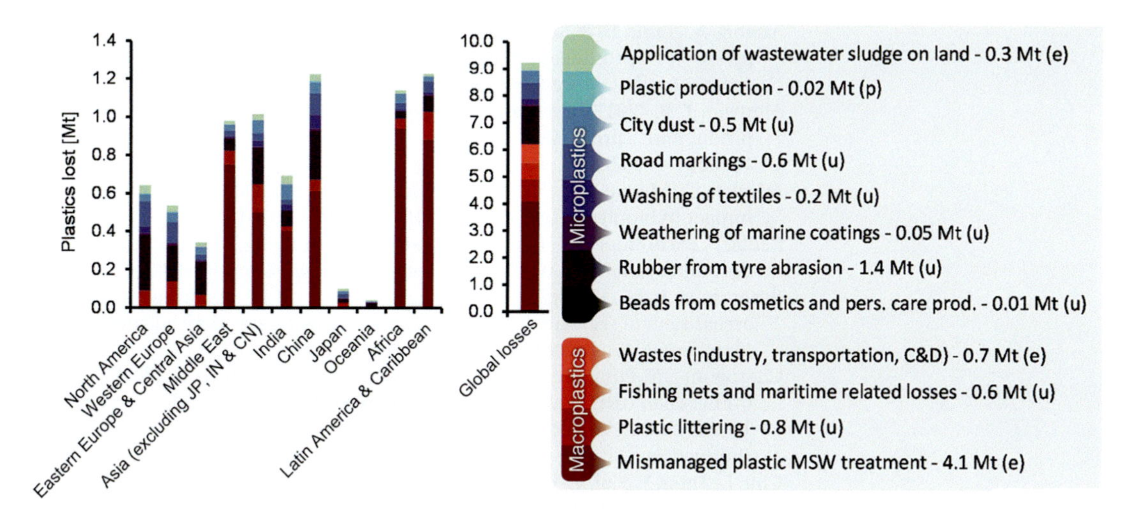

Fig. 2.13 Losses of macroplastics and microplastics to the environment (all marine, freshwater, and terrestrial compartments combined) characterized according to region and loss sources. Losses from maritime activities like fishing or shipping and losses from building industry and the transportation sector could not be assigned to specific regions, and are only indicated in the global estimates. (p) is loss during production stage, (u) is loss during use stage, (e) is loss during end-of-life stage. Reproduced with Elsevier permission from Ryberg, M. W., Hauschild, M. Z., Wang, F., Averous-Monnery, S., Laurent, A., (2019). Global environmental losses of plastics across their value chains. *Resources, Conservation & Recycling, 151*, 104459.

circular economy commitment, which involves recycling and incineration of plastic wastes to transform waste to energy. However, regardless of the extent of advancement of the plastic value chain, the minimization of the environmental impacts of plastic still depends on consumer behavior related to plastic consumption. Consumers are urged to always practice 3R—Reduce, Reuse and Recycle—to minimize waste generation, save the earth, and mitigate the impacts of climate change for the betterment of future generations.

References

Ahuja, N, & Sharma, P. (2017). Life cycle assessment of HDPE and LDPE plastic bags using Simapro 8.3.0 (Yamunanagar, Haryana). *International Journal of Civil Engineering and Technology, 8*(7), 340–345.

Akiyama, M., Tsuge, T., & Doi, Y. (2003). Environmental life cycle comparison of polyhydroxyalkanoates produced from renewable carbon resources by bacterial fermentation. *Polymer Degradation and Stability, 80*, 183–194.

Alfarisi, S., & Primadasa, R. (2018). Carbon footprint and life cycle assessment of PET bottle manufacturing process. In *The 1st international conference on computer science and engineering technology, Universitas Muria Kudus, EAL*.

Alsabri, A., Tahir, F., & Al-Ghamdi, S. G. (2021). Environment impacts of polypropylene (PP) production and prospects of its recycling in the GCC region. In *Material today: Proceedings*. Article in Press.
Altuwair, I. B. (2017). Environmental performance evaluation of polycarbonates production processes. *Industrial Engineering & Management, 6*(3), 1000214–1100225.
Atiwesh, G., Mikhael, A., Parrish, C. C., Banoub, J., & Le, T. T. (2021). Environmental impact of bioplastic use. A review. *Heliyon, 7*(9), E07918.
Bedia, J., Barrionuevo, R., Rodriguez-Mirasol, J., & Cordero, T. (2011). Ethanol dehydration to ethylene on acid carbon catalysts. *Appl Catal B, 103*(3-4), 302–310.
Broeren, M. L. M., Kuling, L., Worrell, E., & Shen, L. (2017). Environmental impact assessment of six starch plastics focusing on wastewater-derived and additives. *Resources, Conservation & Recycling, 127*(246), 255.
Cheroennet, N, Pongpinyopap, S, Leejarkpai, T, & Suwanmanee, U. (2017). A trade-off between carbon and water impacts in bio-based box production chains in Thailand: A case study of PS, PLAS, PLAS/starch, and PBS. *Journal of Cleaner Production, 167,* 987–1001.
Chu, J., Zhou, Y., Cai, Y., Wang, X., Li, C., & Liu, Q. (2022). Life-cycle greenhouse gas emissions and the associated carbon-peak strategies for PS, PVC, and ABS plastics in China. *Resources, Construction & Recycling, 182,* 106295–106306.
Comanita, E. D., Ghinea, C., Rosca, M., Simion, I. M., Petraru, M., & Gavrilescu, M. (2015). Impacts of polyvinyl chloride (PVC) production process. In *The 5th IEEE international conference on E-health and bioengineering. Grigore T. Popa University of Medicine and Pharmacy, Iaşi, Romania, November 19-21.*
Elmowafy, E., Abdal-Hay, A., Skouras, A., Tiboni, M., Casettari, L., & Guarino, V. (2019). Polyhydroxyalkanoate (PHA): Applications in drug delivery and tissue engineering. *Expert Review of Medical Devices, 16*(6), 467–482.
Franklin Associates. (2020). *Cradle-to-gate life cycle analysis low density polyethylene (LDPE) resin.* American Chemistry Council (ACC) Plastics Division.
Franklin Associates. (2021). *Cradle-to-gate life cycle analysis polyvinyl chloride (PVC) resin.* American Chemistry Council (ACC) Plastics Division.
Fukuoka, S., Fukawa, I., Tojo, M., Oonishi, K., Hachiya, H., Aminaka, M., Hasegawa, K., & Komiya, K. (2010). A novel non-phosgene process for polycarbonate production from CO_2: green and sustainable chemistry in practice. *Catalysis Survey from Asia, 14,* 146–163.
Gerngross, T. U. (2001). Polyhydroxyalkanoates: the answer to sustainable polymer production. *ACS Symposium Series, 784,* 10.
Gerngross, T. U., & Slater, S. C. (2000). How green are green plastic? *Scientific American, 283*(2), 36–41.
Gomes, T. S., Visconte, L. L. Y., & Pacheco, E. B. A. V. (2019). Life cycle assessment of polyethylene terephthalate packaging: An overview. *Journal of Polymers and the Environment, 27,* 533–548.
Howell, B. K. (1991). Life-cycle analysis: Polyvinyl chloride packaging. *British Food Journal, 93*(3), 23–29.
Ingarao, G., Licata, S., Sciortino, M., Planeta, D., Lorenzo, R. P., & Fratini, L. (2016). Life cycle energy and Co_2 emissions analysis of food packaging: An insight into the methodology from an Italian perspective. *International Journal of Sustainable Engineering, 10*(1), 31–43.
Kang, D. H., Auras, R., & Singh, J. (2017). Life cycle assessment of non-alcoholic single-serve polyethylene terephthalate beverage bottles in the state of California. *Resources, Conservation and Recycling, 116,* 45–52.
Kim, S., & Dale, B. E. (2005). Life cycle assessment study of biopolymers (polyhydroxyalkanoates) derived from no-tilled corn. *International Journal of Life Cycle Assessment, 10,* 200–210.

Liptow, C., & Tillman, A. M. (2012). A comparative life cycle assessment study of polyethylene based on sugarcane and crude oil. *Journal of Industrial Ecology, 16*(3), 420–435.

Mannheim, V., & Simenfalvi, Z. (2020). The life cycle of polypropylene products: Reducing environmental impacts in the manufacturing phase. *Polymers, 12*, 1–18.

Marten, B., & Hicks, A. (2018). Expanded polystyrene life cycle analysis literature review: An analysis for different disposal scenarios. *Sustainability, 11*(1), 29–35.

Papong, S, Malakul, P, Trungkavashirakun, R., Wenunun, P, Chom-in, T, & Nithitanakul, M. (2014). Comparative assessment of the environmental profile of PLA and PET drinking water bottles from a life cycle perspective. *Journal of Cleaner Production, 65*, 539–550.

Qadir, S. A., Al-Motairi, H., Tahir, F., & Al-Fagih, L. (2021). Incentives and strategies for financing the renewable energy transition: A review. *Energy Reports, 7*, 3590–3606.

Razza, F, Fieschi, M, Innocenti, F. D., & Bastioli, C. (2009). Compostable cutlery and waste managment: An LCA approach. *Waste Management, 29*(4), 1424–1433.

Ryberg, M. W., Hauschild, M. Z., Wang, F., Averous-Monnery, S., & Laurent, A. (2019). Global environmental losses of plastics across their value chains. *Resources, Conservation & Recycling, 151*, 104459.

Schwarz, A. E., Ligthart, T. N., Bizarro, D. G., De Wild, P., Vreugdenhil, B., & van Harmelen, T. (2021). Plastic recycling in a circular economy; determining environmental performance through an LCA matrix model approach. *Waste Management, 121*, 331–342.

Sheehan, J., Cmobreco, V., Duffield, J., Graboski, M., & Shapouri, H. (1998). *Life cycle inventory of biodiesel and petroleum diesel for use in an urban bus (final report)*. Department of Agriculture and U.S. Department of Energy.

Shen, L., & Patel, M. K. (2008). Life cycle assessment of polysaccharide materials: A review. *Journal of Polymer and the Environment, 16*, 154–167.

Simon, B., Amor, M. B., & Földényi, R. (2016). Life cycle impact assessment of beverage packaging systems: Focus on the collection of post-consumer bottles. *Journal of Cleaner Production, 112*(1), 238–248.

Suwanmanee, U., Varabuntoovit, V., Chaiwutthinan, P., Tajan, M., Mungcharoen, T., & Leejarkpai, T. (2012). Life cycle assessment of single use thermoform boxes made from polystyrene (PS), polylactic acid, (PLA), and PLA/starch: Cradle to consumer gate. *The International Journal of Life Cycle Assessment, 18*, 401–417.

Vink, E. T., Davies, S., & Kolstad, J. J. (2010). The eco-profile for current Ingeo polylactide production. *Industrial Biotechnology*, 167–180.

Vink, E. T. H., Glassner, D. A., Kolstad, J. J., Wooley, R. J., & O'Connor, R. P. (2007). The eco-profiles for current and near-future NatureWorks polylactide (PLA) production. *Industrial Biotechnology, 3*(1), 58–81.

Vink, E. T. H., Rabago, K. R., Glassner, D. A., & Gruber, P. R. (2003). Applications of lie cycle assessment to NatureWorks^{TM} polylactide (PLA) production. *Polymer Degradation and Stability, 80*, 403–419.

Zabaniotou, A., & Kassidi, E. (2003). Life cycle assessment applied to egg packaging made from polystyrene and recycled paper. *Journal of Cleaner Production, 11*, 549–559.

Zaki, M. R. M., & Aris, A. Z. (2022). An overview of the effects of nanoplastics on marine organisms. *Science of the Total Environment, 831*, 154757.

Zhao, L., Rong, L., Zhao, L., Yang, J., Wang, L., & Sun, H. (2020). *Microplastics in terrestrial environments* (pp. 423–445). Springer.

3

Plastic wastes and opportunities

3.1 Introduction

Plastic wastes have been an overwhelming global issue for decades. Every year, large quantities of plastics are made, to cater to the market demands for single-use packaging material or short-lived consumer products, which end up as significant amounts of wastes. During the COVID-19 pandemic period, medical wastes consisting of large quantities of plastic materials have been discarded without undergoing a recycling stage, due to hygienic reasons. Hence, managing plastic waste appropriately is crucial for modern society to minimize the environmental impacts, in line with circular economy principles.

In general, plastic recycling involves two major methods: chemical recycling (also known as pyrolysis or gasification processes) and mechanical recycling methods. However, not all plastic wastes can be recycled, as this depends on the extent of soiling when discarded, as well as the ability to segregate the plastic waste, since plastic with multilayer packaging, coatings, adhesives and other features cannot be recycled. In fact, regardless of whether chemical or mechanical recycling is used, segregation of plastic wastes is done according to the Resin Identification Code, as listed in Fig. 3.1. When the waste plastics are sorted properly, the mechanical properties and appearance of the recycled outputs are acceptable.

Commonly, nonrecyclable plastic waste will end up being incinerated or in a landfill. However, with modern incineration technology, the nonrecyclable plastics can be combusted to harvest thermal energy to operate turbines generating electricity. The successful role models are Singapore, Japan, Taiwan, China and the European Union. The emissions can be controlled to a safe level, while the fly ash is used for cement manufacturing or landfills. A study by Jeswani et al. (2021) compared life cycle environmental impacts of chemical recycling, mechanical recycling and energy recovery approaches on mixed plastic waste. As shown

Plastics and Sustainability. https://doi.org/10.1016/B978-0-12-824489-0.00007-6
Copyright © 2023 Elsevier Inc. All rights reserved.

Fig. 3.1 Resin identification code. *PET*, polyethylene terephthalate; *HDPE*, high-density polyethylene; *PVC*, polyvinyl chloride; *LDPE*, low-density polyethylene; *PP*, polypropylene; *PS*, polystyrene.

in Fig. 3.2, they reported that mechanical recycling has the lowest climate change impact (1.99 t CO_2 eq/t plastics) compared to energy recovery, which possesses the highest climate change impact (3.65 t CO_2 eq/t plastics). Meanwhile, there are small differences between chemical and physical recycling. Although the term energy recovery (commonly known as incineration) seems to hold promise, the results of the study show that life cycle energy for incineration (energy recovery) is slightly better than either of the recycling methods, with $<10\%$ difference. The mechanical recycling actually has lower climate change potential, energy use, acidification, freshwater eutrophication, marine eutrophication, photochemical ozone formation and human toxicity impacts than chemical recycling. This could be due to the fact that mechanical recycling is simpler and does not require a complicated catalytic process involving a variety of chemicals to turn the plastic wastes into basic chemicals, such as CO, ethylene or naphtha, for polymerization again. Also, greater energy is needed for the pyrolysis and purification processes than for mechanical recycling. This chapter explores chemical recycling, also known as the pyrolysis process, as well as the incineration of plastic wastes. Both are known as advanced technologies that promise better treatment of plastic wastes; chemical recycling in particular is attracting the attention of industrial players. Finally, chemical recycling and energy recovery are both considered to meet the circular economy principles of today's society.

3.2 Chemical recycling of plastic wastes

The pyrolysis of plastic wastes is also known as chemical recycling, which is to recover valuable chemicals through the application of thermal, catalytic and pressurized effects. The valuable

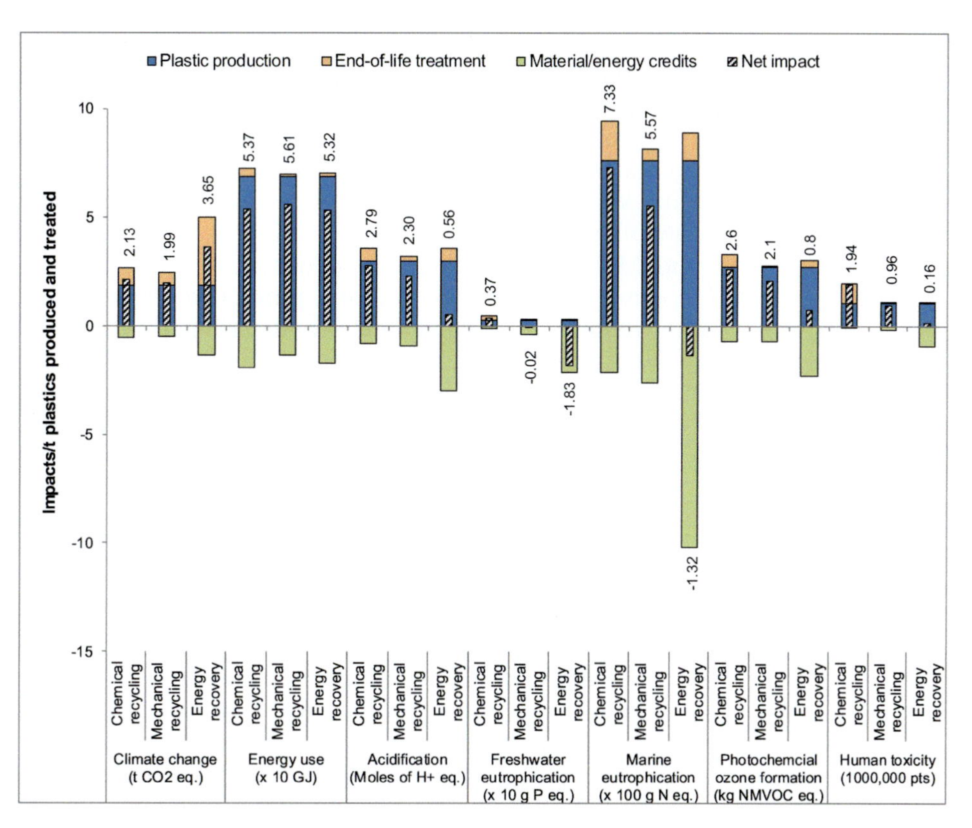

Fig. 3.2 Product and waste perspective: environmental impacts of producing mixed commodity plastics and their end-of-life treatment via chemical and mechanical recycling and energy recovery. 'Energy use' represents the EF 2.0 GaBi Software indicator 'resource use, energy carriers'. End-of-life treatment includes the impacts of collection and treatment of mixed plastic waste via chemical or mechanical recycling or energy recovery. Energy recovery: 30% incineration and 70% refuse-derived fuel. NMVOC: nonmethane volatile organic compounds. Some impacts have been scaled to fit. To obtain the original values, multiply with the factor shown on the *x*-axis against relevant impacts. From Jeswani, H., Kruger, C., Russ, M., Horlacher, M., Antony, F., Hann, S., Azapagic, A. (2021) Life cycle environmental impacts of chemical recycling via pyrolysis of mixed plastic waste in comparison with mechanical recycling and energy recovery. *Science and the Total Environment, 769*, 144483, with permission of Elsevier.

chemicals can then be used as the inputs to produce plastic and reactants for further petrochemical processes. In other words, pyrolysis converts the end-of-life plastic materials into smaller molecules. The heavy chain substances can be used as fuel oil (usually lower in calorific value) and also as bituminous materials for road building. Although chemical recycling offers the advantage of handling a wider spectrum of plastic waste selection with good economics, research and development is still ongoing to achieve better selectivity of product quality, flexibility of plastic waste requirements, and ability to handle the presence of additives and contaminants, to reduce potential impacts on the environment (Dogu et al., 2021).

As mentioned earlier, mechanical recycling has lower environmental impacts compared to chemical recycling and incineration. However, mechanical recycling has limitations, particularly with the following conditions:

(a) Multilayered plastic packaging from different types of plastic materials. This is commonly found in food packaging, which requires prolonged shelf life for resistance of moisture and gas permeation. Separation of multilayer plastic packaging is burdensome and costly.

(b) Unidentified additive contents of plastic materials. For example, plasticized and unplasticized poly(vinyl chlorides) (PVCs) cannot be mixed in mechanical recycling, even if from similar types of PVC. Otherwise, the resulting physical recycled PVC tends to exhibit significant variations in mechanical properties.

(c) Mechanical recycling can be suitable for recycling preconsumer plastic wastes, e.g. plastic waste from a rejected blow moulding process of a beverage factory that produces polypropylene bottles, which can be crushed and re-produced into other beverage bottles. In contrast, postconsumer plastics waste, e.g. plastic bottles found in the municipal waste, can be full of dirt, oils, and other residues. Such postconsumer plastic bottles need to be cleaned with large amounts of water and detergents, causing water pollution as well.

In general, chemical recycling of plastic wastes strongly contributes to the circular economy, by eliminating waste and regenerating the raw materials originating from plastics made earlier. Thus this is inline with the concept of the closing 'loop' of plastic consumption and recycling. Although chemical recycling has a promising process cycle to treat plastic wastes, the technology readiness level remains questionable; the European Union in particular has set a target to recycle all plastics by 2030. A study by Solis and Silveira (2020) commented that the chemical recycling technologies with the highest readiness are pyrolysis, catalytic cracking and conventional gasification. Until now, there have been limited chemical recycling technologies operating on a large scale and available for long-term evaluation. For instance, as reported by Lee et al. (2021), Germany was one of the earliest operators of chemical recycling plants, by Berrenrath and Schwarze Pumpe. Chemical recycling plants Berrenrath in West Germany (1986–97) and SVZ Schwarze Pumpe in East Germany

(1995–2007) used gasification technologies to convert different types and mixtures of carbonaceous waste, including unsorted municipal solid wastes, plastic wastes, tar and oil residues, waste wood, sewage sludge and others. The residues, mainly consisting of carbonaceous waste, were mixed with coal as a co-feedstock into syngas. Eventually, 300–350 tons of methanol per day were supplied as a basic chemical to the chemical industry (Lee et al., 2020). Nevertheless, after about a decade of operation, both plants ceased production, due to the main factors of high operating and maintenance costs for the complex gasification facilities. Moreover, the increasing numbers of waste incineration plants have threatened the economic attractiveness of chemical recycling plants. Nevertheless, chemical recycling is still attracting a great deal of attention from researchers globally. The reason is that chemical recycling technology has been much improved over decades of research and development, and it no longer involves simple thermal cracking. The chemical recycling technology has developed to include plasma pyrolysis, microwave-assisted pyrolysis, catalytic cracking, hydrocracking, plasma gasification and pyrolysis with in-line reforming. The efficiency, selectivity of outputs, energy requirements and environmental friendliness have been significantly improved to cater to investor and policymaker expectations for greener technology.

3.2.1 Pyrolysis process and technology

In general, the pyrolysis processing of plastic wastes involves a complicated operation and requires professional skills. Indeed, most of the time pyrolysis is selected to treat plastic waste mainly due to inability of the plastic waste to undergo mechanical recycling: for instance, degraded plastics, mildly contaminated plastics and municipal solid wastes are difficult to sort with mixed plastics, particularly multilayer packaging. For most of the heavily contaminated plastic wastes and medical plastic wastes, incineration is the safest approach to minimize the hazards and it also offers the energy recovery benefit. Importantly, the main requirement of plastic waste pyrolysis is that the source or type of plastic waste needs to be consistent, because the selection of process parameters and catalysts is tailored to ensure optimum product production and specification. In this section, the main types of pyrolysis processes are discussed to identify the advantages of each approach, together with examples of technology operators, so that readers can obtain better understanding of pyrolysis technology development and operation.

(1) Thermal cracking

(a) *Process description*

Thermal cracking is the conventional pyrolysis process that manipulates parameters of temperature, pressure and residence time to depolymerize the waste plastics in the absence of oxygen. Pyrolysis in general is a complicated reaction process and requires a high amount of energy. The thermal cracking of waste plastics produces gas, char and liquid oil. In most of the cases, pyrolysis oil is the preferable product, with its good calorific value, to be blended with petroleum fuel oil to operate industrial boilers. Meanwhile, the produced gases are also useful as they can be recycled into the pyrolysis stream again to fuel the heating process. A simple thermal cracking process of polypropylene and polystyrene has been demonstrated by Angyal et al. (2007), seen in Fig. 3.3. It was found that the changes of the polystyrene portion in the thermal cracking process can lead to variation of product composition (carbon number), as shown in Fig. 3.4. In fact, this is the limitation of pyrolysis when being fed with multilayer plastic packaging and plastics with impurities (e.g. additives, colourants, fillers and others). The pyrolysis process can be disrupted when chlorine compounds are

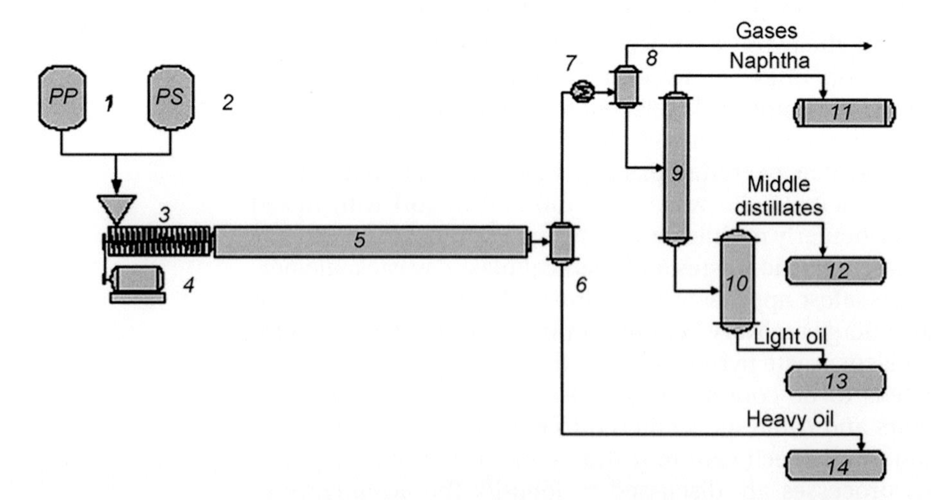

Fig. 3.3 Cracking setup: (1) polypropylene storage tank, (2) polystyrene storage tank, (3) extruder, (4) electric motor, (5) reactor, (6) separator, (7) cooler, (8) liquid-gas separator, (9) distillation column, (10) vacuum distillation column, (11–14) product storage. From Angyal, A., Miskolczi, N., Bartha, L. (2007). Petrochemical feedstock by thermal cracking of plastic waste. *Journal of Analytical and Applied Pyrolysis, 79*, 409–414, with permission from Elsevier.

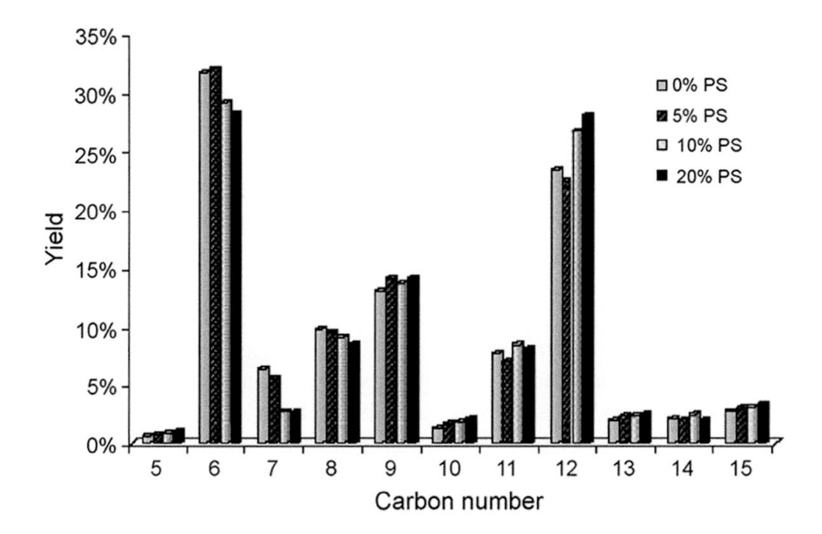

Fig. 3.4 The composition of aliphatic hydrocarbons of the naphtha-like fraction. From Angyal, A., Miskolczi, N., Bartha, L. (2007). Petrochemical feedstock by thermal cracking of plastic waste. *Journal of Analytical and Applied Pyrolysis, 79*, 409–414, with permission from Elsevier.

added, which can generate corrosive substances that damage the equipment. In recent years, a wider range of pyrolysis technology, such as catalysis pyrolysis, plasma pyrolysis and microwave pyrolysis, has been developed to overcome the shortcomings of the thermal cracking process.

(b) *Advantages and disadvantages*

Advantages: Simple operation with product yield being highly dependent on the selection temperature, pressure, residence time and heating rate.

Disadvantages: The product quality is highly dependent on the feedstocks. Difficult to handle halogen compounds present in waste plastics, which are corrosive.

(c) *Example of technology operator*

Plastic Energy is the technology provider for a thermal cracking process that targets low-density polyethylene, high-density polyethylene, polystyrene and polypropylene. They operate two plants located in Sevilla and Almeria, Spain (Solis & Silveira, 2020). In addition, in Japan a thermal cracking plastic waste plant can be found at Sapporo Plastic Recycling and Mogami Kiko.

(2) Plasma pyrolysis

(a) *Process description*

Plasma pyrolysis is an improvement to the thermal cracking process in which the operation temperature can be as high as 2000–9000°C at an ultrafast heating rate of 0.1–0.5 s. This is compared to thermal cracking, which usually occurs at a temperature of 1000°C or below; when the waste plastics are heated at several thousand degree Celsius, they decompose into monomers or small molecules with limited high hydrocarbon composition present. As a result, plasma pyrolysis is suitable for use to treat multilayer waste plastics and most of the additives can be decomposed except for mineral types. The resulting good quality syngas is suitable for direct use to run gas turbines for power generation. A simple laboratory scale plasma pyrolysis process is shown in Fig. 3.5. Graphite and carbon electrodes are located in the gasifier and supplied with high voltage to generate ultrahigh heat to transform the waste plastic into combustible gas. Meanwhile, slag removal is installed at the bottom to conduct nongaseous product removal from waste plastics. The remainder consists of the separation process and gas collection steps, depending on requirements. The analysis of the products indicates the presence of CH_4, C_2H_6, CH_3H_8, C_4H_{10}, C_2H_2 and traces of higher hydrocarbons.

(b) *Advantages and disadvantages*

Advantages: When the waste plastic is heated to an ultrahigh temperature, the plastic is decomposed into monomers or simple molecules with good calorific values. Theoretically, this pyrolysis method is suitable to handle a variety of waste plastics from different types of plastics, particularly multilayer packaging. Moreover, this is not limited to plastic molecules but also includes a wide range of additives, particularly halogen compounds. Examples of halogen additive compounds can usually be found in the plastic products with flame retardants added.

Disadvantages: This pyrolysis process can be costly to operate due to the electricity required to operate the ultrahigh temperature electric arc pyrolyser. In addition to the syngas produced from the process, there are also slag side products. The composition of the slag is dependent on the nondecomposable minerals, such as calcium carbonate, kaolin clay, transition metal oxides, etc. The waste handling of the slag can be troublesome, as slags are categorized as scheduled wastes in many countries.

(c) *Example of technology operator*

The Facilitation Centre for Industrial Plasma Technologies of the Institute of Plasma Research, Gandhinagar,

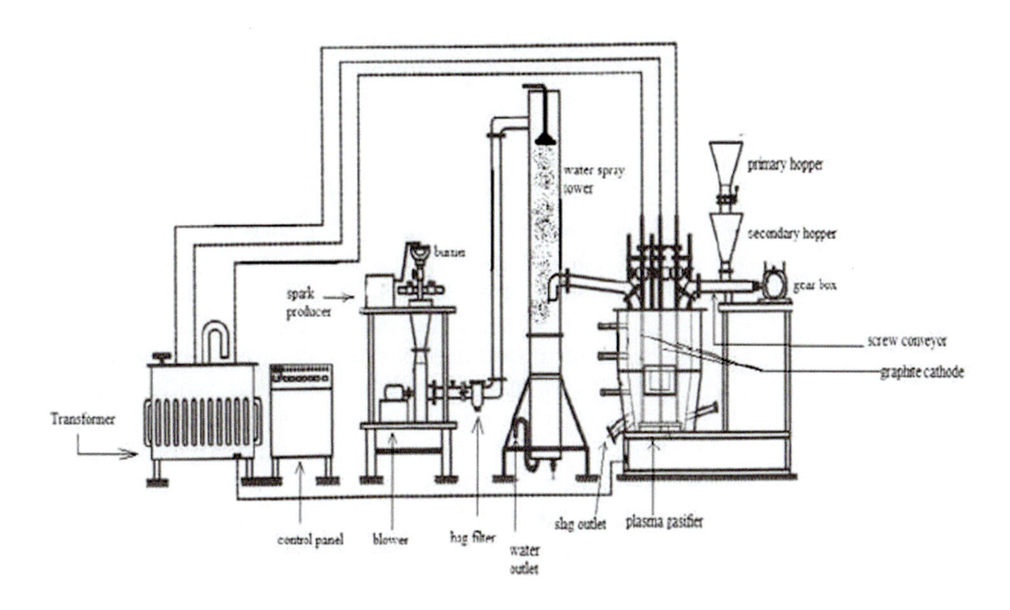

Fig. 3.5 Waste plastic plasma pyrolyzer (Top); Schematic representation (Bottom). Actual laboratory scale waste plastic plasma pyrolyzer. From Punčochář, M., Ruj, B., Chatterjee, P. K. (2012). Development of process for disposal of plastic waste using plasma technology and option for energy recovery. *Procedia Engineering, 42,* 420–430, with permission from Elsevier.

Gujarat, India operates a plasma pyrolyser to handle the disposal of hospital wastes, including PVC blood bags, rubber gloves, catheters and others.

(3) **Hydrocracking**

(a) *Process description*

On top of the thermal cracking pyrolysis previously described, hydrocracking involves the addition of hydrogen during the cracking process, so that a better quality of products can be obtained. The hydrogen reacts during the plastic liquefaction stage to reduce unsaturated hydrocarbon compounds, providing reduction of oxygen-rich compounds while improving the quality and calorific values of the petroleum products. In order to promote the reactivity of hydrogen to the liquefaction, the hydrogen is added at a high pressure of 70 atm and a temperature of 375–400°C in the presence of various acidic catalysts, such as Ni/S or NiMo/S supported catalysts and zeolite (Ding, Liang, & Anderson, 1997) and bifunctional platinum catalysts (González-Marcos, Fuentes-Ordóñez, Salbidegoitia, & González-Velasco, 2021; Jumah, Anbumuthu, Tedstone, & Garforth, 2019; Munir, Abdullah, Piepenbreier, & Usman, 2017; Takashi, 2009).

(b) *Advantages and disadvantages*

Advantages: High-quality liquid oil is produced for electricity power generation.

Disadvantages: A local hydrogen generation plant needs to be set up to handle the processing. It is not suitable to handle halogen-rich inputs and is easily deactivated by compounds of nitrogen, sulphur and impurities in waste plastics, which can cause poisoning of the catalyst and are corrosive to the processing environment.

(c) Example of technology operator

Laboratory scale operation.

(4) **Microwave-assisted pyrolysis**

(a) *Process description*

Microwave-assisted pyrolysis was developed mainly to overcome the shortcoming of low efficiency of the heating process in the conventional pyrolysis process. In fact, the controlling of heating steps is of the utmost importance to ensure that good quality products can be obtained. When microwaves are being applied, they induce vibrations of the plastic molecules, causing the heat to be absorbed and allowing better conversion of electrical energy into the targeted components under a proper heat rate and high temperature. Nevertheless,

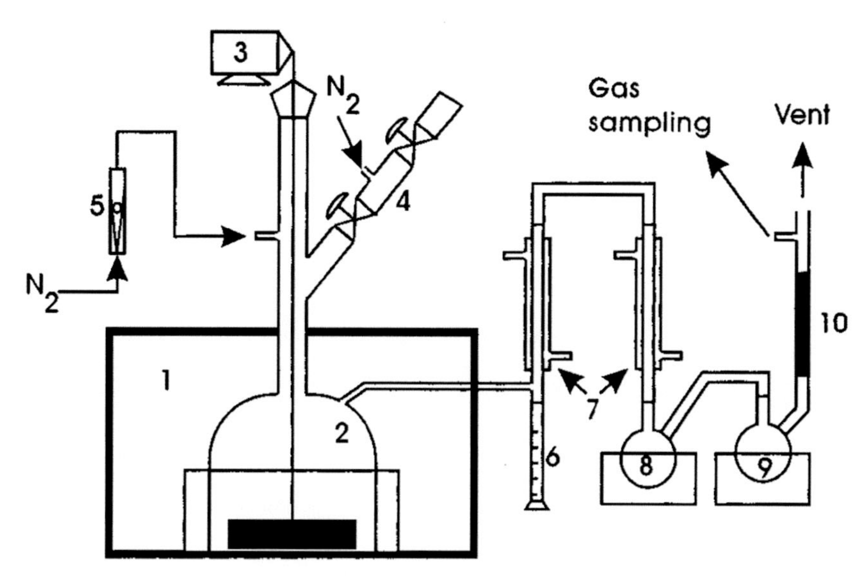

Fig. 3.6 Microwave-assisted pyrolysis apparatus: (1) Modified catering microwave oven, (2) quartz vessel of 180 cm diameter, (3) motor with maximum speed of 6 rpm, (4) top feeder, (5) purging/carrier nitrogen gas using rotameter, (6) main collection flasks, (7) two water-cooled Liebig condensers, (8 and 9) cold traps the uncondensable gas flow through and (10) cotton wool filter to collect any aerosol present. From Ludlow-Palafox, C., Chase, H. A. (2001). Microwave-induced pyrolysis of plastic wastes. *Industrial & Engineering Chemistry Research, 40*, 4749–4756, with permission of ACS Publications.

microwave effects can only work on chemically polar types of plastics and are limited for polyethylene and polypropylene, which have poor dielectric strength. Subsequently, different absorbents can be added, for instance, mixing with polystyrene, which can absorb heat and transfer it to polyethylene. As reported by Ludlow-Palafox and Chase (2001), microwave-induced pyrolysis can deal with laminate types of waste plastic: for example, toothpaste tubing with aluminium layers can be recovered together with the hydrocarbons. The setup for microwave-assisted pyrolysis is shown in Fig. 3.6, where the pyrolysis system is able to separate aluminium from the plastic laminate. The solid aluminium traces can be sieved out from the leftover carbon in the quartz reactor, while other traces of white pigment, denoted as titanium dioxide, can also be found adhered to the wall of the reactor. The recovered hydrocarbons from the laminate and HDPE pellets showed they produced almost identical syngas (refer Fig. 3.7). This indicates that microwave-assisted pyrolysis is a suitable targeted pyrolysis approach to treat multilayer waste plastics.

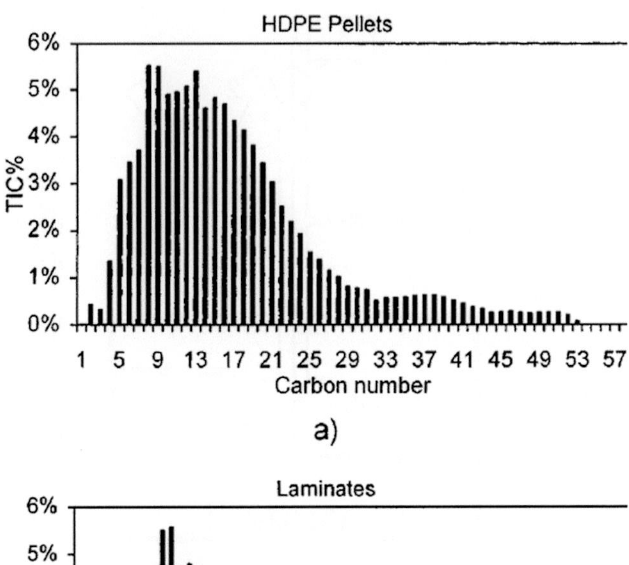

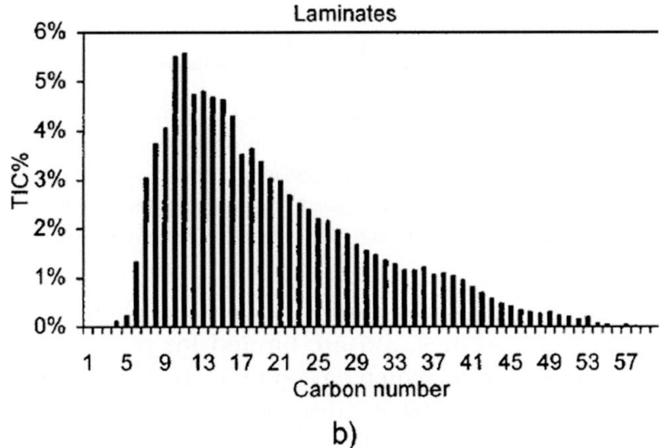

Fig. 3.7 Carbon number distribution for pyrolysis at 500°C without carrier nitrogen gas of (A) High-density polyethylene pellets and (B) aluminium/polymer laminate (toothpaste tube). From Ludlow-Palafox, C. and Chase, H. A. (2001). Microwave-induced pyrolysis of plastic wastes. *Industrial & Engineering Chemistry Research, 40,* 4749–4756, with permission of ACS Publications.

(b) *Advantages and disadvantages*

Advantages: Good heat distribution, faster reaction and suitable to treat multilayer waste plastics with high-temperature effects.

Disadvantages: Highly dependent on the microwave-sensitive types of plastics and required absorbents to increase the thermal effect on the waste plastics.

(c) *Example of technology operator*

Laboratory and pilot scale operation.

(5) Catalytic cracking

(a) *Process description*

Catalytic cracking is one of the most popular research areas in pyrolysis, since the catalyst can accelerate the decomposition reaction, with better yield and selectivity and, most importantly, reduction in the energy cost. Almeida and Fátima Marques (2016) listed the benefits of catalysts in pyrolysis:

- Lower energy required to initiate the decomposition by using lower temperature
- Increased competitiveness with reduced cost of operation
- Good yield and selectivity of high-value products
- Shortened residence time, so the volume of the reactor can be reduced with lower capital investment.
- Inhibited formation of cyclic hydrocarbons, and aromatic and branched hydrocarbon products

A detailed comparison of the products of thermal cracking and catalytic cracking was made by Seo et al. (2003) with the apparatus setup as shown in Fig. 3.8; the products are listed in Table 3.1 for high-density polyethylene. An obvious observation is that the gaseous

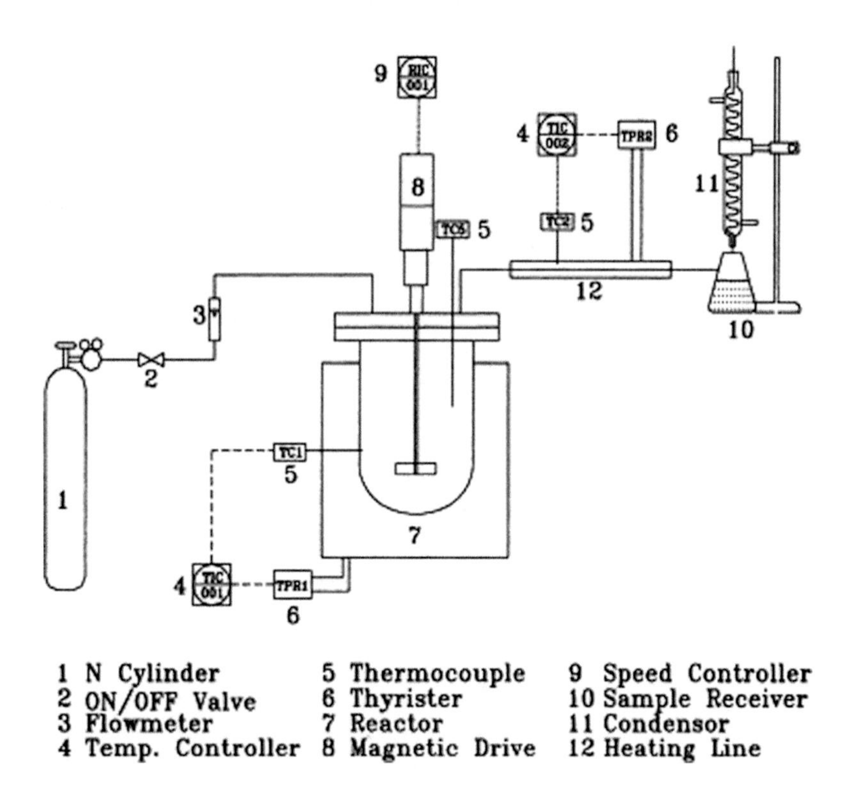

Fig. 3.8 Schematic diagram of apparatus setup for catalytic pyrolyser by Seo et al. (2003). From Seo, Y.-H., Lee, K.-H., Shin, D.-H. (2003). Investigation of catalytic degradation of high-density polyethylene by hydrocarbon group type analysis. *Journal of Analytical and Applied Pyrolysis, 70*, 383–398, with permission of Elsevier.

Table 3.1 Products of liquid, gas, coke and liquid when HDPE undergoes various types of catalysts at 450°C.

Yield of products	Liquid (%)	Gas (%)	Coke (%)	Liquid		
				C_6–C_{12} (%)	*C_{13}–C_{23} (%)*	*$>C_{24}$ (%)*
Thermal cracking only	84.00	13.00	3.00	56.55	37.79	5.66
ZSM-5 (powder)	35.00	63.50	1.50	99.92	0.08	0
Zeolite-Y (powder)	71.50	27.00	1.50	96.99	3.01	0
Zeolite-Y (pellet)	81.00	17.50	1.50	86.07	11.59	2.34
Mordenite (pellet)	78.50	18.50	3.00	71.06	28.67	0.27
Silica-alumina (powder)	78.00	18.50	3.00	71.06	28.67	0.27
Alumina (powder)	82.00	15.90	2.10	53.02	43.27	3.71

Zeolite catalyst

ZSM-5 (powder): BET surface area 455 m^2/g, pore size (Å) 5.4 × 5.6
Zeolite-Y (powder): BET surface area 873 m^2/g, pore size (Å) 7.4
Zeolite-Y (pellet): BET surface area 586 m^2/g, pore size (Å) 7.4
Mordenite (pellet): BET surface area 443 m^2/g, pore size (Å) 6.7 × 7.0

Nonzeolite catalyst

Silica-alumina (powder): BET surface area 455 m^2/g
Alumina (powder): BET surface area 455 m^2/g

products for catalyst-added pyrolysis are in larger quantities compared to thermal cracking only. Also, in the liquid composition, low hydrocarbons C_6–C_{12} make up a major part of the composition except for alumina (powder). Selecting highly acidic types of catalysts is preferable to achieve a high composition of gaseous components; zeolite in particular exhibits excellent catalytic efficiency on cracking, isomerization and aromatization due to a strongly acidic property and a micropore crystalline structure. Sahu et al. (2014) conducted a feasibility study on the catalytic cracking of waste plastic to produce fuel oil with amorphous silica-alumina selected, to achieve a conversion of 94.36%. The economic analysis showed that a typical plant with an annual feed rate of 120,000 tons required a capital cost of USD $58,591,260, whilst the rate of return could be 35.97%. Nevertheless, the feasibility of catalytic cracking is not merely dependent on the profitability, since the type of waste plastic supply can affect products. For instance, when the waste plastic is polyvinyl chloride or polyethylene terephthalate, additional processing is needed to remove hydrogen chloride and benzoic acids, respectively, which tend to cause a deficiency in the process (Fukushima et al., 2009).

(b) *Advantages and disadvantages*

Advantages: Better in selectivity of low carbon number products, energy cost savings and high production with lower operating temperature and residence time.

Disadvantages: High value of catalysts for operation. Requires further care for waste plastic containing halogen compounds that can cause poisoning of catalysts.

(c) *Example of technology operator*

One of the best-known operators of catalytic cracking of waste plastic technology is Sapporo Plastics Recycling Co., Ltd., who started waste plastics liquefaction in 2000 using the fluid catalytic cracking (FCC) catalyst for their pyrolysis process. The major component of the FCC catalyst is crystalline zeolite. Sapporo Plastics Recycling operates two commercial plastic liquefaction facilities in Japan, which are Niigata Plastics Liquefaction Centre (6000 tons/year) and Sapporo Waste Plastics Liquefaction Plant (14,800 tons/year) (Fukushima et al., 2009). One of the main focuses of this technology is the ability to process municipal waste plastics that contain polyvinyl chloride (PVC). The waste plastics are transformed into pellet form and fed into the dichlorination process by initially heating to 300°C to melt and then the temperature is further increased to 1200°C. The generated hydrochloric gas is captured by dissolving in the water and is sold for alkali neutralization, while the hydrocarbon gases are collected and distilled accordingly for selling as fuel, as shown in Fig. 3.9 for the pyrolysis process of Sapporo Plastic Recycling Plant. However, an obstacle of this pyrolysis process is that, when cracking polyethylene terephthalate (PET) waste plastics, the equipment is susceptible to corrosion and clogging. The reason for this is that benzoic acid $C_6H_5(COOH)$ and terephthalic acid $C_6H_4(COOH)_2$ are formed during the thermal decomposition of PET. The solution for this problem is to add hydrated lime in the process in order to neutralize the acidic condition.

(6) Gasification and plasma gasification

(a) *Process description*

Gasification has been studied for a long time for various types of solid matter, such as biomass, organic waste, sludges, and waste plastics. The gasification process converts the solid into a simple gaseous mixture of

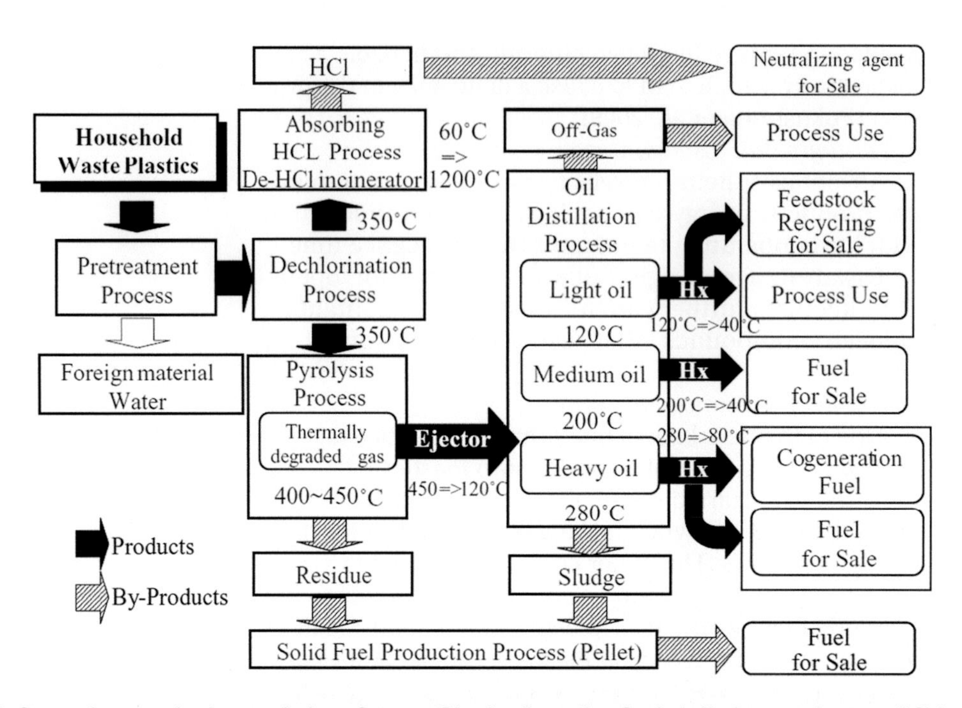

Fig. 3.9 Steps of waste plastics pyrolysis at Sapporo Plastics Recycling Co. Ltd. *Hx*, heat exchanger; *HCl*, hydrogen chloride. From Fukushima, M., Shioya, M., Wakai, K., Ibe, H. (2009). Toward maximizing the recycling rate in a Sapporo waste plastics liquefaction plant. *Journal of Material Cycles Waste Management, 11*, 11–18, with permission from Springer Nature.

carbon monoxide (CO), hydrogen (H_2), carbon dioxide (CO_2) and methane (CH_4) through a partial oxidation reaction. In fact, this is also the main difference between pyrolysis and gasification, with pyrolysis being conducted in a totally inert environment, while gasification is carried out in an environment with a low concentration of oxygen, which is very much below the stoichiometric requirement for combustion to occur. The gasification of waste plastics is always treated as the most effective solution for highly contaminated waste or unsorted plastic waste. At a temperature of 700–1500°C for conventional gasification or up to 15,000°C for plasma gasification, a majority of the organic compound decomposes into syngas of H_2/CO. The syngas is used to produce basic products like methanol, gasoline, gasoline additives, formaldehyde, acetic acid and ethylene, among others.

(b) *Advantages and disadvantages*

Advantages: The conventional gasification process is well-established technology and it is suitable for the production of high-quality hydrogen and methane from a wider spectrum of plastic wastes, including multilayered packaging. By using steam or a supply of high purity oxygen gas, gasification can produce high calorific-value syngas (with the absence of atmospheric nitrogen). However, the cost of operation is higher with the use of oxygen. Meanwhile, plasma gasification uses low-quality plastic wastes to produce high-quality gases using the ultrahigh temperature of plasma torches.

Disadvantages: The supply of steam and oxygen requires an on-site facility and the cost of consumption is high. Also, the plasma torches require high electrical voltage, which is expensive. In return, the gases produced are more feasible to be used as feedstock for petrochemical production after a separation process to achieve high-purity gases and transport to the nearest petrochemical plants for utilization. In other words, this is a more complicated process. On the other hand, for energy generation using the gas products of conventional gasification, a localized power generation turbine can be set up and the generated electricity fed into the national electrical grid to generate revenue.

(c) *Example of technology operator*

The gasification technology developed by Enerkem is made up of a bubbling fluidized bed reactor with a front-end feeding system. There are currently four Enerkem gasification facility sites, with on-going construction or plans for more operations: Varennes, Canada (Input: 200,000 metric tons of waste to produce 125 million litres of biofuels and renewable chemicals (under construction)); Edmonton, Canada (100,000 dry tons of waste per year to produce 38 million litres of biofuels and renewable chemicals per year (under commissioning)); Rotterdam, The Netherlands (360,000 tons of waste to produce 80,000 renewable products (in planning stage)); Tarragona, Spain (400,000 metric tons of waste to produce 220 million litres of biofuels (in planning stage)). In addition, Showa Denko, Kawasaki, Japan has operated a gasifier with a daily treatment capacity of 195 tons since 2003. This pressurized gasification technology was developed by EBARA coupled fluidized-bed low-temperature gasification furnace, with Ube Industries' high-temperature gasification furnace (EBARA, 2021). The advantages of this technology include the ability to thermally decompose plastic wastes together with other high-calorific value wastes to produce syngas of CO and H_2. This setup has minimal concerns about generation of dioxins from a high-temperature gasification process. The by-products of fly ash are recovered as molten slag for cement manufacturing purposes. Other operators include Rheinbraun, Berrenrath (1985–97) which treated 700 tons per day (tpd) coat-waste mixture to produce 350 tpd methanol; SVZ, Schwarze Pumpe (1991–2007) treated 1300 tpd coal (75%)-waste mixture to produce 300 tpd methanol (Lee et al., 2018).

3.3 Incineration of plastic wastes and energy recovery

According to the European Waste Framework Directive 2008/98/EC, the hierarchy of waste management is Reduce, Reuse, Recycle, Recover, and finally Landfill (Voss et al., 2021). In other words, the incineration of plastic wastes is deemed to be a last option before going to landfill. Hence, inline with waste management hierarchy, the policymakers are focusing on enacting policies related to recycling of plastic waste as the priority, although they have studied the difficulties and obstacles in both mechanical and chemical recycling. The recycling approach is suitable primarily when the waste plastics are nicely segregated, with low to moderate contamination, whereas incineration is intended mainly to handle those waste plastics with high contamination or medical wastes that can cause hygiene problems if recycled. In fact, during the COVID-19 pandemic, the capacities of worldwide incineration facilities have been stretched to the limit (Jing, 2020). For instance, during the COVID-19 outbreak, municipal solid wastes in Wuhan, China from designated hospitals, Fangcang shelter hospitals, isolation locations and residential areas (e.g. face masks, test kits) were collected and categorized as healthcare waste due to high infectiousness and strong survivability of COVID-19. According to Yang, Lu, et al. (2021) and Yang, Yu, et al. (2021), the average production of healthcare wastes per 1000 persons in Wuhan increased tremendously from 3.64 to 27.32 kg/day. Stationary facilities, mobile facilities, coprocessing facilities (incineration plants for municipal solid waste) and nonlocal disposal were consecutively utilized to improve the disposal capacity, which was from 50 to 280.1 tons/day. The waste treatment facilities, particularly incineration, were in high demand during the COVID-19 outbreak period (refer to the chronology in Fig. 3.10). This is because

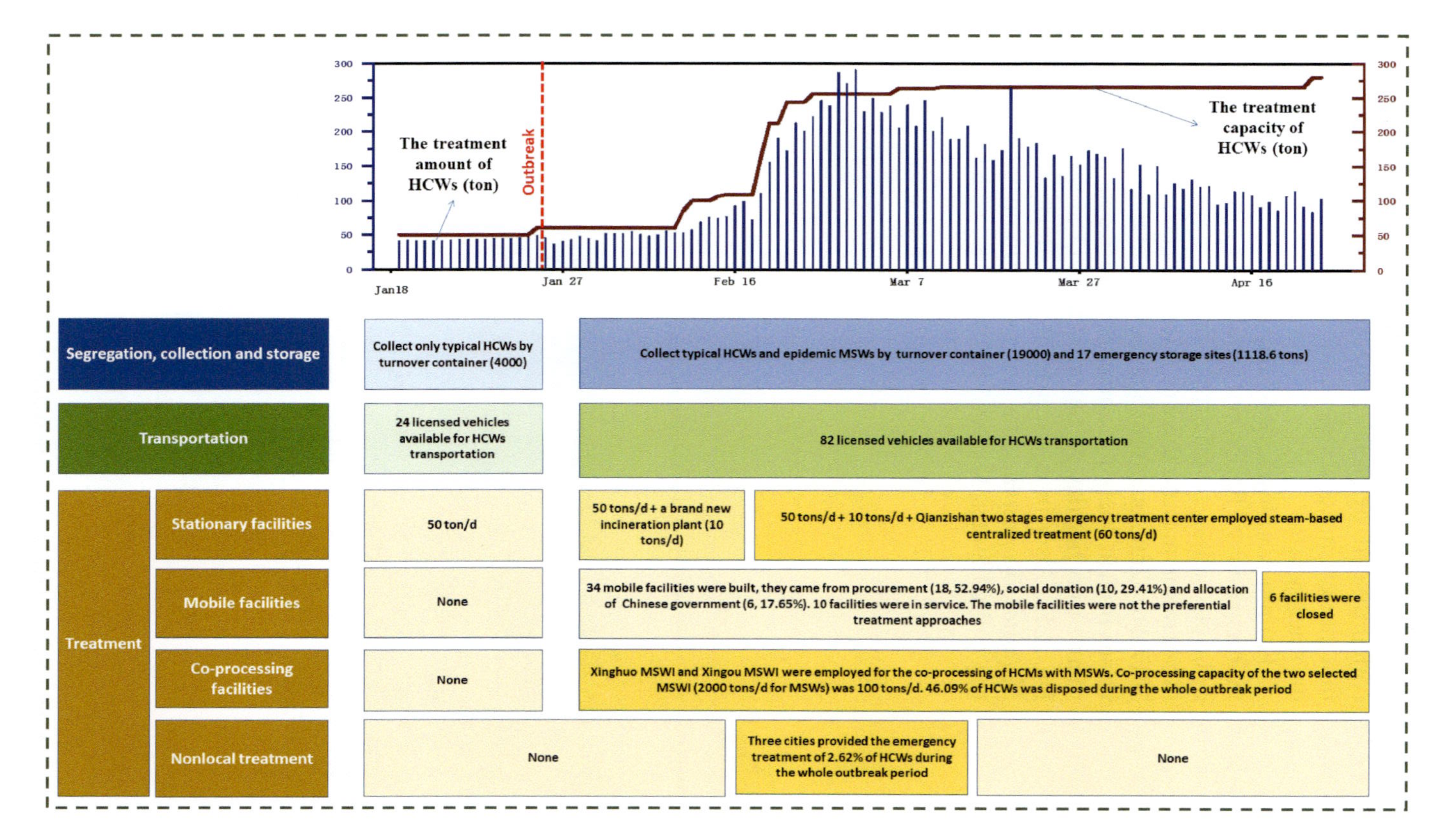

Fig. 3.10 The chronology of healthcare waste management related to incineration during COVID-19 outbreak in Wuhan. Note: *HCW*, health care waste. Reproduced with Elsevier permission from Yang, Z., Lu, F., Zhang, H., Wang, W., Shao, L., Ye, J., He, P. (2021) Is incineration the terminator of plastics and microplastics. *Journal of Hazardous Materials, 401*, 123429; Yang, L., Yu, X., Wu, X., Wang, J., Yan, X., Jiang, S., Chen, Z. (2021). Emergency response to the explosive growth of health care waste during COVID-19 pandemic in Wuhan, China. *Resources, Conservation and Recycling, 106*, 105074.

incineration is the safest technology to destroy the deadly virus and germs, by burning off at high temperature. Otherwise, the contaminated personal protection equipment waste, if buried in a landfill, poses a risk of environmental exposure to the virus or other germs.

From a worldwide perspective, the incineration technology of municipal waste has grown rapidly. For instance, the studies done by Lu et al. (2017) revealed that the total capacity of municipal waste incineration around the world was 707,805 metric tons per day in 2017. For China only, the total capacity of incineration of municipal waste, both in operation and under construction, was about 439,000 metric tons per day, as reported in 2017 at the location shown in Fig. 3.11. Also, in a news report by Royte (2019) published by *National Geographic*, the European Union incinerated 42% and the United States 12.5% of their wastes, which is inline with the

Fig. 3.11 Current and ongoing project for municipal waste incineration in China, 2017 (Lu et al., 2017).

initiative to restrict landfilling of organic wastes. Compared to the incineration process of many decades ago, in which the burning of plastics could lead to emissions of harmful gases and compounds such as dioxins and furans in flue gas, many techniques are available nowadays for the treatment of flue gas emissions, depending upon the type of feed stocks. For instance, sulphur compounds can be used, e.g. $(NH_4)_2SO_4$ and pyrite (FeS_2), to treat the flue gases, or the operating conditions of incineration can be changed (Mukherjee et al., 2016) to control the quality of the emissions. Incineration technology is undeniably a popular method to treat waste plastics, as compared to landfilling methods.

3.3.1 Incineration equipment and technology

In its general principle of operation, incineration is about the combustion process, which requires a supply of air to assist in complete combustion of the wastes. There is variety in incineration technology, such as moving grate incinerators, rotary kiln incinerators and fluidized bed incinerators, and they can be found widely in the market from technology providers. The main difference in the technology is the technique of air supply to combusted wastes at ultrahigh temperatures, to minimize generation of toxic flue gases. The energy generated will be used for boilers to produce steam to operate turbines to generate electricity. It is important to mention that the incineration of waste is governed by the Waste Incineration Directive, which is known as Directive 2000/76/EC of the European Parliament and of the Council of 4 December 2000 on the incineration of waste. This Directive states that the operating temperature needs to be kept to at least 850°C. Article 6 of the Directive says:

> *Incineration plants shall be designed, equipped, built and operated in such a way that the gas resulting from the process is raised, after the last injection of combustion air, in a controlled and homogeneous fashion and even under the most unfavourable conditions, to a temperature of 850°C, as measured near the inner wall or at another representative point of the combustion chamber as authorised by the competent authority, for two seconds. If hazardous wastes with a content of more than 1% of halogenated organic substances, expressed as chlorine, are incinerated, the temperature has to be raised to 1100°C for at least two seconds.*
>
> *Each line of the incineration plant shall be equipped with at least one auxiliary burner. This burner must be switched on automatically when the temperature of the combustion gases after the last injection of combustion air falls below 850°C or 1100°C as the case may be.*

It shall also be used during plant start-up and shut-down operations in order to ensure that the temperature of 850°C or 1100°C as the case may be is maintained at all times during these operations and as long as unburned waste is in the combustion chamber.

During start-up and shut-down or when the temperature of the combustion gas falls below 850°C or 1100°C as the case may be, the auxiliary burner shall not be fed with fuels which can cause higher emissions than those resulting from the burning of gasoil as defined in Article 1(1) of Council Directive 75/716/EEC, liquefied gas or natural gas.

The Directive also includes other stringent requirements for operation of incinerators (Directive 2000/76/EC, 2000).

3.3.1.1 Moving grate incinerator

The moving grate incinerator is the common incinerator to treat municipal waste. The moving grate, as shown as Item 5 in Fig. 3.12, is mainly to move the wastes, reducing the size of the waste while enabling the air to distribute well in the waste

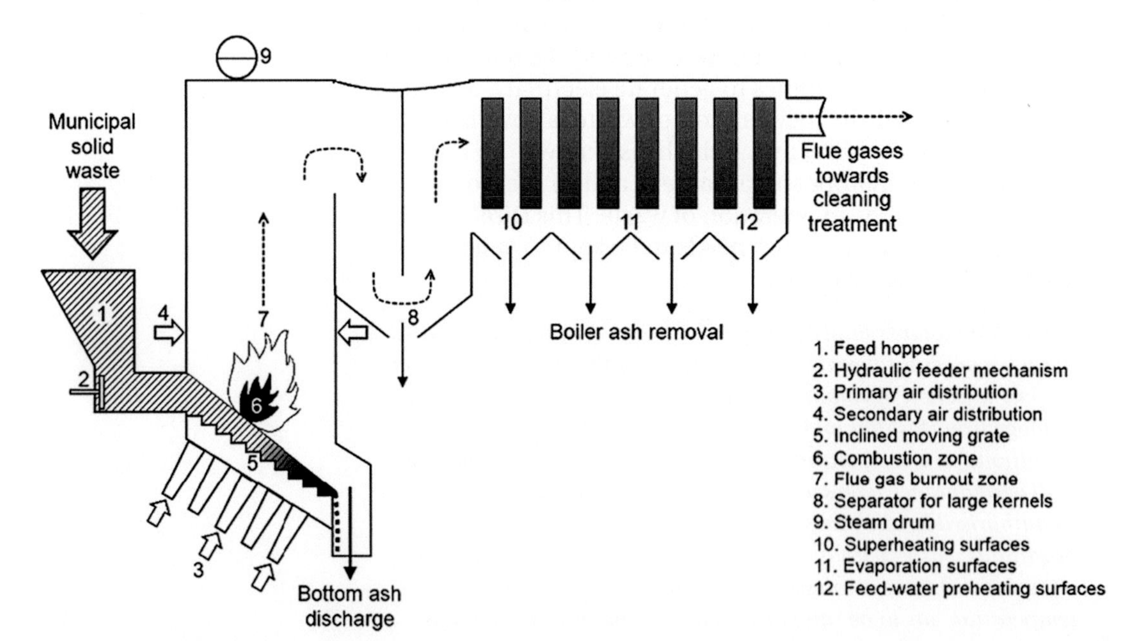

Fig. 3.12 Workflow of moving grate incinerator. From Malinauskaite, J., Jouhara, H., Czajczynska, D., Stanchev, P., Katsou, E., Rostkowski, P., Thorne, R. J., Colon, J., Ponsa, S., Al-Mansour, F., Anguilano, L., Krzyzynska, R., Lopez, I. C., Vlasopoulos, A., Spencer, N. (2017). Municipal solid waste management and waste-to-energy in the context of a circular economy and energy recycling in Europe. *Energy, 141*, 2013–2044, with permission of Elsevier.

particles to achieve complete combustion. The waste is being fed through the hopper, followed by the wastes moving along the descending grate, combusting and finally reaching the ash pit. During the movement of the wastes on the grates, the air is pumped through the grate. In addition to the air being used for the combustion process, the supply of the air is also being used to cool and maintain the temperature of the combustion chamber. In order to achieve complete combustion and conversion of flue gas from unfavourable gases, a secondary combustor is added with excess air supply to achieve a highly turbulent mixing state, so that proper breakdown of toxic substances can occur effectively. After the combustion, the ash and slag are collected for cement manufacturing.

3.3.1.2 Rotary kiln incinerator

In general, the rotary kiln incinerator (refer Fig. 3.13) has a similar operating principle as the moving grate incinerator, where the waste is fed from the top of the kiln while air is pumped to the lower end of the kiln to assist combustion of the wastes. The kiln is a rotating cylinder that adds turbulence to the process with good mixing, to achieve complete combustion. There are two combustion chambers for this incinerator: the primary chamber

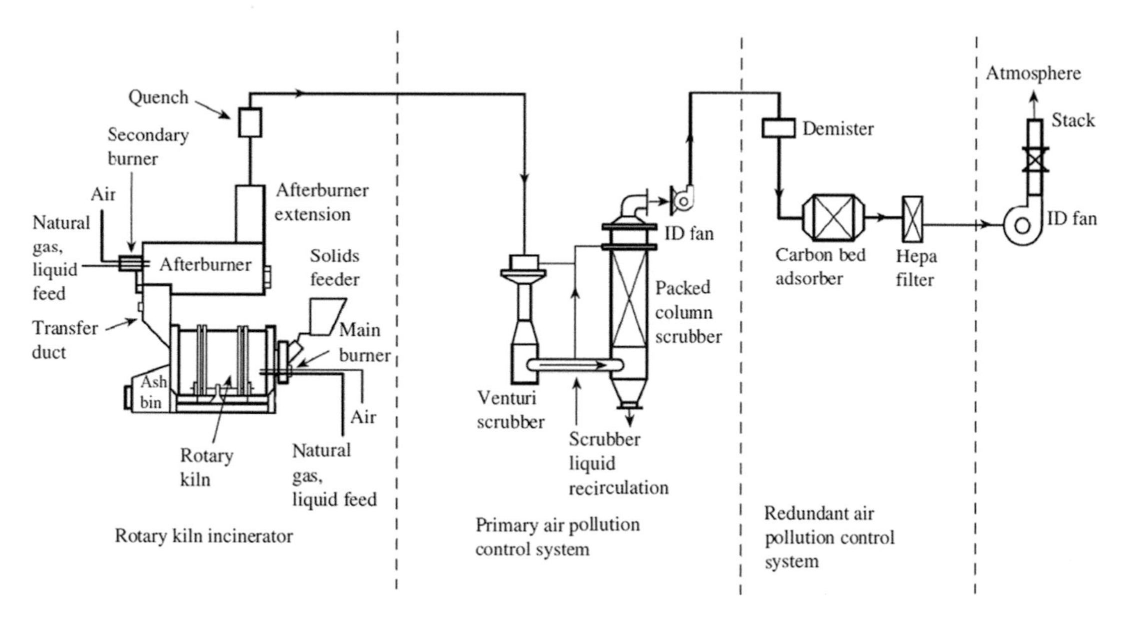

Fig. 3.13 Workflow of rotary kiln incinerators. From Al-Salem, S. M. (2019). Energy production from plastic solid waste (PSW). In *Plastics to energy* (pp. 45–64). Elsevier (chapter 3),. with permission from Elsevier.

breaks down the waste into volatile flue gas, while the secondary combustor ensures full breakdown of hazardous flue gases. In some designs, an air pollution control system such as a scrubber and carbon filter are installed to meet the stringent air emissions quality of local authorities.

3.3.1.3 Fluidized bed incinerator

A fluidized bed incinerator is constructed in a vertical orientation, as shown in Fig. 3.14. Commonly, the outer shell is made of a steel enclosure that is refractory lined. A powerful air blower is used to assist the combustion and the firing air is passed to the air nozzle to a fluidizing sand or limestone bed, where the wastes are fed through the inlets to induce turbulence. The efficiency of combustion is improved with better surface area thermal contact with the waste particles, as well as the air bubbling mechanism.

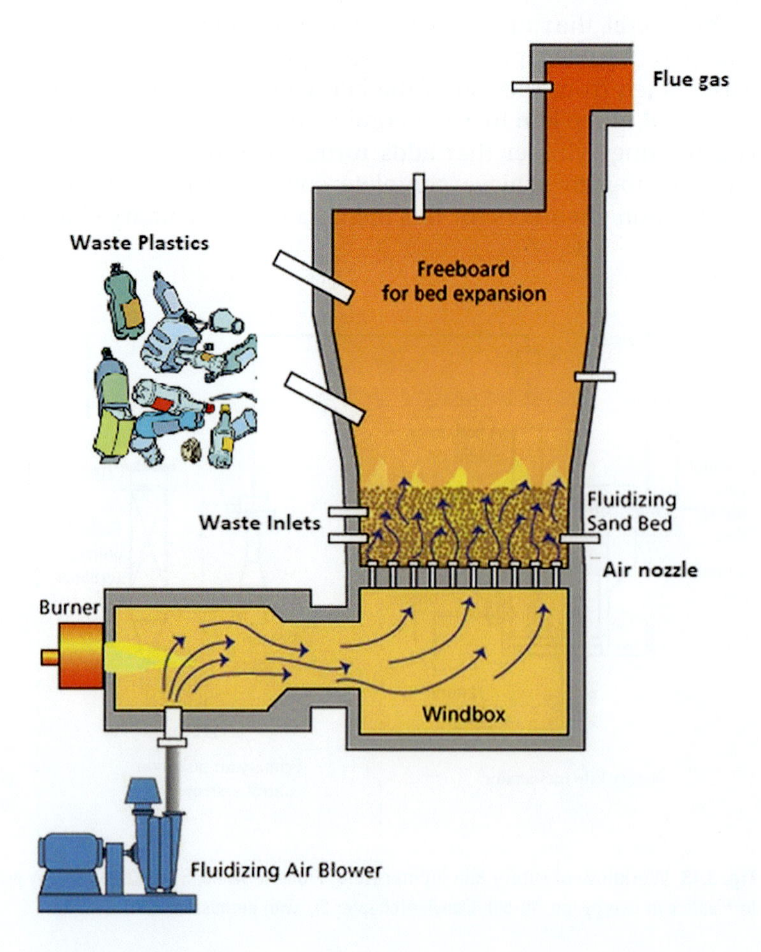

Fig. 3.14 Schematic diagram of fluidized bed incinerator.

The bed of sand, usually at a height of about 0.75–1 m is filled on top of the grid. Basically, the combustion occurs in two zones, which are the zone of the sand bed itself when the temperature rises in contact with the organic compounds, followed by the zone of the freeboard area to break down the remaining volatile compounds in order to control the quality of the emissions.

3.3.2 Challenges of plastic incineration and improvements

3.3.2.1 *Microplastics*

While incineration is known to be able to mitigate the impacts of plastic wastes, there are still traces of unburned material existing in the fly ash, bottom ash and soil, as the residues from the incineration plant. As found by Shen et al. (2021), incineration contains an abundance of microplastics with 23, 171 and 86 particles/kg in fly ash, bottom ash and soil, respectively. The presence of microplastics is mainly due to incomplete combustion of the plastic wastes. Among the microplastics found in the residue, the fibre type of microplastics is mainly found in the fly ash, while bottom ash and soil consist of fragments (43.0% and 29.3%), film (26.3% and 25%), foam (13.0% and 25.1%) and fibre (17.7% and 20.7%). Heavy metals such as Cr, Cu, Zn and Pb were observed to adsorb on the surface of the microplastics. The analysis of Shen et al. (2021) found that the content of heavy metals in bottom ash, fly ash and soil samples was significantly different, as shown in Table 3.2. Moreover, the presence of heavy metal elements also highly depends on the source of the waste plastics.

Table 3.2 Heavy metal content of bottom ash, fly ash and soil from incineration plant (Shen et al., 2021).

Element	Bottom ash (mg/kg, dry weight)	Fly ash (mg/kg, dry weight)	Soil (mg/kg, dry weight)
Cr	8.69	2.15	3.67
Cd	5.63	10.06	3.48
Pb	9.67	92.58	7.27
Zn	528.02	88.99	248.50
Cu	210.41	10.99	43.56
Mn	1152.76	3.76	802.3

For instance, for fire-resistant types of plastic materials, the residue of incineration contains heavy metal elements of aluminium and antimony, sourced from the flame retardants aluminium trihydrate and antimony trioxide added to plastics. Another study by Yang, Lu, et al. (2021) and Yang, Yu, et al. (2021) analysed the microplastics extracted from bottom ash in mass burn incinerators, bottom ash disposal centres and four fluidized bed incinerators. The results revealed that there was an abundance of microplastics, at 1.9–565 particles/kg of bottom ash. This means that every metric ton of wastes produced 360–102,000 microplastic particles after incineration. Microplastics at the sizes of 50 μm to 1 mm made up 74% of the total particles found, while granules, fragments, film and fibres made up 43%, 34%, 18% and 5% of the microplastics, respectively, as shown in Fig. 3.15.

In the reviews of current incineration challenges by Cui, Liu, Xia, Jiang, and Skitmore (2020) and Wielgoshinski (2011), the following suggestions were made in order to improve the combustion efficiencies in incineration.

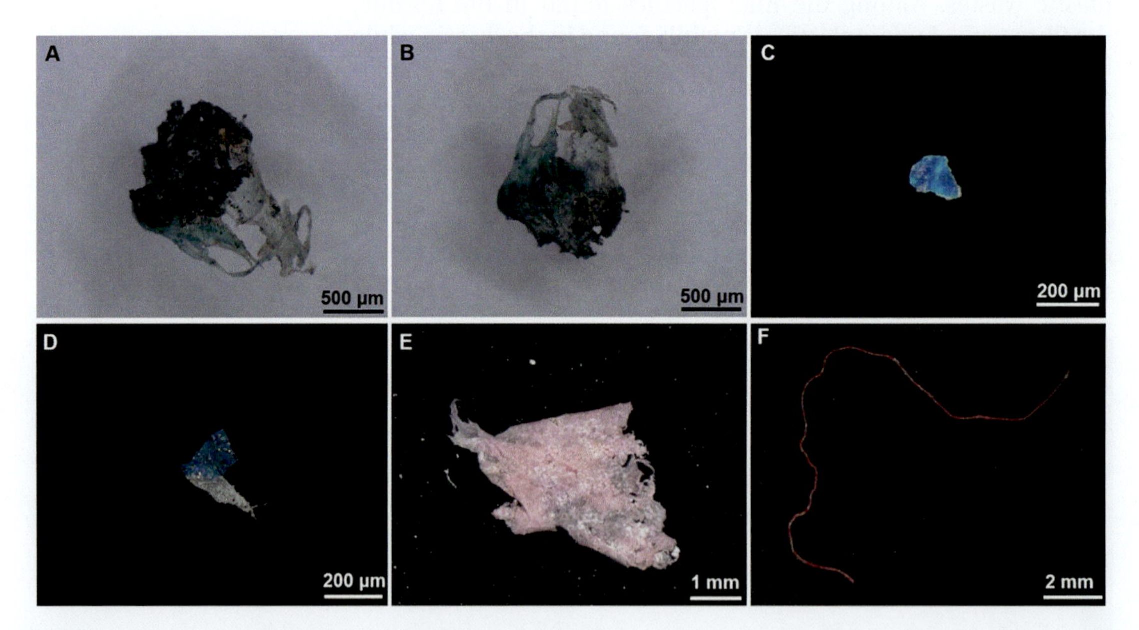

Fig. 3.15 Structure of microplastics in bottom ash. (A) and (B) are granules, (C) and (D) are fragments, (E) is film and (F) is fibre. From Yang, Z., Lu, F., Zhang, H., Wang, W., Shao, L., Ye, J., He, P. (2021) Is incineration the terminator of plastics and microplastics. *Journal of Hazardous Materials, 401*, 123429; Yang, L., Yu, X., Wu, X., Wang, J., Yan, X., Jiang, S., Chen, Z. (2021). Emergency response to the explosive growth of health care waste during COVID-19 pandemic in Wuhan, China. *Resources, Conservation and Recycling, 106*, 105074.

Combustion at high temperature

- All incinerators need to operate to ensure the gas resulting from the process is at a temperature of 850°C for 2 s to ensure full decomposition of toxic substances to CO_2 and harmless substances.

Low-pollution combustion

- Improvements that can be done include recirculation of flue gases to reduce NO_x and using oxygen-enriched combustors. Luizzo et al. (2007) reported that NO_x can be reduced with the recirculation of flue gas, while oxygen-enriched combustion can reduce it by 23% when the oxygen concentration increases from 21% to 30% (Chang et al., 2015). High NO_x emission is undesirable because NO_x can lead to acid rain as well as corrosion of the boiler and incinerator.

Dechlorination in furnace

- Chlorine compounds and fumes of hydrochloride are formed during the combustion process of polyvinyl chloride (PVC) or halogen-based flame-retardant plastic compounds. The main problem with the formation of HCl in the flue gas is that it can cause corrosion of the boiler tubes and incinerator structures. In addition, HCl can cause the formation of polychlorinated dibenzo-*p*-dioxins and dibenzo-furans (PCDD/Fs) and dioxinlike compounds.
- Powdered sorbents based on naturally available alkaline metal or alkaline earth metal substances can be used to react with HCl in flue gas and reduce the concentration. For instance, dolomitic sorbent can be added to the combustion chamber to reduce HCl, HF and SO_2 at a range of 7%–37%, 39%–80% and 34%–95%, respectively (Biganzoli et al., 2015).

Adsorbent with high sorption capacity

- Removal of PCDD/Fs and dioxinlike compounds can be accomplished using activated carbon.
- There are two approaches to be employed for adsorption of activated carbon, the first type being a solid bed (replaced by batch) and the second type being continuous stream adsorption using a monolithic honeycomb bed with powdery activated carbon.

Catalytic conversion of toxic compounds

- Nippon Shokubai and Shell Dioxin Destruction System have been involved in development of catalysts to enhance decomposition of dioxin into harmless carbon dioxide, water and other compounds.
- Vanadium-tungsten catalyst on a titanium oxide carrier effectively decomposes PCDD/Fs with the release of CO_2, water and HCl. Moreover, this catalyst can also be used to reduce nitrogen oxide (Wielgoshinski, 2011).

3.4 Conclusion

Plastic wastes have been known to cause environmental pollution. Hence, proper management of plastic wastes is of the utmost importance. Waste plastics can undergo mechanical recycling, chemical recycling and incineration to generate electricity to achieve the circular economy in the community. Chemical recycling is mainly to convert the waste plastics into basic chemicals, so they can be used to reproduce other materials and fuels. Chemical recycling consists of either pyrolysis or gasification processes. Both processes are considered to be advanced technologies that require expertise in operation, with large capital investments for profitable operation. Nevertheless, the type of waste plastics needs to be consistent (segregation is needed) so that the quality is guaranteed. Meanwhile, the incineration process is suitable for highly contaminated waste plastics or medical wastes. Upon undergoing the combustion process of incineration, thermal energy is used to generate steam to operate a turbine and produce electricity. The main concern about incineration is the emission from the flue gas, fly ash and bottom ash. Thus stringent regulations on the operation of incinerators have been rolled out by authorities, to minimize toxic emissions to the environment.

In conclusion, there are a variety of ways to handle plastic wastes, yet success in mitigating the impacts of plastic wastes still depends on the extent of cooperation from communities. It is important to discard wastes properly and segregate accordingly, to ease the waste-handling work performed by relevant parties. Finally, waste plastics can be further transformed to generate benefits to the community.

References

Almeida, D., & Marques, M. F. (2016). Thermal and catalytic pyrolysis of plastic waste. *Polímeros*, *26*(1), 44–51.

Angyal, A., Miskolczi, N., & Bartha, L. (2007). Petrochemical feedstock by thermal cracking of plastic waste. *Journal of Analytical and Applied Pyrolysis*, *79*, 409–414.

Biganzoli, L., Racanella, G., Rigamonti, F., Marras, R., & Grosso, M. (2015). High temperature abatement of acid gases from waste incineration. Part I: Experimental tests in full scale plants. *Waste Management*, *36*, 98–105.

Chang, Y., Wu, K., Che, Y., & Che, C. (2015). Atomized air of oxygen-enriched combustion for a 4500000 kcal/h industrial furnace. *Energy Fuels*, *29*, 3476–3482.

Cui, C., Liu, Y., Xia, B., Jiang, X., & Skitmore, M. (2020). Overview of public-private partnerships in the waste-to-energy incineration industry in China: Status, opportunities, and challenges. *Energy Strategy Reviews*, *32*, 100584.

Ding, W., Liang, J., & Anderson, L. L. (1997). Hydrocracking and hydroisomerization of high-density polyethylene and waste plastic over zeolite and silica – alumina-supported Ni and Ni – Mo sulfides. *Energy Fuels, 11*, 1219–1224.

Directive 2000/76/EC. (2000). Available at https://eur-lex.europa.eu/legal-content/EN/TXT/PDF/?uri=CELEX:32000L0076&from=EN. (Accessed 30 March 2022).

Dogu, O., Pelucchi, M., Hijver, R. V., Steenberge, P. H. M. V., D'hooge, D. R., Cuoci, A., ... Geem, K. M. V. (2021). The chemistry of chemical recycling of slid plastic waste via pyrolysis and gasification: State-of-the-art, challenges, and future directions. *Progress in Energy and Combustion Science, 84*, 100901.

EBARA. (2021). *Gasification technologies.* Available at https://www.eep.ebara.com/en/business_technology/technology_3.html. (Accessed 30 March 2022).

Fukushima, M., Shioya, M., Wakai, K., & Ibe, H. (2009). Toward maximizing the recycling rate in a Sapporo waste plastics liquefaction plant. *Journal of Material Cycles Waste Management, 11*, 11–18.

González-Marcos, M. P., Fuentes-Ordóñez, E. G., Salbidegoitia, J. A., & González-Velasco, J. R. (2021). Optimization of supports in bifunctional supported Pt catalysts for polystyrene hydrocracking to liquid fuels. *Topics in Catalysis, 64*, 224–242.

Jeswani, H., Kruger, C., Russ, M., Horlacher, M., Antony, F., Hann, S., & Azapagic, A. (2021). Life cycle environmental impacts of chemical recycling via pyrolysis of mixed plastic waste in comparison with mechanical recycling and energy recovery. *Science and the Total Environment, 769*, 144483.

Jing, Y. (2020). *How Wuhan copes with mountains of medical waste.* Available at https://news.cgtn.com/news/2020-03-17/How-Wuhan-copes-with-its-mountains-of-medical-waste-OUxhr4jW1i/index.html. (Accessed 20 March 2022).

Jumah, A., Anbumuthu, V., Tedstone, A. A., & Garforth, A. A. (2019). *Catalyzing the hyrocracking of low density polyethylene, 58*, 20601–20609.

Lee, R. P., Laugwitz, A., Mehlhose, F., & Meyer, B. (2018). *Waste gasification: Key technology for closing the carbon cycle.* Available at https://tu-freiberg.de/sites/default/files/media/professur-fuer-energieverfahrenstechnik-und-thermische-rueckstandsbehandlung-16460/publikationen/10-04-waste_gasification_ifc2018.pdf. (Accessed 22 March 2022).

Lee, R. P., Meyer, B., Huang, A., & Voss, R. (2020). Sustainable waste management for zero waste cities in China. Potential, challenges and opportunities. *Clean Energy, 4*, 169–201.

Lee, R. P., Tschoepe, M., & Voss, R. (2021). Perception of chemical recycling and its role in the transition towards a circular carbon economy: A case study in Germany. *Waste Management, 125*, 280–292.

Lu, J.-W., Zhang, S., Hai, J., & Lei, M. (2017). Status and perspectives of municipal solid waste incineration in China: A comparison with developed regions. *Waste Management, 69*, 170–186.

Ludlow-Palafox, C., & Chase, H. A. (2001). Microwave-induced pyrolysis of plastic wastes. *Industrial & Engineering Chemistry Research, 40*, 4749–4756.

Luizzo, G., Verdone, N., & Bravi, M. (2007). The benefits of flue gas recirculation in waste incineration. *Waste Management, 2007*, 106–116.

Mukherjee, A., Debnath, B., & Ghosh, S. K. (2016). A review of technologies of removal of dioxins and furans from incinerator flue gas. *Procedia Environmental Sciences, 35*, 528–540.

Munir, D., Abdullah, Piepenbreier, F., & Usman, M. R. (2017). Hydrocracking of a plastic mixture over various micro-mesoporous composite zeolites. *Powder Technology, 316*, 542–550.

Royte, E. (2019) Is burning plastic waste a good idea? Available at: https://https://www.nationalgeographic.com/environment/article/should-we-burn-plastic-waste [Online]. National Geographic. Accessed on 11 October 2022.

Sahu, J. N., Mahalik, K. K., Nam, H. K., Ling, T. Y., Woon, T. S., Rahman, M. S. A., … Jamuar, S. S. (2014). Feasibility study for catalytic cracking of waste plastic to produce fuel oil with reference to Malaysia and simulation using ASPEN Plus. *Environmental Progress & Sustainable Energy, 33*, 298–307.

Seo, Y.-H., Lee, K.-H., & Shin, D.-H. (2003). Investigation of catalytic degradation of high-density polyethylene by hydrocarbon group type analysis. *Journal of Analytical and Applied Pyrolysis, 70*, 383–398.

Shen, M., Hu, T., Huang, W., Song, B., Qin, M., Yi, H., Zeng, G., & Zhang, Y. (2021). Can incineration completely eliminate plastic waste? An investigation of microplastics and heavy metals in bottom ash and fly ash from an incineration plant. *Science of the Total Environment, 779*, 146528.

Solis, M., & Silveira, S. (2020). Technologies for chemical recycling of household plastics—A technical review and TRL assessment. *Waste Management, 105*, 128–138.

Takashi, Y. (2009). *Current recycling technologies for plastic waste in Japan*. Osaka: Plastic Waste Management Institute.

Voss, R., Lee, R. P., Seidl, L., Keller, F., & Frohling, M. (2021). Global warming potential and economic performance of gasification-based chemical recycling and incineration pathways for residual municipal solid aste treatment in Germany. *Waste Management, 134*, 206–219.

Wielgoshinski, G. (2011). The reduction of dioxin emissions from the processes of heat and power generation. *Journal of the Air & Waste Management Assocition, 61*, 511–526.

Yang, Z., Lu, F., Zhang, H., Wang, W., Shao, L., Ye, J., & He, P. (2021). Is incineration the terminator of plastics and microplastics. *Journal of Hazardous Materials, 401*, 123429.

Yang, L., Yu, X., Wu, X., Wang, J., Yan, X., Jiang, S., & Chen, Z. (2021). Emergency response to the explosive growth of health care waste during COVID-19 pandemic in Wuhan, China. *Resources, Conservation and Recycling, 106*, 105074.

4

Blending technology to improve ecofriendliness of plastics

4.1 Introduction

Many of the commodity plastics are nondegradable. To be more accurate, plastics require more than 50 years to degrade to substances harmless to the environment. However, when plastics are exposed to the external environment, they are susceptible to surrounding environmental attacks. For instance, ultraviolet radiation from sunlight, oxygen from the atmosphere, moisture, heat, forces (like wind or rain) or animals (like fish, worms, or insects) can attack the plastic, and eventually the waste plastics are degraded into small fragments. At that time, the fragments are called microplastics, and they can endanger the ecosystem, particularly soil and aquatic organisms, with impacts that are widespread throughout the entire food chain. Hence, consumers are urged to reduce, reuse and recycle plastic materials while being responsible in handling of plastic wastes, for instance not simply dumping them into the open environment, drainages, rivers or oceans. Plastic wastes should be physically or chemically recycled by competent industries, or they should be incinerated instead of disposed of in landfills, so that the waste can be transformed into electrical energy and returned to the economic loop for the benefit of the community.

Since commodity plastics are widely used and are nondegradable, the blending of a variety of fillers, from biomass, minerals, specialty chemical additives and others, are used to reduce the weight of the plastic content per piece of plastic products. After the weight reduction, the material cost of the plastics is reduced and without compromising the properties of applications. Although such a move can be promising, there is another concern that consumers are not well informed about the type of additives being included, so the recycling process may lead to unexpected

Plastics and Sustainability. https://doi.org/10.1016/B978-0-12-824489-0.00001-5
Copyright © 2023 Elsevier Inc. All rights reserved.

results, inasmuch as the properties of the plastic can vary based on the sources of waste plastics, particularly postconsumer products. Thus the industrial producers should provide more information about the plastic product recyclability so that consumers can segregate the plastic wastes according to industry requirements.

4.2 Blending of biomass

The blending of biomass with plastics was done many decades ago with a variety of biomass types, depending on the local availability and harvesting season. For instance, rice husk and straw have been widely added to thermoplastics to produce wood-like natural fibre composite materials. Such thermoplastic composites can be used to produce profiles to substitute for wooden building materials such as floor decking, gazebos, or small loaded wooden bridges in cottages or parks. The natural fibre composite possesses good weathering resistance, termite resistance, water resistance and durability. Table 4.1 tabulates the number of research publications on biomass blending with commodity plastics: polyethylene, polypropylene, polystyrene, polyvinyl chloride and polyethylene terephthalate extracted from the Scopus database. It is obvious that the blending of biomass and natural fibres from agricultural sources has attracted a great deal of attention from researchers while development has been moving towards positive growth.

Table 4.1 Number of publications on types of agricultural biomass blended with plastics from Scopus database (until January 2022 searching documents algorithm biomass type and plastic—e.g., Wood flour and Poly).

Type of biomass	Number of publications
Flax	1298
Rice husk	1198
Jute	1155
Coconut fibre from shell	1097
Wood flour	1060
Kenaf	974
Sisal	716
Empty fruit bunch	462

4.2.1 Filler and reinforcement agents

The most common application of biomass is being added as a filler to reduce the composition percentage of plastics. In fact, the use of biomass from lignocellulosic sources is huge, particularly after the harvest seasons. Many countries' policymakers have actually treated the use of lignocellulose as a waste-to-wealth initiative, instead of using the agricultural wastes to be combusted as fuel for a drying process, or merely letting them rot in the agricultural fields after harvesting. Another common biomass is starch, or cellulosic type of biomass. When a comparison is made between lignocellulosic and cellulosic biomasses, the most obvious difference is that lignocellulose possesses a coarser structure compared to cellulose, resulting in the lignin portion being more durable, which causes lignocellulose to require a further milling process to reduce the particle sizes so the blending process can be done more easily. On the other hand, both lignocellulose and cellulose possess biodegradability upon being added to thermoplastics. However, this is limited to the portion of the lignocellulose or cellulose to be degraded by microorganism activity. The leftover plastic skeleton will still take a long time to degrade, leading to controversy over the microplastic pollution problems. Nonetheless, when biomass is blended with biodegradable plastic such as polylactic acid, the resulting composite is fully biodegradable and this composite material is compostable upon burial in soil. In summary, blending of biomass with thermoplastics has advantages and disadvantages as follows:

Advantages

(1) Reduce the nonrenewable composition of fossil-based thermoplastics.
(2) Reduce the cost of plastic compounds after substitution with biomass.
(3) Improve degradability of plastic compounds due to biodegradability of lignocellulose and cellulose.
(4) Value-added products of biomass for advanced applications, such as construction materials with natural fibre reinforcement.

Disadvantages

(1) The biodegradation is limited to the lignocellulosic or cellulosic portion, with the remaining nondegradable thermoplastic fragments possibly leading to microplastics pollution.
(2) Limited recyclability of biomass thermoplastic composites due to thermal degradation of biomass occurs when undergoing melt processing for a second time. Visible burning marks

and odour can be detected when reprocessing under high temperature and shearing rate in blending of biomass composite materials.

4.2.1.1 Blending process of biomass with thermoplastics

Stage 1: Premixing and biomass/natural fibre conditioning

Prior to the blending process of biomass in thermoplastics, the biomass requires a size reduction process to transform the large particle size and aspect ratio to a smaller particle size, for ease of compounding by twin screws extruder. The extent of particle size reduction and selection of machineries for size reduction depends on the rigidity of the structure of the biomass and application. Referring to Table 4.2 showing the composition of selected biomass, when the amount of lignocellulose is higher, the rigidity of the biomass is higher. Rice husk having silica content is abrasive, which can cause the surface contact of the machinery to easily wear out when the particle sizes of the biomass are large. Commonly the size reduction is done initially by a mechanical machine milling/crushing process, followed by a pulverizer to minimize the size using the principle of impact/coalition among the particles and classification according to particle sizes and aspect ratio as well. A large aspect ratio means that biomass has transformed into a fibre form with a reinforcing effect provided the surface treatment is done properly. The examples of milling and the pulverizer can be seen in Fig. 4.1. On the other hand, the biomass from kenaf, jute or flax sources are naturally in long fibre forms. Hence, many studies have reported that jute, hemp,

Table 4.2 Composition of biomass.

Types	Cellulose/ hemicellulose	Lignin	Pectin
Flax	65–85	1–4	5–12
Kenaf	45–57	8–13	3–5
Sisal	50–64	–	–
Jute	45–63	12–25	4–10
Hardwood	40–50	20–30	0–1
Softwood	40–45	36–34	0–1
Rice husk	25–35	17–20	–
Rice straw	20–50	15–20	–
Empty fruit bunch	60–70	20–28	–

Fig. 4.1 Hammer mill and high-speed pulverizer for crushing and reducing the size of biomass to particle forms for mixing with thermoplastics.

sisal and coir can be used to replace glass fibre for plastic material composites (Wambua, Ivens, & Verpoest, 2003). Long natural fibres can be made into woven mats by laying them on a mould for epoxy spraying to become composite articles. In addition, the long fibres can also be used to produce profile products via a pultrusion process.

Due to the moisture affinity of agricultural biomass or fibres, a drying process is crucial to minimize the defects such as pinholes, decolourization, shrinkage and warpage resulting from the vaporization of the moisture content of biomass. Moreover, excessive moisture can cause depolymerization when blending with nylon or polyethylene terephthalate. This is because the vaporized moisture can trigger a hydrolysis reaction to occur from the reversible reaction of condensation polymerization. In order to improve the drying process for agricultural biomass, the process operator should ensure installation of a proper dehumidifier prior to the compounding process, together with a vacuum system being attached to the extruder to remove the excessive water vapour when processing at high temperature. In other words, after the drying process, the moisture content should be $<1\%$ from the hopper dryer while a remaining 1% can be removed during the extrusion compounding process. Nevertheless, the moisture content should be $<0.1\%$ when the compounding is to blend nylon, polyethylene terephthalate or polylactic acid, for which the

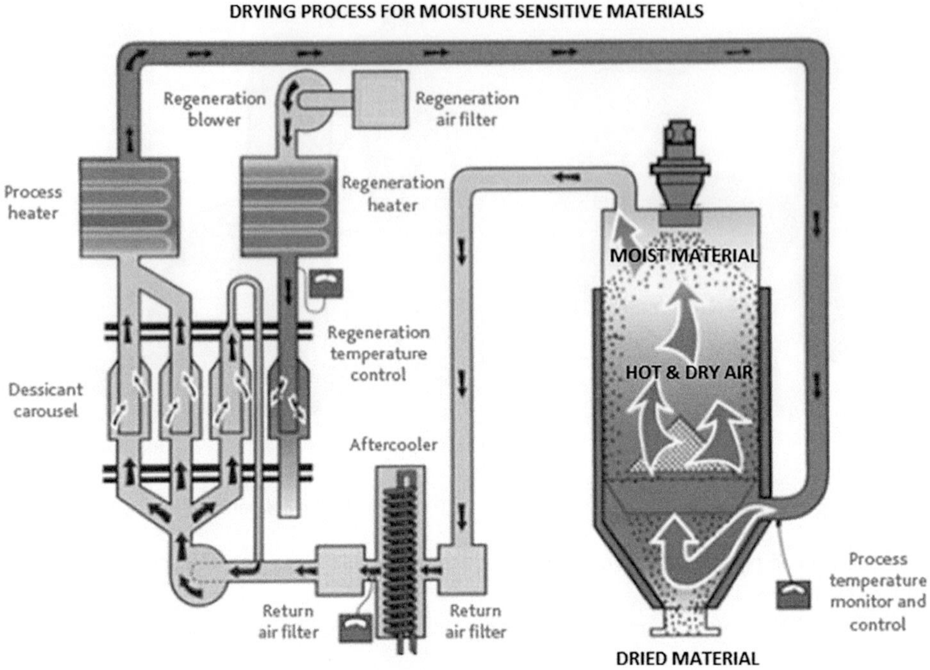

Fig. 4.2 Dryer and dessicant for moisture-sensitive materials to remove excess moisture prior to blending process.

plastics are made by a condensation polymerization reaction. As mentioned earlier, the excessive moisture in contact with these types of plastics can cause depolymerization or a chain scissioning reaction. When extruded nylon composite is too brittle, it can be due to chain scissioning and this is an irreversible process. Consequently, the process operator should have crucial quality control in the premixing process. A typical dryer and dehumidifier can be found in Fig. 4.2.

Stage 2: Compounding of biomass/natural fibre with thermoplastics

The compounding of biomass/natural fibre with thermoplastics can be accomplished using a common twin screw compounder or an internal mixer compounder. The twin screw compounder is preferable because it has better dispersion and is a continuous process, whereas the internal mixer is a batch kneading process. Inherently, blending of natural fibre with thermoplastics can be as simple as adding natural fibre and thermoplastics together. However, such mixing composition is only suitable for adding up to 10 wt.% of biomass/natural fibre using high melt flow index plastics and extruding at a temperature of

140°C or below. This is because most of the agricultural fibre degrades at a temperature of 150°C or even lower. The natural fibre can undergo early degradation when subjected to both thermal and high shear effects. Hence, a mere blending of natural fibre and plastics can only be done using polyethylene, which has a melting temperature starting at 120°C.

In order to blend with high melting point thermoplastics such as polypropylene, polystyrene, polyvinyl chloride, etc., additives packages are needed. For instance, internal and external lubrication are needed in order to promote the dispersion and minimize the internal friction among the components, particularly when the biomass particles are large. For instance, internal lubrication of calcium stearate is added to polyvinyl chloride at a dosage of about 2% to reduce premature degradation caused by the internal friction when blending with filler. Without the addition of calcium stearate internal lubricant, high shearing and friction can cause the release of corrosive hydrogen chloride radicals to attack fibres and polymer chains, and subsequently a brownish colour is visible after the extrusion. For optimal effect, a mixture of fatty acid soap and amides is recommended for formation of chemical bonding between polymer and filler to achieve homogeneous blends. An example of a lubricant system made especially for natural fibre is provided by the Performance Additives ULTRA-PLAST series, which can promote the formation of homogeneous blends by secondary bonding, such as hydrogen bonding between natural fibres and plastic matrix.

From a chemical perspective, the blending of biomass/natural fibres with thermoplastics is actually an incompatible mixture, because natural fibre is characteristically polar, while thermoplastics such as polyethylene and polypropylene are nonpolar. When the blending is carried out under high shear and thermal effects from the extruder, the compounds may seem to be homogeneous. But the fact is that such an incompatible blend has weaker mechanical properties and limited reinforcing effects. In other words, the interfacial adhesions between fibres and plastics are weak, as shown in Fig. 4.3, resulting in an ineffective force transfer to the fibre from the polymer matrix. Also, the durability of the composite is lower and easily broken within a shorter period of time. In order to mitigate the incompatibility, a compatibilizer can be added to the composites. Maleic anhydride compatibilizer is commonly blended into the natural fibre and thermoplastics system to improve interfacial interaction between the components as shown in Fig. 4.4. For instance, linear low-density polyethylene-grafted maleic anhydride can form secondary bonding with the cellulose of wood particles consisting of an -OH functional group via

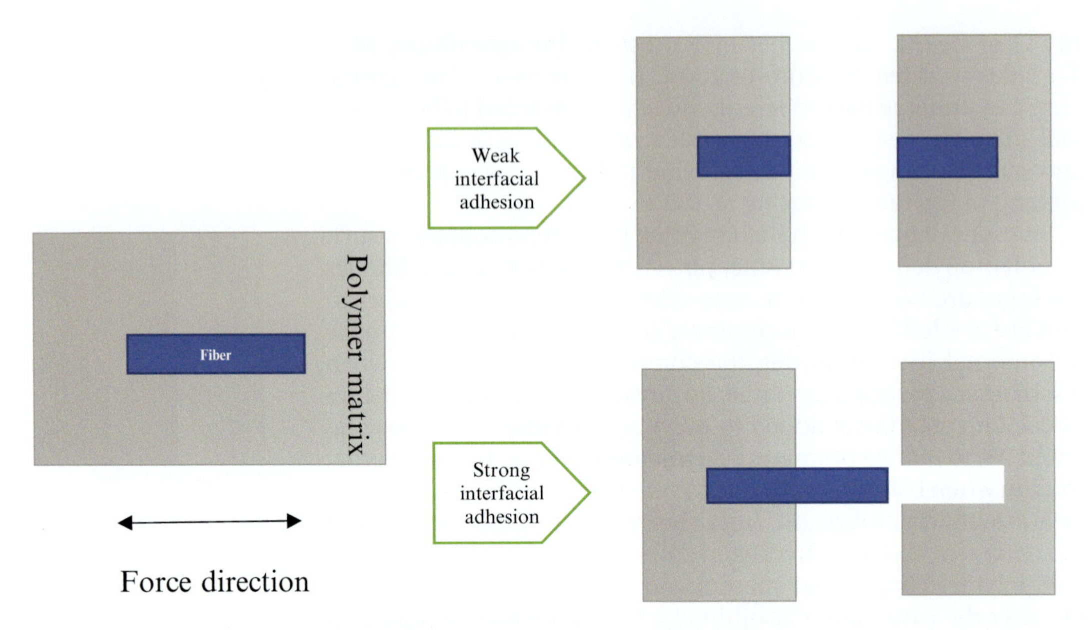

Fig. 4.3 Mechanism of weak and strong interfacial adhesion between fibre and plastic matrix. Addition of a coupling agent is needed in order to improve the interfacial adhesion so that the applied forces can transfer to the fibre effectively.

hydrogen bonding. When there is no compatibilizer, the polar functional group of -OH from the cellulose has low interaction with the nonpolar -CH group of polyethylene. One of the well-established compatibilizers is Dow Fusabond, which can be added to polyethylene, polypropylene, polyamide and others at a dosage of <5 wt.%. The selection of grade is highly dependent on the plastic matrix in use as well as the processing methods.

Stage 3: Extrusion using twin screws extruder

The process of compounding biomass/natural fibre with thermoplastics is recommended to be done using a twin screws co-rotating extruder to achieve the best homogeneous mixing of thermoplastic and natural fibre. Although biomass/natural fibre can be premixed before filling to the feeding zone for easy processing, there is a risk of overheating along the pathway of metering and mixing zones. This results in an extruded compound that is seriously degraded, discoloured and smelly. An alternative is to use a side feeder to push feeding of the biomass/natural fibre nearly to the middle (between the metering and mixing) zones, so that the residence time of biomass/natural fibre can be shorter in order to minimize the possibility of

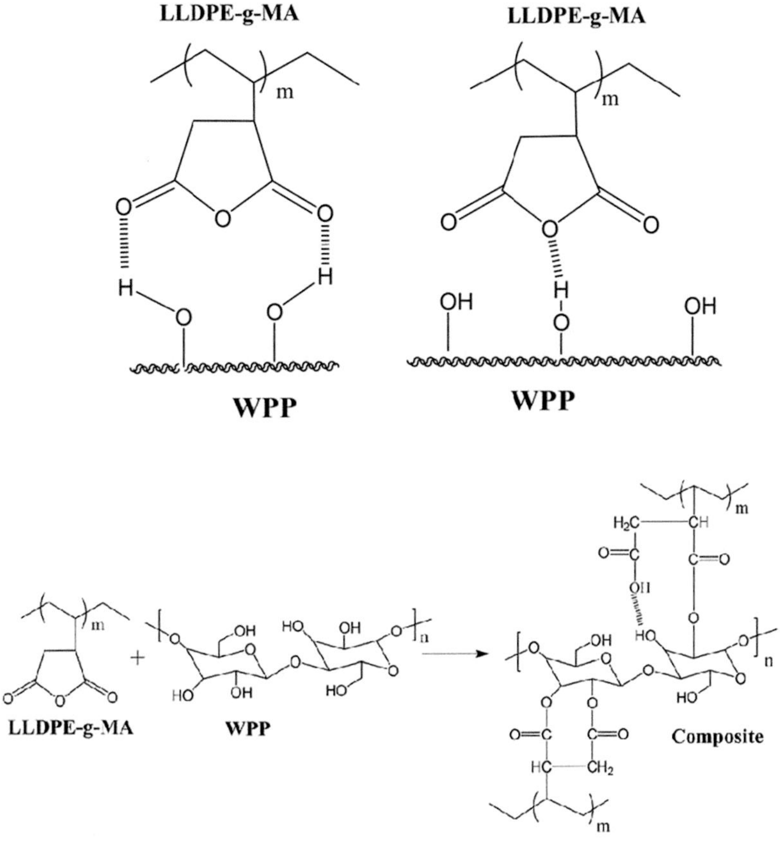

Fig. 4.4 Formation of secondary bonding (top) esterification reaction (bottom) between linear low-density polyethylene-grafted-maleic anhydride (LLDPE-g-MA) with paper powder (WPP). The LLDPE attaches to thermoplastic size while the MA is attached to the wood particles. From Saini, A., Yadav, C., Bera, M., Cupta, P., Maji, P. K. (2017). Maleic anhydride grafted linear low density polyethylene/waste paper powder composite with superior mechanical behavior. *Journal of Applied Polymer Science, 134*, 45167, with permission of Wiley.

overheating. In other words, the plastic resin, wax, coupling agents and additives can be at the beginning of the feeding zone to achieve a preliminary mixing before reaching the middle pathway where the natural fibre is fed. With such an operating procedure, a better quality and more durable composite can be produced. After the extrudate is out from the die, the solidification and pelletizing stages follow. Although a water bath cooling process can be used to solidify the hot molten compounds, the drawback of this is that the compound can absorb water when immersed in a water bath. The resulting compound after pelletizing still requires another drying process before undergoing injection moulding and the other remaining processes. A better approach is to engage a high-speed air-cooling system to solidify the hot molten compounds. With proper storage, only a light drying is required prior to undergoing the next plastic processing step. A typical type of twin screw co-rotating extruder can be seen in Fig. 4.5, which is made by Dr. Collin GmBH, Germany. The complete set of this twin screw extruder costs >€500,000 including the extruder, side feeder and pelletizing system, but excluding the hopper dryer.

4.3 Blending of recycled plastics

There are two main purposes of blending recycled plastics with virgin plastics: the first is to minimize the weakness of recycled plastics with virgin plastics and the second is to reduce the cost of virgin plastics. Most of the time, postconsumer recycled plastics are unlikely to be used alone due to variation of quality resulting from the wide spectrum of collection sources. For instance, polyethylene terephthalates (PETs) to produce beverage bottles and thermoforming food packaging are similar type of plastics. However, both applications use a different grade of PET. Moreover, it is difficult to identify the history of use of postconsumer waste plastics, where longer exposure to sunlight, thermal or mechanical effects can greatly degrade the quality of the recycled plastics. As such, it is advisable that basic quality control should be carried out during receiving of products from recyclers, with the simplest methods and tests as follows:

(a) Visual examination: Clear and brighter colour of recycled plastic resin is preferable compared to yellowish or darker colour. The yellowish or darker colour indicates degradation may have occurred.

(b) Odour: A good recycled resin is odourless or smells like virgin plastic.

(c) Hand feel: A good recycled resin should feel rigid/firm and seldom powdery. For certain types of plastics such as PET and nylon, the handheld feeling should be dry. Otherwise, the resin may undergo degradation or depolymerization.

(d) Melt flow rate: Two main observations can be found from the melt flow rate. First is checking the consistency of melt flow rate from batch to batch. Inconsistency of melt flow rate indicates change of quality, due to degradation, different types of additives or processing aids or source/grades of plastic. Secondly, the melt flow rate is the indication of suitability of processing. For optimum mixing and processing, the melt flow rate should be as close to the virgin plastics processing as

(A)

(B)

(C)

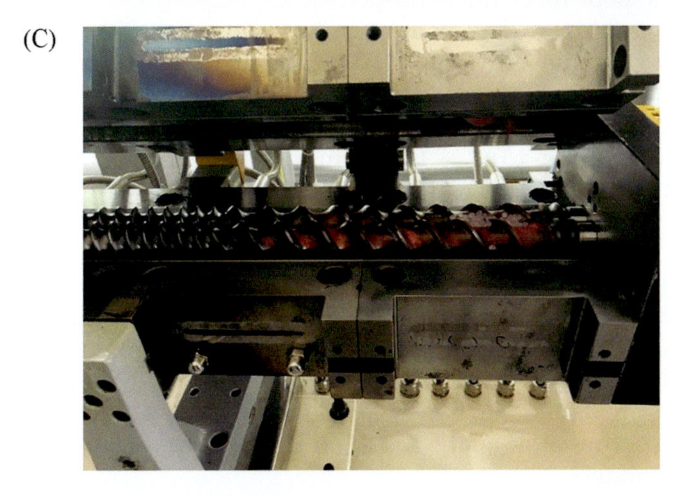

Fig. 4.5 (A) Co-rotating twin screw extruder; (B) control panel and motor of extruder; (C) screw of extruder. This extruder is built by Dr Collin, GmBH model OMega 20PC with screw diameter 26.00 mm, screw minor diameter 15.20 mm, barrel diameter 26.30 mm, screw functionality length L/D 42. Photo courtesy Ant Spirits Sdn. Bhd.

Fig. 4.6 International Resin Identification Code for commodity plastics. *PET*, polyethylene terephthalate; *HDPE*, high-density polyethylene; *PVC*, polyvinyl chloride; *LDPE*, low-density polyethylene; *PP*, polypropylene; *PS*, polystyrene; *OTHER*, other types of plastics.

possible, to avoid excessive tuning of the processing parameters.

The best approach is to blend a plastic with similar types of plastics. In fact, the International Resin Identification Codes (refer to Fig. 4.6) should be labelled on the plastic products, so the segregation and recycling can be done effectively. Blending similar types of plastics together is important for the following reasons.

(a) To avoid inferior mechanical properties due to incompatibility, particularly when blending different polarities of plastic together. For instance, when polystyrene is mixed with polyethylene it results in a dual-phase condition, as shown in Fig. 4.7. When forces are applied to the blends, the forces are unable to be transferred and distributed evenly, due to the poor mechanical properties.

(b) Even though low-density polyethylene (LDPE) and high-density polyethylene (HDPE) are chemically similar types of plastics, the blending of these plastics can cause microstructure changes. As shown in Fig. 4.8, the blend comprising 15wt.% HDPE in an LDPE matrix apparently has a reduction of the average size of spherulites, with even distribution of spherulites visible (Li et al., 2020). This means that after adding HDPE in LDPE, the elongation property of LDPE is lowered, while at the same time the strength of the LDPE is higher. In other words, unless both are blended with the purpose of improving elongation or mechanical strength of the plastics, the extent of improvement is hardly predictable and requires further experimental analysis in order to achieve the desired products.

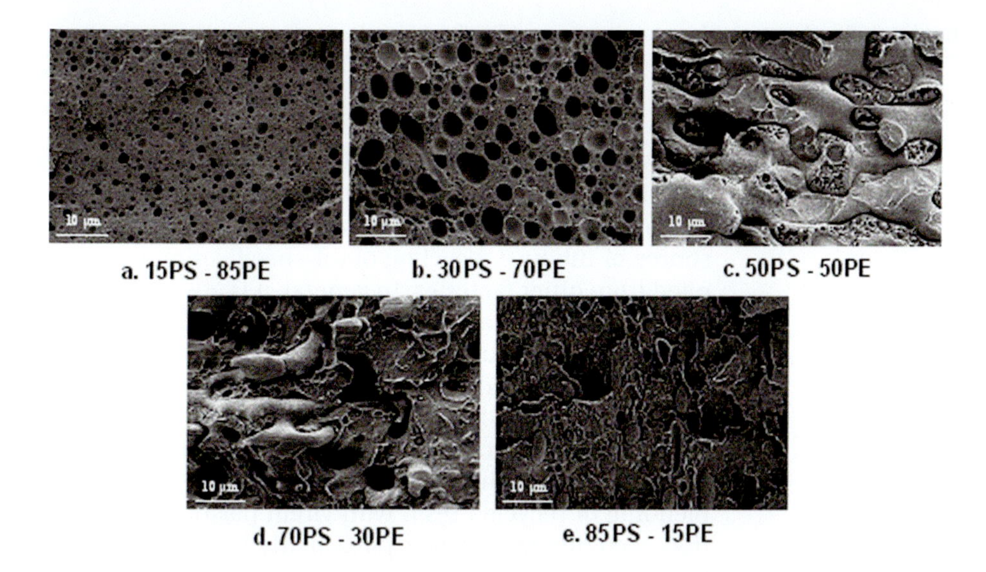

Fig. 4.7 Scanning electron micrograph of polystyrene-polyethylene blends with dual phase microstructure visible. From Thritha, V. M., Lehman, R. L., Nosker, T. J. (2004). Morphological effects on glass transition in immiscible polymer blends. *Material research society symposium proceeding, online proceedings library, 856*, 1112. Open Access Resources.

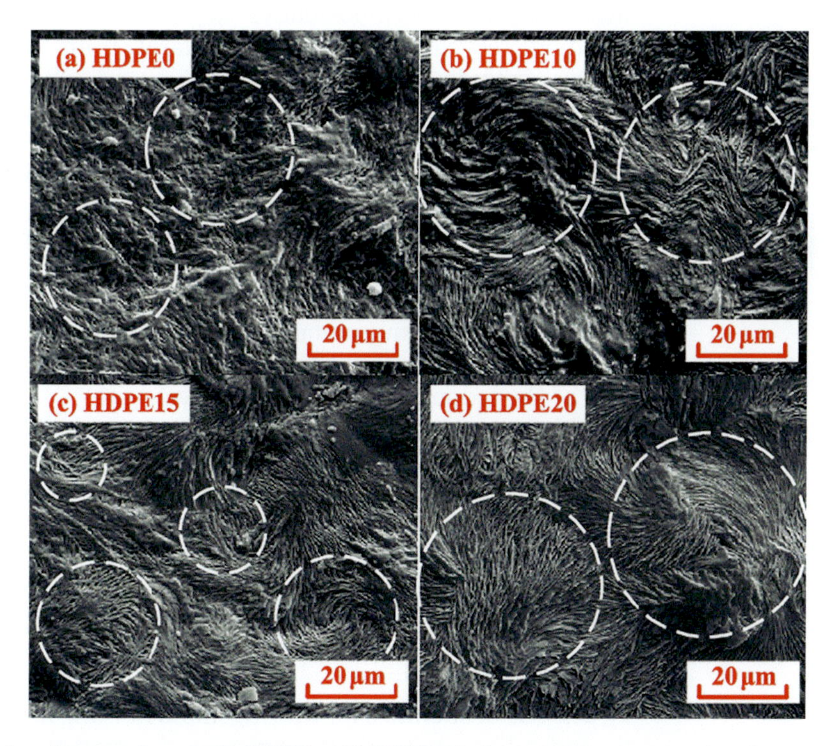

Fig. 4.8 Scanning electron micrograph of HDPE and LDPE blends. It can be observed that the microstructure gradually changed, with visible distribution of spherulites throughout the microstructure when HDPE increased. From Li, Z., Zhou, S., Xu, B., Fan, M., Su, J. (2020). Effect of crystalline morphology on electrical tree growth characteristics of high-density and low-density polyethylene blends. *IEEE Access, 8*, 114413. Open access source.

Compared to postconsumer recycled plastics, preconsumer recycled plastics can be used without hesitation regarding quality. Preconsumer recycled plastics can be obtained either from a source of inhouse recycling of off-specification products or supplied from other factories. In general, it is advised to use postconsumer recycled plastics by blending with virgin plastics at the amount of maximum 20 wt.%, while preconsumer recycled plastics can be blended with virgin plastics up to 80%. The recycled resin can be added during the plastic products processing process, such as injection moulding, blow moulding, blown film, etc. through the hopper/feeder. In certain circumstances, the blending of recycled resin with virgin plastics can cause a variation of colour appearance, particularly for whitish colours, where the recycled plastic can experience degradation with mild yellowish colour visible. Thus operators should make adjustments to the colouring package as well as adding antioxidants to minimize degradation. An example of an antioxidant and thermal stabilizer is Irganox produced by BASF. Irganox is a sterically hindered primary phenolic antioxidant stabilizer that protects organic substrates against thermooxidative degradation. Other issues of blending recycled plastics, particularly from postconsumer sources, include:

(a) Presence of plasticizers in the recycled PVC, which can cause fire risk.
(b) Presence of lubricants in recycled plastics, which can affect the flow of molten plastic during processing.
(c) Incompatibility processing aids in the recycled plastics, which can cause an oily surface from the migration of additives.
(d) Presence of heavy metals or flame retardants in the recycled plastics, which may affect the RoHS compliance (Restriction of Hazardous Substances Directive).
(e) Moisture content of recycled plastics, such as PET and nylon.

These issues require targeted troubleshooting by experienced persons so they do not cause substantial off-specification in the production.

4.4 Blowing agent as weight reduction approach

The use of chemical blowing agents is nothing new in the plastic and rubber industry. In particular, the market has been very familiar with expandable polystyrene, or styrofoam. This plastic material is very lightweight and possesses excellent insulation performance. Expandable polystyrene uses the physical foaming/ blowing agent pentane to expand. But the drawback is that expanded polystyrene is very cheap and consumers tend to use it for single-use plastic applications, such as food and beverage containers, representing a huge challenge in waste management. From another perspective, expandable polystyrene is at least 40 times lighter than original polystyrene resin. This has undeniable savings in material consumption, to minimize the use of fossil-based plastics. However, expandable polystyrene is accused of being a major cause of serious pollution due to the fact that collection and recycling of expandable polystyrene is not profitable; the material is so bulky and the cost of transportation can be very high, even overtaking the value of expandable polystyrene for recycling. Thus expandable polystyrene is widely discarded after one use without recycling.

As with the physical blowing agents, there are two main purposes for adding chemical blowing agents to plastics and rubbers. The first is to produce a foamed or cellular structure for thermal insulation and acoustic applications, and the second is to reduce the material usage, with the cellular structure being filled with gases to achieve lightweight products. Indeed, the second purpose is identical to the objective of adding fillers (calcium carbonate, talc, kaolin clay, wood flour and others) to reduce material cost. But the drawback of adding a mineral type of filler is the weight of the products increases significantly, due to the high density of mineral constituents. Thus a chemical blowing agent is used to produce a lightweight product with good insulation properties as well.

There are several types of chemical blowing agents available on the market for plastic and rubber applications, as listed in Table 4.3. The selection for types of blowing agents is highly dependent on the processing temperature of the plastic and rubber. This is because each type of chemical blowing agent possesses a unique decomposition temperature. If the plastic processing temperature is below the decomposition temperature of the blowing agent, the gas release is less or none, resulting in poor expansion of the plastic products. On the other hand, in the selection of blowing agent, consideration needs to be given as to whether an open cell or closed cell structure is desired. The open cell structure is mainly for acoustics application, while the closed cell structure is suitable for insulation purposes. As such, the open cell structure is usually more flexible, whereas closed cell products are more rigid.

The analysis conducted by Petchwattana and Covavisaruch (2011) of zinc-modified azodicarbonamide with different particle

Table 4.3 Type of chemical blowing agent

Chemical structure	Name of substance	Decomposition temperature in °C	Gas yield in mL/g and gas types
H_2N–C(=O)–N=N–C(=O)–NH_2	Azodicarbonamide (AZO, ADCA) Yellow-orange powder	205–215 (activated grade: 150–190)	270 N_2, CO, NH_3, CO_2
H_2NHN–SO_2–C6H4–O–C6H4–SO_2–$NHNH_2$	Oxybis (benzenesulphonyl hydrazide) (OBSH) White powder	155–165	160 N_2, H_2O
H_3C–C6H4–SO_2–$NHNH_2$	*p*-Toluenesuphonyl hydrazide (TSH) White powder	120–130	120 N_2, H_2O
H_2N–C(=O)–NH–NH–SO_2–C6H4–CH_3	Toluenesulphonyl semicarbozide (TSS) White powder	230–250	140 N_2, H_2O
C6H5–(1H-tetrazol-5-yl)	5-Phenyltetrazole (5PT)	215–225	220 N_2

sizes and content in rice-hull polyvinyl chloride composites found that the foam density (or composite density) dropped when the amount of modified azodicarbonamide increased at 2.0 wt.%, as shown in Fig. 4.9. Beyond this content, the foam density increased. A similar trend can be observed in ultimate flexural strength as well. The foam density was reduced to a minimum before it started to increase, mainly due to the capability of the PVC composite structure to withstand the liberation of gas when decomposition occurred. When too many large cells formed within the structure during the decomposition period, it causes a weakening effect to the structure, and collapse. As shown in Fig. 4.10, the cells are more visible and well distributed at 1.0% and 1.5% of 5 μm modified azodicarbonamide (mAC5) and a subsequent increment of blowing agent exhibits a low porosity structure. When the porosity is reduced, the strength of the composite

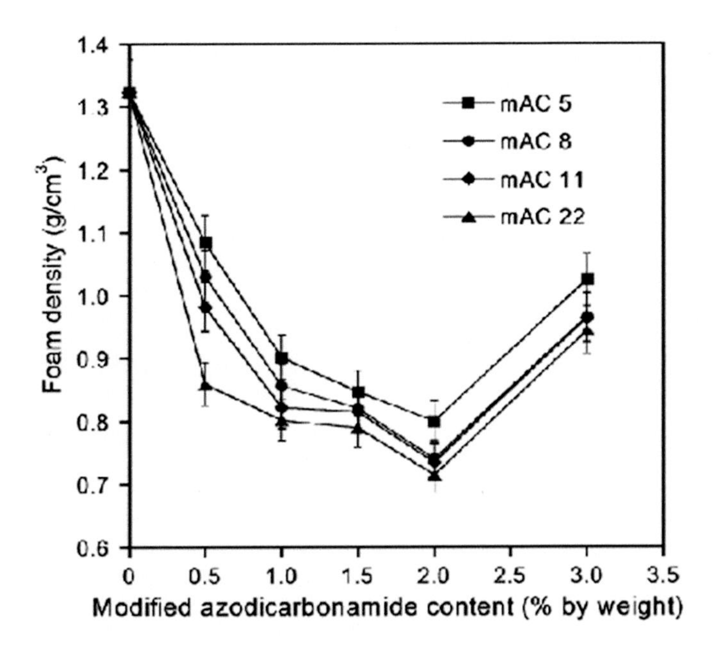

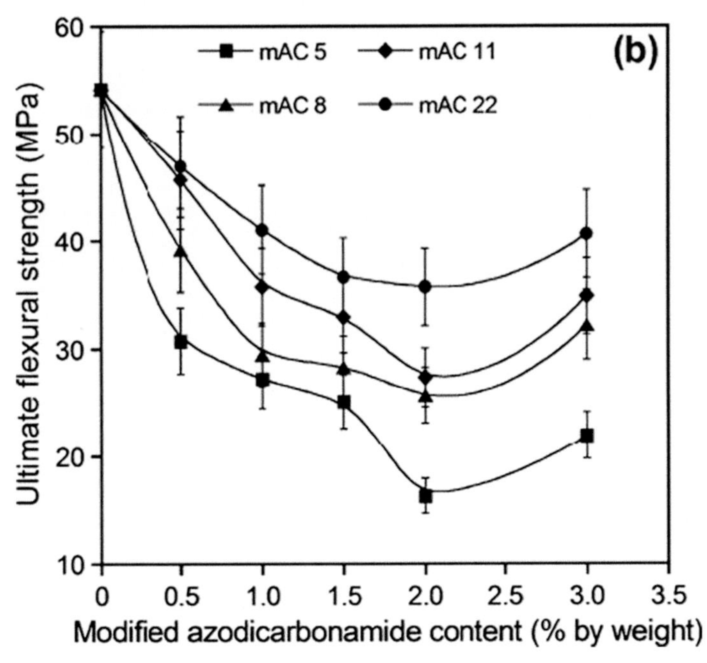

Fig. 4.9 (Top) Foam density corresponds to modified azodicarbonamide content in PVC-rice hull composite (mAC5, mAC8, mAC11 and mAC22 corresponds to a particle size of 5, 8, 11 and 22 μm) (Bottom) ultimate flexural strength (MPa) of foamed PVC-rice hull composite. From Petchwattana, N., Covavisaruch, S. (2011). Influences of particle sizes and contents of chemical blowing agents on foaming wood plastic composites prepared from poly(vinyl chloride) and rice hull. *Materials and Design*, *32*, 2844–2850, with permission of Elsevier.

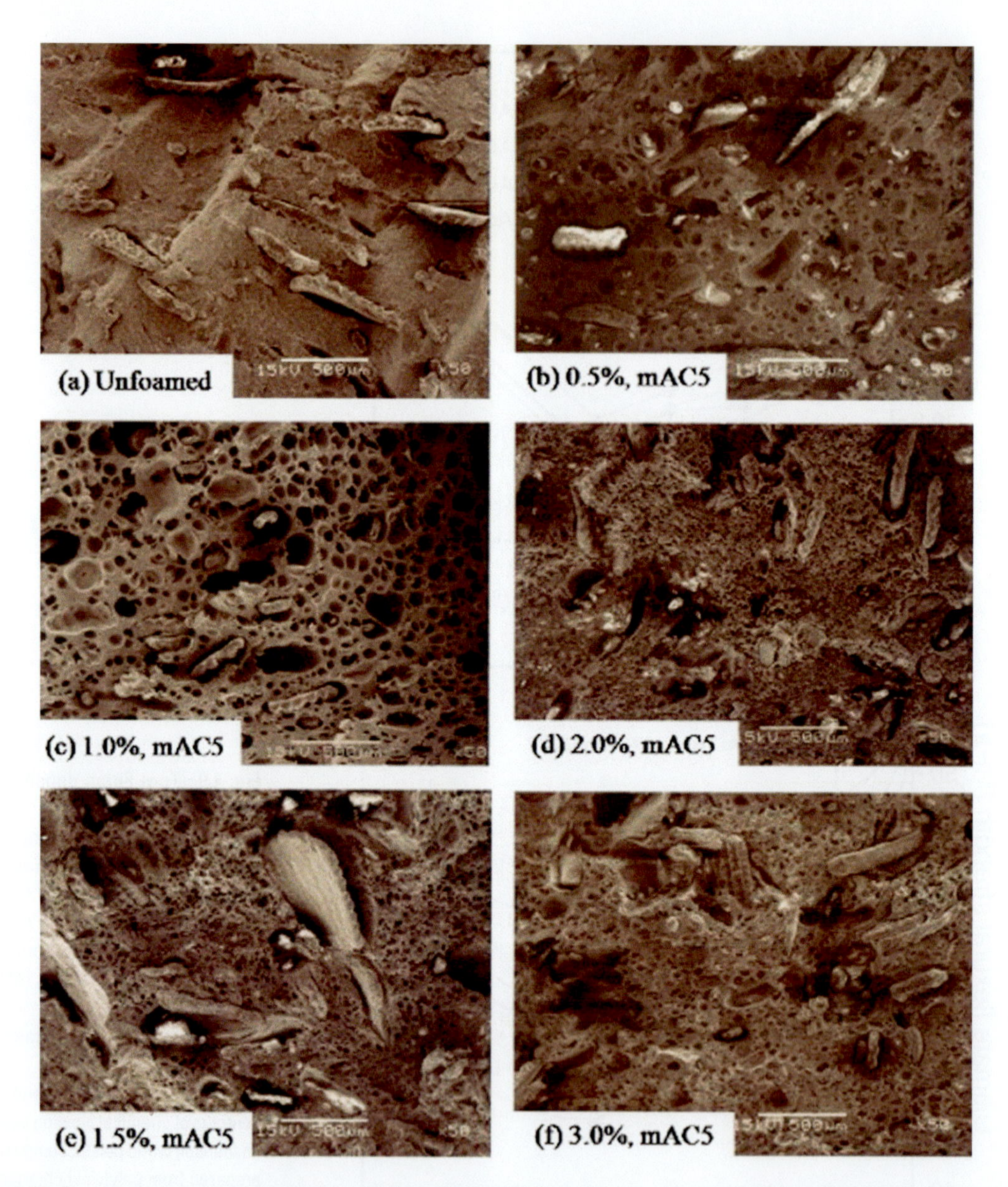

Fig. 4.10 Scanning electron micrograph for surface of foam PVC-rice hull composites (mAC5, mAC8, mAC11 and mAC22 correspond to the particle size of 5, 8, 11 and 22 μm). From Petchwattana, N., Covavisaruch, S. (2011). Influences of particle sizes and contents of chemical blowing agents on foaming wood plastic composites prepared from poly(vinyl chloride) and rice hull. *Materials and Design, 32,* 2844–2850, with permission of Elsevier.

is increased. In short, although the usage of a blowing agent is able to reduce cost and weight significantly, there is a caution of sacrificing the mechanical properties. Thus the formulation developer should initially look into laboratory experimental works in order to obtain the optimum combination to meet the product requirements.

4.5 MuCell and IQ Foam technology for weight reduction

MuCell is also known as microcellular injection moulding. This technology was developed by Massachusetts Institute of Technology (MIT) and licenced by Trexel Inc. It involves injecting of a supercritical blowing agent into the molten plastic during the injection moulding process (or, more accurately, during the metering or plasticizing period). During the moulding stage, the molten plastic dissolves blowing agent and experience pressure drops significantly where cell nucleation and growth happens, subsequently foaming takes place in the mould cavity. As seen in Fig. 4.11, the MuCell process mimics a conventional injection moulding machine with the addition of supercritical gas, N_2 or CO_2, injectors to introduce it into the barrel when the plastic is

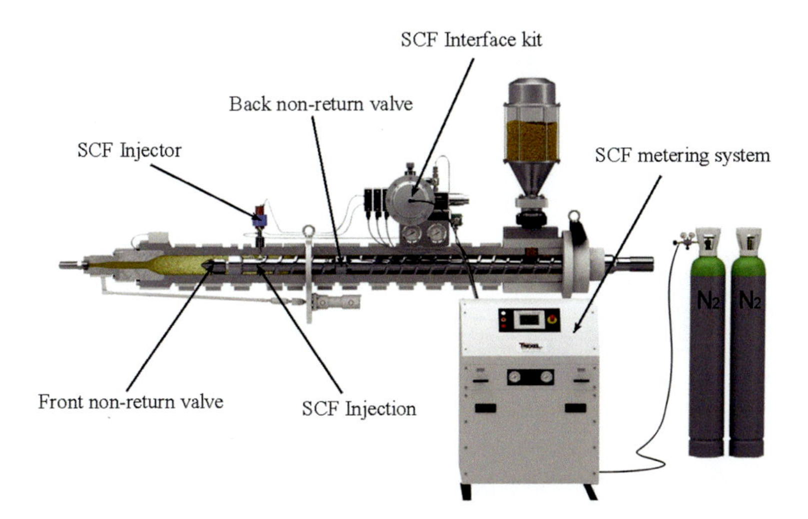

Fig. 4.11 MuCell technology assembly. From Gomez-Monterde, J., Hain, J., Sanchez-Soto, M., Maspoch, M. L. (2019). Microcellular injection moulding: A comparison between MuCell process and the novel micro-foaming technology IQ Foam. *Journal of Material Processing Technology, 268,* 162–170, with permission of Elsevier.

in the molten stage. In order to ensure that good mixing and homogenizing occurs in the molten plastic together with the supercritical fluid, a special reciprocating screw is engaged. A robust control of pressure is crucial to avoid foaming inside the barrel. Such a pressure control system is known as a microcellular plasticizing pressure (MPP) monitoring system. The main advantages of MuCell is weight reduction, good dimensional stability, energy and clamping force decrease, and shorter cycling time. However, in return, a higher operating cost is needed for the supercritical fluid consumption, surface quality is compromised and mechanical properties are lowered.

IQ Foam is a technology developed by Volkswagen AG in which the plastic resin is impregnated with gas before the melting process (see Fig. 4.12). The technology uses two units of gas injectors to introduce blowing agents with valves, which are used to regulate the flow of gas and lock it within the chambers before feeding into the barrel. It is worth noting that only a moderate to low gas pressure supply is needed compared to MuCell, which requires high-pressure gas. The step-by-step operation of IQ Foam technology is illustrated in Fig. 4.13. According to Gomez-Monterde et al. (2019), polypropylene-glass fibre reinforced composite for MuCell and IQ Foam can achieve densities as low as 0.82 and 0.81 g/cm^3, respectively, as compared to the density of the original compound of 1.04 g/cm^3. This indicates both technologies are about saving material use for the commitment to produce more environmentally friendly products.

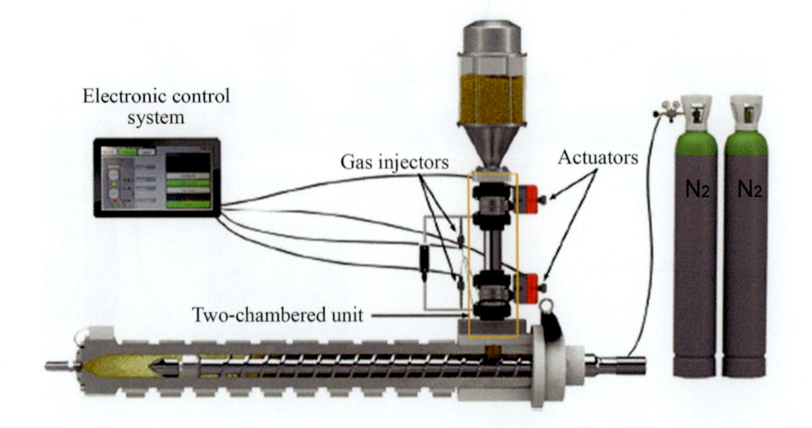

Fig. 4.12 IQ Foam technology assembly. From Gomez-Monterde, J., Hain, J., Sanchez-Soto, M., Maspoch, M. L. (2019). Microcellular injection moulding: A comparison between MuCell process and the novel micro-foaming technology IQ Foam. *Journal of Material Processing Technology, 268*, 162–170, with permission of Elsevier.

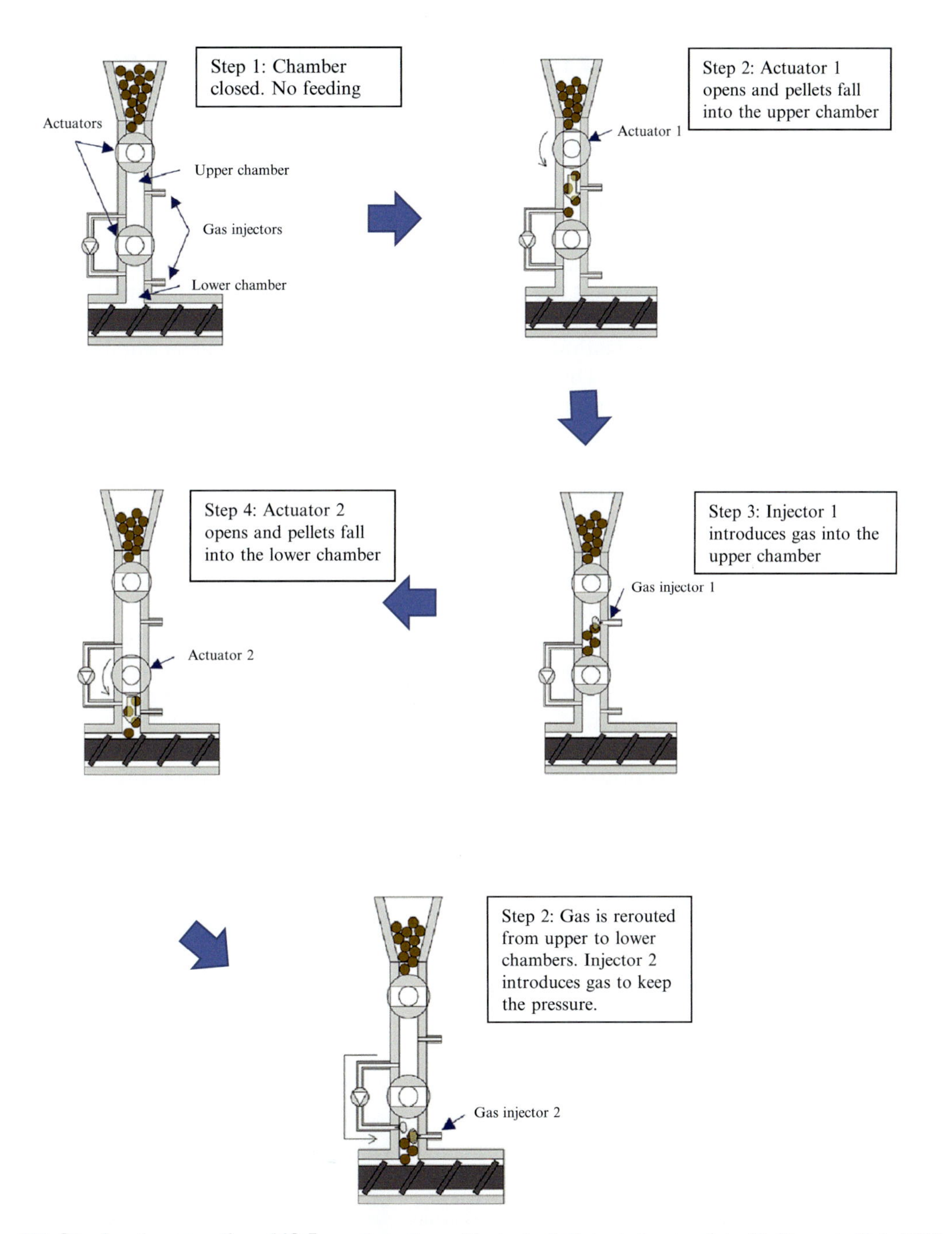

Fig. 4.13 Step-by-step operation of IQ Foam. From Gomez-Monterde, J., Hain, J., Sanchez-Soto, M., Maspoch, M. L. (2019). Microcellular injection moulding: A comparison between MuCell process and the novel micro-foaming technology IQ Foam. *Journal of Material Processing Technology, 268*, 162–170, with permission of Elsevier.

4.6 Pro-degradant/oxo-degradable additive

4.6.1 Background

As is known, the commodity plastics like polyethylene, polypropylene, polystyrene, etc. are all nondegradable, or degradation can only happen after a long period of time. In fact, the inertness of plastics towards many of the degradants like moisture, sunlight, acidic/basic conditions and microbial activities is the factor that prolongs the occurrence of plastic degradation. Nonetheless, plastics are formed through a polymerization reaction and the polymerization is a reversible process that needs catalysts to take place. Thus, based on this principle, when the plastics are brought to a suitable and catalyzed condition, degradation or chain scissioning can be accelerated, causing degradation to happen earlier by breakdown of the macromolecules to smaller fragments (ketones, aldehydes, carboxylic acids, alcohols, alcohols, etc.). As such, pro-degradants are additives being developed and used to serve the purpose of being added to plastic products so they can degrade earlier. The pro-degradant/oxo-degradable additives are made of transition metal salts or oxides such as cobalt, manganese or iron, as well as organic ligand complexes. There are two well-known producers of these additives, namely EPI Environmental TDPA and Symphony Environmental d2W additives. Both have been marketing their products for about a decade.

The most common plastic materials that involve the use of pro-degradant additives are plastic bags (commonly supermarket grocery bags and garbage bags), as shown in Fig. 4.14. They are also known as oxo-degradable plastic bags. When the plastic bags

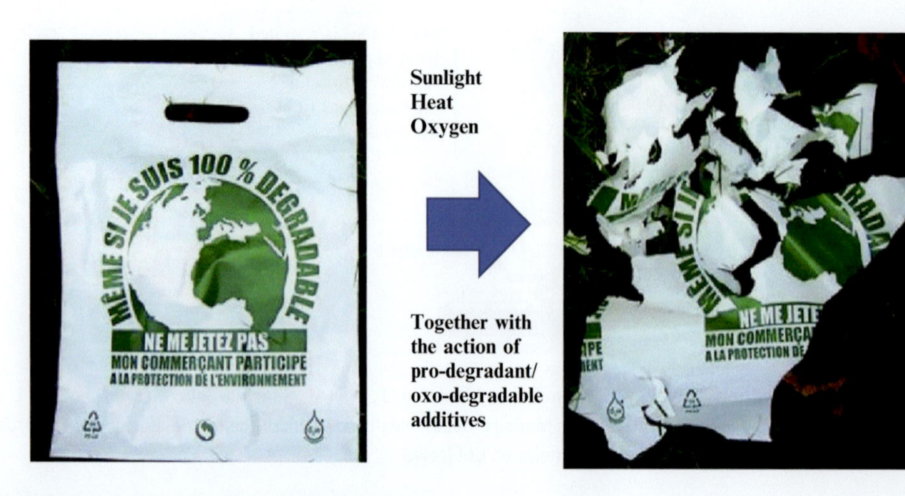

Fig. 4.14 Oxo-degradable plastic fragmentation process.

are exposed to sunlight and heat, they tend to weaken and finally break down into fragments. Such pro-degradant additives are also added to single-use food trays and containers; it can be noticed that the containers turn into a yellowish colour and become fragile at a faster rate. From the perspective of appearance, oxo-degradable plastic products can decrease in size and finally disappear.

4.6.1.1 Controversy about oxo-degradable plastics

Although oxo-degradable plastics can accelerate the breakdown of nondegradable plastics, during the fragmentation process, microplastics form that can still remain in the environment for some time, posing a pollution risk to the soil and aquatic organisms. Moreover, the oxo-degradable additives are made up of heavy metals as well, and the traces of such heavy metals together with the leftover microplastics can lead to environmental toxicity. A recent publication by Shruti et al. (2020) reports that such oxo-degradable single-use plastic exceeds the standard of international norms (94/62/EC; EN 13432; 2016/1416; ASTM D6400-04) for Cu, Cr, Mo, Zn, Fe and Pb with maximum concentrations of 1898 mg/kg, 1586 mg/kg, 95 mg/kg, 1492 mg/kg, 1900 mg/kg and 7528 mg/kg, respectively. Subsequently, the microplastics enter the food chain, which can lead to serious toxicity. In addition, oxo-degradable plastics are unable to be recycled, and can contaminate other nonoxo-degradable plastics to be recycled. The presence of pro-degradant additives can cause the neat polyethylene to decrease in mechanical strength due to the imminent thermal degradation during the plastic processing, which runs at melting temperature (Aldas et al., 2018).

In light of the seriousness of pollution caused by oxo-degradable plastics, the European Union has banned oxo-degradable plastic through Article 5 of the Single-Use Plastic Directive (EU) 2019/904. A news item in the Bangkok Post (2021) reported that Thailand is planning to ban oxo-degradable plastics together with single-use plastic bags by 2025. In the latest development, South Australia prohibited production and manufacture of oxo-degradable products starting March 1, 2022. All these countries have thoroughly debated and finally decided to ban oxo-degradable plastic to minimize its impact on the environment.

4.7 Conclusion

The application of blending technology is able to improve the ecofriendliness of plastic products, such as by blending with biomass and specialty additives to minimize the plastic composition of the products, as well as to reduce the overall weight of the products. In addition, the blending of recycled plastic with virgin plastics is another alternative to overcome the drawback of inferior properties of recycled plastic that results from undesirable degradation during first-time usage. As an advancement in current polymer processing technology, the addition of physical blowing agents is used in MuCell and IQ Foam to produce cellular/foamed plastic goods. Pro-degradants can be added to the plastics to improve their degradability, to reduce the burden of landfills in handling nondegradable plastic waste. All the approaches mentioned are mainly to mitigate and minimize the impacts of plastic wastes. Reduce, reuse and recycle are always the basic steps that need to be carried out first, followed by other technological efforts in order to effectively protect the earth from plastic pollution for the betterment of future generations.

References

Aldas, M., Paladines, A., Valle, V., Pazmino, M., & Quiroz, F. (2018). Effect of the prodegrandant-additive plastics incorporated on the polyethylene recycling. *International Journal of Polymer Science, 2018,* 2474176.

Bangkok Post. (2021). *Oxo-degradable plastics worry firms.* 7 Jun 2021.

Gomez-Monterde, J., Hain, J., Sanchez-Soto, M., & Maspoch, M. L. (2019). Microcellular injection moulding: A comparison between MuCell process and the novel micro-foaming technology IQ Foam. *Journal of Material Processing Technology, 268,* 162–170.

Li, Z., Zhou, S., Xu, B., Fan, M., & Su, J. (2020). Effect of crystalline morphology on electrical tree growth characteristics of high-density and low-density polyethylene blends. *IEEE Access, 8,* 114413.

Petchwattana, N., & Covavisaruch, S. (2011). Influences of particle sizes and contents of chemical blowing agents on foaming wood plastic composites prepared from poly(vinyl chloride) and rice hull. *Materials and Design, 32,* 2844–2850.

Shruti, V. C., Pérez-Guevara, F., Roy, P. D., Elizalde-Martínez, I., & Kutralam-Muniasamy, G. (2020). Identification and characterization of single use oxo/biodegradable plastics from Mexico City, Mexico: Is the advertised labeling useful? *Science of The Total Environment, 739,* 140358.

Wambua, P., Ivens, J., & Verpoest, I. (2003). Natural fibres: Can they replace glass in fibre reinforced plastics? *Composites Science and Technology, 63,* 1259–1264.

5

Effective plastic design and packaging

5.1 Multilayer packaging

Polymers (also known as plastics) are the most commonly used materials in packaging applications. Packaging is applied to protect products such as food materials by separating the products from external influences, including physical, chemical and biological effects. The packaging material must be able to protect the packaged products from external effects, be safe to consumers (especially for food packaging), and be easy to obtain and produce at a reasonable cost (Pandey et al., 2020). Polymers such as polyethylene (PE), polystyrene (PS), polyethylene terephthalate (PET), polyvinyl chloride (PVC), polyamide (PA), etc. are commonly selected as packaging materials due to their excellent mechanical, gas permeability and moisture barrier properties, and their low cost (Asgher et al., 2020; Sharanyakanth & Radhakrishnan, 2020). The use of these polymers in packaging applications can also be attributed to their easy availability, low material costs, and light weight (Nguyen et al., 2020). The use of polymers as packaging material is very important in various industries, especially in food processing (Mahmoudi & Parviziomran, 2020).

In general, the types of packaging are classified into three main categories based on their functionality: primary packaging, secondary packaging and tertiary packaging. Primary packaging refers to packaging that is in direct contact with the product, and is usually the first direct protection of the product (Chung et al., 2018). For instance, bottles in the food industry that hold products such as beverages, cooking oil, peanut butter, etc. are classified as primary packaging. It is very important to select suitable polymer materials as primary packaging, especially for the food processing and preservation industries. These selected polymer materials must be nontoxic and nonreactive with the contact

Plastics and Sustainability. https://doi.org/10.1016/B978-0-12-824489-0.00004-0
Copyright © 2023 Elsevier Inc. All rights reserved.

food, beverage or other product, and be able to effectively protect the products from external physical, chemical and biological detrimental effects (Ferrara et al., 2020; Wan et al., 2020). Secondary packaging refers to a layer of packaging outside the primary packaging (Mahmoudi & Parviziomran, 2020). In other words, secondary packaging is used as a protection layer to the primary packaging. Secondary packaging can be used to pack several primary packages together for better storage efficiency, ease of product distribution and convenience. In addition, the application of secondary packaging could also preserve the quality of the product in primary packages during transportation: for example, protect the product from small damage caused by mechanical factors, such as change of shape and weight losses during transportation activities (Albaar et al., 2016). Tertiary packaging (also known as transit packaging) is used for bulk handling in warehouse storage and transportation/shipping. Tertiary packaging is very important, especially for transport and shipping in any production and distribution network, where it could enhance the logistical efficiency (Chung et al., 2018; Mahmoudi & Parviziomran, 2020).

Packaging is crucial in various types of goods production, especially in the food industry. In general, the application of packaging in manufacturing industries is mainly to provide suitable protection of products or goods from external factors. For example, packaging is very important for electrical and electronic components or devices in the electronics industry, serving as protection from external effects such as scratching or contamination from dirt, moisture and other hazards. As mentioned, the use of packaging is also very important in the food and beverage industries, where it can be treated as a storage container to store food products. In addition, it is used in the food industry to protect food products from the penetration or invasion of insects, mite pests and other external bicontaminants (Stejskal et al., 2017). Suitable and effective packaging can help to reduce significant losses of stored food from insect infestations or other biocontaminants (Hagstrum & Sbbramanyam, 2009).

5.1.1 Permeability and compatibility

The selection of materials having excellent resistance ability against the diffusion of gases, water, organic liquids and vapours is very important in the packaging industry. The excellent barrier behaviour and low cost of polymer materials are the main reasons why the polymer materials are widely used in the packaging industry. The usage of polymer packaging is very important, especially in the food industry, biomedical industry, and others, in protecting the products against the exposure of biocontaminants, air, moisture and other contaminants that might be harmful to consumers (Rudmann et al., 2020). Numerous research studies have been conducted to improve the mechanical properties and barrier behaviour of polymeric packaging materials (Thellen, Schirmer, Ratto, Finnigan, & Schmidt, 2009). Currently, the packaging films most commonly used in industries are monolayer packaging materials, made of polyethylene, polypropylene, polyester, etc. (Stejskal et al., 2017). However, some modification methods for polymer packaging materials have been conducted to improve the barrier and mechanical properties of these films. The incorporation of montmorillonite layered silicates into polymer has been found to improve the barrier and mechanical properties by dispersing the layered MMT silicates in the polymer matrix (Bee et al., 2014; Thellen et al., 2009). The good dispersion of montmorillonite in a polymer matrix could create an increased travelling path length of the solutes to diffuse through the polymer matrix (Bharadwaj, 2001; Thellen et al., 2009). On the other hand, the use of multilayer packaging has become increasingly important in the food industry due to its excellent gas barrier property and its improved mechanical properties. Multilayer packaging can better control the release of active compounds from the packaged food and prevent the loss of active substances through diffusion to the environment. Also, the multilayer structure is important in the food packaging industry, as the pristine polymer layer laminates the outer recycled polymer layer to prevent the passage of unwanted substances to the food (Chytiri et al., 2006).

On the other hand, a number of research studies have been conducted on laminating polymer films with cellulose nanomaterials such as cellulose nanocrystals, cellulose nanofibrils, etc., to induce barrier performance of polymer packaging materials (Wang, Yao, Wang, & Li, 2017). The use of multilayer packaging with cellulose nanomaterials was found to display excellent oxygen barrier behaviour in comparison to other common polymer materials packaging such as vinyl alcohol (EVOH). The polarity difference between the cellulose nanomaterials and oxygen significantly reduces the solubility of oxygen when contacting with oxygen gas (Wang et al., 2020). This is because the oxygen gas molecules are blocked from passing the narrow gap of cellulose nanomaterials by strong H-bonding (Aulin et al., 2010). Furthermore, the random arrangements of cellulose nanomaterials create a tortuous and lengthy path that further reduces the permeability of oxygen gas through the cellulose nanomaterials, as shown in Fig. 5.1 (Wang et al., 2020). As seen in Fig. 5.1, the very random arrangement of cellulose nanofibrils can induce a travelling path

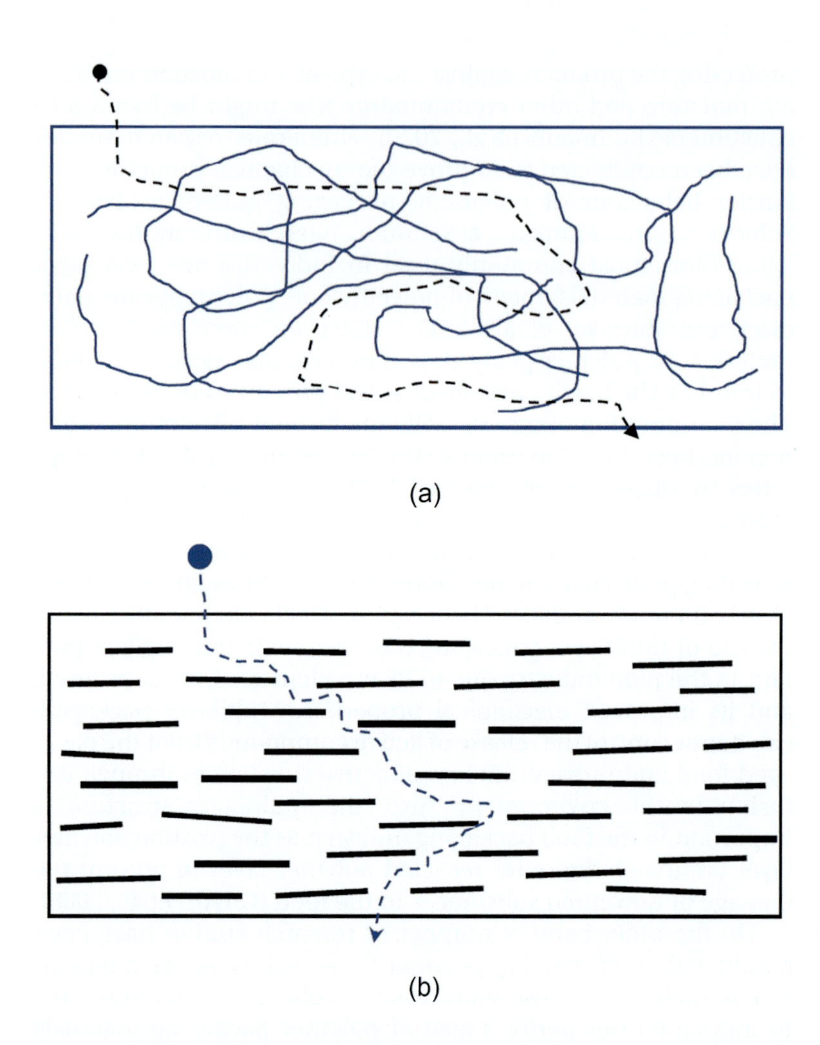

Fig. 5.1 The diffusion path of gases through the polymer of (A) cellulose nanofibrils and (B) cellulose nanocrystals.

for the oxygen gas to diffuse through the polymer matrix, thus improving the gas barrier effect of the matrix. On the other hand, the nanofibres of cellulose nanocrystals are mainly uniform and rigid in their structure, which causes them to be arranged in a more ordered structure in the polymer matrix. The higher ordered structure of the nanocrystals can significantly increase the density. The high density of cellulose nanocrystals also indicates the limited presence of voids in the polymer matrix, which could restrict the diffusion of oxygen (Wang et al., 2020). According to various research studies conducted by Shamini and Yusoh (2014) and Visanko et al. (2015), the oxygen diffusion barrier effect of multilayers of PP laminated with cellulose nanomaterials such as cellulose nanofibrils or cellulose nanocrystals is highly increased due to the longer diffusion path. Multilayer packaging laminated with cellulose nanofibrils was also found to provide a better oxygen barrier effect than multilayer packaging laminated with cellulose nanocrystals. In addition, multilayer packaging with cellulose nanofibril and cellulose nanocrystal films was also found to have excellent water vapour barrier properties. According to Wang et al. (2020), an adhesive layer is necessary between the PP layer and cellulose nanomaterials to reduce the penetration of water vapour through the cellulose nanomaterials. This is because the high numbers of pores on the surface of cellulose nanomaterials indicate a greater surface area. This can further cause water vapour to easily weaken the adhesion effect between the interface of the PP layer and the layer of cellulose nanomaterials by diffusing into the cellulose nanomaterials (Wang et al., 2020). An adhesive layer is applied between the PP layer and cellulose nanomaterials during lamination to reduce the contact area available for water vapour diffusion. Therefore the presence of the adhesive layer could limit the swelling effect of the cellulose nanomaterials layer by resisting the water vapour.

Various methods have been investigated and conducted to reduce the use of polymer materials by recycling the polymer materials, due to restrictions on the disposal of polymer materials in landfills in some countries. Most of the polymer materials are widely used for packaging purposes, especially in the food processing and manufacturing industries due to the good resistance ability of polymer materials against water vapour and oxygen gas, and their considerable mechanical properties. In order to reduce the disposal of polymer wastes in landfills, some researchers have suggested to recycle the used polymer in various industries including the food packaging industry. However, the recycled polymer matrix may contain some hazardous chemical species or degraded chemical residues that might migrate from the packaging layer to the food surface. Because of this, multilayer packaging is widely employed in the food industry where the outer layer of the multilayer packaging is made of recycled polymer materials and the inner layer, which is in contact with food, is made of pristine polymer materials (Pauer et al., 2020). On the other hand, biodegradable polymers such as polylactic acid, polyvinyl alcohol, etc. have also been used to replace nonbiodegradable polymer materials as the polymer packaging materials in the food packaging industry. However, the cost of the biodegradable polymers is significantly higher than that of common nonbiodegradable polymers. In order to overcome the cost of the biodegradable materials, multilayer packaging of biodegradable polymers and recycled polymer materials is applied in food packaging industries. The inner layer that contacts the food is made of thin pristine biodegradable polymer, while the outer layer is made of recycled polymer materials separated from the food by the biodegradable polymer.

The preparation of multilayer packaging with active packaging technology can be achieved by coating the antimicrobial agents directly onto the surface of the polymer. This type of active packaging in multilayers is efficient in protecting packaged food from recontamination by microbials by controlling the release of an active antimicrobial compound from the multilayer packaging to the prepacked food. The active packaging with multilayers consists of an outer layer of with high permeability resistance properties to prevent the loss of chemical substances to the environment and an inner layer to control the mass flux of active antimicrobial agent release. The thickness, structure and chemical composition of the inner layer are important factors that could control the release of the active antimicrobial agent. The selection of active antimicrobial additives is highly dependent on the application's requirements. In a study conducted by Gherardi, Becerril, Nerin, and Bosetti (2016), they incorporated an antimicrobial agent into the adhesive layer that sticks the polyethylene layer to the aluminium, as illustrated in Fig. 5.2. They found that the growth of microorganisms was only exhibited in the polyurethane layer contacting with the adhesive without an antimicrobial agent. This indicates that the addition of an active antimicrobial agent could effectively eliminate the growth of microorganisms. According to this research, the addition of an active antimicrobial agent (essential oils of *Cinnamomum zeylanicum* and *Origanum vulgaris*) into the adhesive layer showed the antimicrobial activity of the prepared plastic packaging against the growth of three

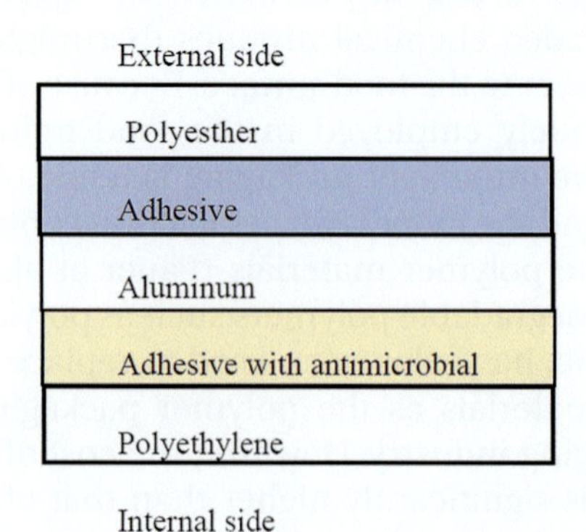

Fig. 5.2 The structure of multilayer antimicrobial packaging materials for tomato puree.

different microorganisms (bacteria, mould and fungi). The protective action of active packaging with multilayers is achieved through the release of antimicrobial substances from the adhesive layers of multilayer packaging to the contacted food. In this research, they also mentioned that the active multilayer packaging can be prepared without a significant increase in production cost with effective antimicrobial activity. In addition, it is important for the chemical composition of the adhesive layer (with active antimicrobial agent added) in active multilayer packaging to reach a critical antimicrobial activity. The incorporation of an antimicrobial agent in the adhesive layer of multilayer packaging is a new technology that is effective and efficient without greatly increasing the cost of production.

In the food packaging industry, various research studies have been conducted to prolong the storage shelf life of food by improving the antioxidant packaging materials. Researchers investigated the effect of using different types of antioxidant additives, polymer packaging matrices, the selection of processing techniques and efficient controlled release methods. Packaging with antioxidant additives can release antioxidants slowly and continuously from the packaging film to the food surface at a certain range of release rate. Thus the concentration of the active components will slowly control the migration and diffusion of active components around the food products inside the packaging. The uses of multilayer packaging in food packaging are also important in the raw food processing industry, such as poultry, meat, spices, frozen shrimp, frozen fish, etc. In the food processing industry, the prepacked food will be subjected to irradiation to eliminate the possibility of microbial contamination (Chytiri et al., 2006). However, the irradiation action on the prepacked food might cause the polymer packaging to be degraded by releasing unwanted substances. The use of multilayer packaging could prevent permeability of chemical substances released by the outer polymer layer due to irradiation treatment to the packed food by laminating a layer of polymer materials with higher resistance ability against the low irradiation dosage (Chytiri et al., 2006). The use of low irradiation on the prepacked food is necessary and important in avoiding recontamination by microbials. To prevent the generation of chemical substances during the irradiation process, the packaging materials that are in contact with food must be virgin polymers with higher irradiation resistance ability. However, pristine polymer with high irradiation resistance effects poses a higher material cost. An outer layer of polymer packaging made of recycled polymer resin can be coated on the outer surface of the virgin polymer packaging materials to reduce the total packaging cost. The layer of pristine polymer resin with higher irradiation resistance ability laminated with a recycled polymer layer could reduce the material cost and avoid the transfer of unwanted chemical substances to the food. In a study conducted by Pauer et al. (2020), a multilayer of polyethylene and polyamide packaging provided good resistance towards water vapour and oxygen permeability with considerable mechanical properties. These results have led to multilayers of polyethylene and polyamide becoming commonly used in the food industry. Besides, the transparency of this multilayer packaging of polyethylene and polyamide is also attractive to be used as food packaging, especially for smoked meat products such as bacon. The common packaging systems used in bacon in blocks involve thermoforming the multilayers of PP and PA into films, vacuum bags, etc.

A study conducted by Chytiri et al. (2006) investigated the effect of replacing virgin LDPE with recycled LDPE on mechanical properties and gas permeability of LDPE multilayer packaging materials. Referring to Fig. 5.3A, increasing the recycled amount in the multilayer packaging slightly reduced the tensile strength. However, increasing the recycled LDPE amount in the multilayer packaging was found to provide higher rigidity at 50% recycled LDPE in comparison to others. On the other hand, the replacement of virgin LDPE with recycled LDPE was found to provide a less significant effect on the gas and water vapour transmission. By referring to Table 5.1, use of recycled LDPE was found to provide a similar permeability effect of gases and water vapour to virgin LDPE. This indicates that the use of recycled LDPE in multilayer packaging did not have any significant effect on the permeability resistance of the multilayer packaging materials. The permeability of gases such as oxygen and carbon dioxide and water vapour through the multilayer LDPE packaging is highly dependent on the degree of crystallinity and the molecular structure of the polymer matrix. This indicates that the replacement of virgin LDPE with recycled LDPE does not change or affect the crystallinity and molecular structures of LDPE in polymer matrix. On the other hand, the application of irradiation on all the LDPE packaging (100% virgin LDPE, 50% virgin LDPE + 50% recycled LDPE and 100% recycled LDPE) was also observed to have a less significant effect on the permeability of all samples.

The use of irradiation on LDPE packaging could induce the formation of crosslinking, chain scissioning and free radical recombination, among other effects, in LDPE multilayer packaging. Referring to Fig. 5.3, increasing irradiation doses gradually increased the tensile strength of the LDPE multilayer packaging with 100% virgin LDPE added and the samples with 50% virgin

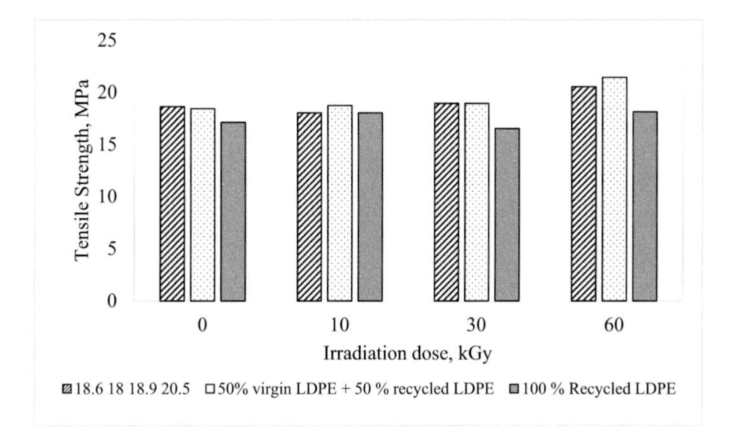

(a) Tensile strength

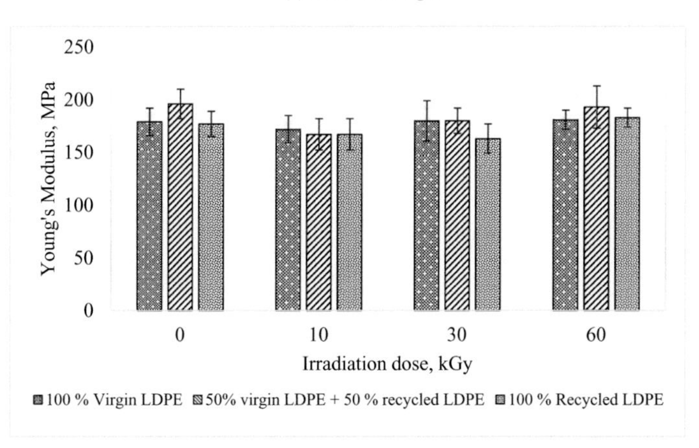

(b) Young's modulus

Fig. 5.3 The effect of replacing the virgin LDPE with recycled LDPE on (A) tensile strength and (B) Young's modulus of LDPE multilayer packaging materials. Data collected from Chytiri, S., Goulas, A. E., Riganakos, K. A., Kontominas, M. G. (2006). Thermal, mechanical and permeation properties of gamma-irradiated multilayer food packaging films containing a buried layer of recycled low-density polyethylene. *Radiation Physics and Chemistry, 75*, 416–423.

LDPE and 50% recycled LDPE added. This might be attributed to the formation of a crosslinking network that could slightly increase the tensile strength of LDPE multilayer packaging with 100% virgin LDPE and LDPE multilayer packaging with 50% virgin LDPE + 50% recycled LDPE added. This indicates that increasing the recycled polymer amount did not restrict the generation of free radicals by irradiation to form the crosslinking networks. However, the formation of crosslinking networks was observed to have a less significant effect on the permeability resistance of gases and water vapour. The formation of crosslinking networks in the polymer matrix of all types of LDPE multilayer packaging did not change the molecular structure of LDPE very much, so the permeability to gases and water vapour was thus not affected by the irradiation treatment.

Table 5.1 The effect of irradiation doses on the permeability to oxygen, carbon dioxide and water vapour of LDPE multilayer packaging with 100% virgin LDPE, 50% virgin LDPE + 50% recycled LDPE and 100% recycled LDPE.

LDPE multilayer packaging	Irradiation doses (kGy)	Oxygen[a] ($cm^3/m^2/day$)	Carbon dioxide[a] ($cm^3/m^2/day$)	Water vapour ($g/m^2/day$)
100% virgin LDPE	0	2693 ± 71	$10,572 \pm 235$	1.4 ± 0.1
	10	2893 ± 176	$11,268 \pm 685$	1.1 ± 0.2
	30	2617 ± 77	$10,359 \pm 396$	1.1 ± 0.1
	60	2799 ± 48	$11,226 \pm 320$	1.1 ± 0.1
50% virgin LDPE + 50% recycled LDPE	0	2814 ± 52	$11,156 \pm 308$	1.0 ± 0.2
	10	2747 ± 162	$10,998 \pm 689$	1.1 ± 0.1
	30	2888 ± 136	$11,252 \pm 659$	1.3 ± 0.1
	60	2883 ± 48	$11,632 \pm 320$	1.5 ± 0.1
100% recycled LDPE	0	2640 ± 38	$10,364 \pm 212$	1.4 ± 0.2
	10	2704 ± 109	$10,850 \pm 489$	1.4 ± 0.1
	30	2769 ± 101	$11,030 \pm 456$	1.6 ± 0.1
	60	2747 ± 89	$10,912 \pm 398$	1.4 ± 0.2

[a]Tested under atmospheric conditions.
All the data were collected from Chytiri, S., Goulas, A. E., Riganakos, K. A., Kontominas, M. G. (2006). Thermal, mechanical and permeation properties of gamma-irradiated multilayer food packaging films containing a buried layer of recycled low-density polyethylene. Radiation Physics and Chemistry, 75, *416–423.*

5.1.2 Weight and material savings

Flexible multilayer polymer packaging is popular for food and medical packaging and its use has increasingly gained attention in these industries. This is because multilayer polymer packaging has lower energy consumption during production, and its use leads to lower packaging weight and reduced material usage. Multilayer packaging with different types of materials has led to packaging films with considerable functional properties, such as high resistance to water vapour permeability, effective oxygen barrier and good mechanical properties, among other benefits (Soares et al., 2022), because different types of materials have different functional properties. Multilayer packaging can achieve the performance requirements of the packaging industry by using suitable combinations of various polymer materials with the required functional properties. Table 5.2 shows that different types of polymer materials offer different functional properties, causing the selection of suitable polymer materials to become very important in multilayer packaging.

Table 5.2 The functionality and application of different type of polymers.

Polymers	Functionality	Application
Polyethylene	Heat resistance; moisture barrier; good resistance to gas or aroma diffusion effects	Beverage
Polypropylene	Good mechanical properties; good resistance against gas or aroma permeation; moisture barrier	Microwaveable food packaging
Polyethylene terephthalate (PET)	Considerably good mechanical properties; moisture barrier; good resistance against gases and aroma	Food packaging, especially bottles, packaging for meat, etc.
Polystyrene	Considerably good mechanical properties; good resistance effect against gases or aroma diffusion; printable	Outside layer of food packaging for printing purposes
Polylactic acid (PLA)	Considerably good mechanical properties; moisture barrier; good resistance effect against gases or aroma	Disposable packaging, cups and container

In the fresh and raw food packaging industry, multilayer packaging films can consist of various layers of different polymer materials with different functional properties. In this application, the packaging materials are required to provide a good oxygen and water vapour barrier with good mechanical properties. The types of polymer materials used for producing multimaterial multilayers are basically dependent on the specific industry's requirements and the development techniques of the multilayer packaging. In order to achieve these requirements, the polymer materials selected to produce multilayer packaging materials must be able to be fabricated together using the same development techniques. The use of thin multimaterial multilayer packaging materials can minimize the thickness of the packaging to avoid wastage of the polymer materials used as multilayer packaging films, while offering various effective functional properties, such as resistance to oxygen and aromas as well as suitable mechanical properties (Mieth, Hoekstra, & Simoneau, 2016). The reduction of the total polymer volume used in fabricating multimaterial multilayer packaging limits the overuse of polymer materials in packaging applications. Thus the waste disposal of multimaterial multilayers can be significantly decreased due to the reduction of the polymer volume. However, multimaterial multilayer packaging causes the recycling of the polymer materials to be more difficult, because polymer waste consisting of multiple types of polymer materials in layers is hard to separate.

Currently, the packaging industry is also required to meet sustainability criteria according to the requirements of packaging uses (Pauer et al., 2020). Sustainable packaging materials play a major role as product protection, indicating the high priority of packaging in various industries, such as the food industry, in securing or protecting the product. Thus the environmental impact of a prepacked product is much greater due to the environmental impact of the packaging itself. Due to the environmental impact of polymer packaging waste, the recyclability of polymer materials as packaging is one of the main approaches in the packaging industries.

In the food and beverage industries, a recycled polymer materials layer is used as a base to prepare the multimaterial multilayers. In the uses of multimaterial multilayer packaging, the functional properties of the multimaterials are highly prioritized over the recyclability of the polymer multilayer packaging. Recycling of postmultilayer polymer packaging for certain specific polymer materials has already been achieved and applied by some companies. Basically, the multilayer polymer packaging consists of various layers with different thicknesses. This homogeneous multilayer polymer packaging is produced through a coextrusion or lamination process. According to Pauer et al. (2020), reducing the thickness of the nonrecyclable, lightweight flexible polymer multilayer packaging can maintain and even improve the functional properties. However, the use of different types of polymer materials basically has restricted the recyclability of the waste from multimaterial multilayer polymer packaging and thus increased the environmental burden.

According to Pauer et al. (2020), food loss prevention is the priority in food packaging design, as food loss could result in a higher environmental burden. Then, lightweight designs are also prioritized for food packaging over the recyclability due to the environmental burden. This is due to the fact that the separation of different polymer materials from multilayer packaging films requires a complex and difficult process. The separation process requires knowledge of chemical reactions to separate the different contacted polymer materials. This delamination step is an important step in the recycling process and is required to separate the multimaterials in the multilayer packaging before proceeding to the rest of the recycling process. However, the delamination process of multilayer polymers is not commonly used to separate the polymer layer before proceeding to the main recycling action.

The difficulties of recycling the multilayer polymer packaging with immiscible polymers can be overcome by the use of compatibilizers. However, these recycling methods are not commonly used on an industrial scale. Another possible solution for recycling the multilayer flexible polymer packaging is using chemical recycling. This process consists of chemolysis and pyrolysis. Chemolysis of condensation polymers such as PET and PA produces valuable monomers, while pyrolysis of mixture polymers produces waxes and gaseous and liquid fuels (Ragaert, Delva, & Geem, 2017).

As mentioned, two general methods can be used to recycle multilayer packaging items: separation of different components for recycling purposes in other industries and processing the used multilayer packaging using a compatibilizer (Kaiser, Schmid, & Schlummer, 2018). Delamination of the multilayer packaging system is one of the separation methods for recycling purposes. The delamination method can be conducted by performing a chemical decomposition of the adhesive layer or adhesive between the multilayers. According to Zhang et al., 2015, the delamination of multilayers is currently carried out using a postconsumer liquid packaging board through a bonding damage concept. In the delamination process for food packaging materials made of polyethylene/aluminium/PET multilayer packaging, an acetone solvent is used to dissolve the adhesive between the multilayers for reducing the bonding strength between the layers (Soares et al., 2022). However, the application of physical methods in the delamination process is unrealistic, attributed to the presence of bonding strength. The physical processing methods of delamination for multilayer packaging waste can be carried out by using various methods like shedding and washing (Kaiser, Schmid and Schlummer, 2018). The delamination process has involved a shredding method followed by a microemulsion process with the application of a surfactant. This delamination technique is able to reduce the interlayer forces caused by the adhesion effect and thus separate the multilayer structure of PE-PET, PE-aluminium, PP-aluminium, etc.

Another technique, a selective dissolution separation technique, uses a solvent to separate the layers of multimaterial multilayer packaging. The selective dissolution of thermoplastic polymers such as polyethylene, polypropylene, polystyrene, polyvinyl alcohol and PET is used as an important recycling technique by dissolving a selective polymer. Because different types of polymers pose different solubility behaviour, a range of solvents will be selected to dissolve the targeted polymers and keep the other polymers in the solid phase. This selective dissolution technique can be used to selectively recover a specific polymer for further purposes at a high purity. By using this technique, the end products produced from the separated recycled polymers can be higher quality, comparable to the end products produced from virgin polymer resins.

On the other hand, the polymer wastes of multimaterial multilayers are relatively unstable in terms of mechanical stability. The second basic recycling method is conducted without performing multilayer separation, by using a compatibilizer. Compatibilizers are chemical substances used to recycle multilayers by increasing their mechanical stability as a compatibilized single stream without separation (Soares et al., 2022). The application of compatibilizers can increase the miscible ability of all polymer layers by incorporating another chemical substance to reduce the interlayer bonding. The selection of the suitable compatibilizer is directly dependent on the types of polymer components inside the multilayer polymer packaging, due to the blocky structure of the compatibilizer.

On the other hand, emerging treatment methods are being applied to recycle multilayer polymer wastes that are difficult to be recycled in normal ways. The recycling of multilayer polymer waste through emerging treatments might release some monomers and hydrocarbon gases. This process is normally performed through gasification, pyrolysis and/or hydrocracking and is conducted under high temperatures in the absence of oxygen gas.

5.2 Food safety requirements and specialty packaging

Packaging is a very important coordinated system in the manufacturing sector involving processing, transportation, goods distribution, retailing and protecting goods to fulfil market demands. In the food industry, food packaging is divided into two main categories, which are primary packaging and secondary packaging. The selection of food packaging materials, especially primary packaging, is very important due to the fact that direct contact of packaging materials with the food is unavoidable. The main role of packaging in the food industry is to act as a protective layer to the food by separating the food from the external environment. This packaging layer can protect the food from any external contaminants and ensure the packed foods maintain a good appearance and condition (preventing any shape distortion of the foods due to vibration effects) during transportation. In addition, the use of food packaging can also ease the food-handling process during the transportation and product distribution process. Food packaging also helps to preserve the food for a longer storage period due to the time loss during transportation. Packaging also plays another important role in the food industry, by providing customers with information such as the ingredients, production date, expiry date, instructions for use, tracking and tracing of the foods products, etc. Furthermore, the food packaging can help in marketing the food products, as fanciful and beautiful packaging designs can attract more customers to spend money on the products. According to the Food and Agriculture Organization of the United Nations, food packaging materials must meet main requirements: have good mechanical properties (especially tensile strength); must not release any harmful substances that might contaminate the contact food; recycled polymer materials are prohibited from being in direct contact with food; and regulations affect the usage of polyvinyl chloride as packaging material, with excess chloride monomers.

According to the Food Safety Practices and General Requirements for Food Standards Australia New Zealand (2021), the selection of packaging materials must fulfil three basic requirements, as shown in Table 5.3. The selection of food packaging materials is very important, especially when the packaging could make the food unsafe due to the potential of food contamination by the packaging or during the packaging process. Improper food packaging materials, such as dirty packaging or damaged packaging, could also cause food to be contaminated by harmful

Table 5.3 Regulations regarding food packaging requirements for various countries.

Regulations	Food packaging requirements
Food Standards Australia New Zealand Standard 3.2.2: Food Safety Practices and General Requirements (2000)	• Only use suitable packaging materials that are fit for their intended purpose • The materials used in food packaging are not likely to contaminate the food • Ensure there is no likelihood that the food could be contaminated during packaging • No potential of packaging that could cause the food to be unsafe • No chemicals that can leach from the packaging materials under certain conditions
Malaysia Food Regulations 1985 P.U. (A) 437/85 Incorporating P.U. (A) 200/2017	• Harmful packaging is prohibited as food packaging • The usage of polyvinyl chloride polymer is prohibited due to the fact that a release of vinyl chloride monomer could contaminate the contact food • The use of recycled packages is prohibited for food packaging • The use of damaged packages is also prohibited

Continued

Table 5.3 Regulations regarding food packaging requirements for various countries—cont'd

Regulations	Food packaging requirements
US Food & Drug Administration (FDA) Section Food Ingredients & Packaging	• Only chemical substances intended for use in food packaging are allowed to be used in food packages • The use of BPA (bisphenol A) and melamine as additives in plastic packaging is not allowed • The food packages used for food packaging must pass the migration test to avoid the migration of chemical substances from the plastic packaging to food
Law of People's Republic of China on Agricultural Product Quality Safety (2006)	• The additives in food packages, such as food containers, to improve the quality and properties of the package materials must meet the expected purpose during the packaging preparation • Harmful additives must not be used in food packaging that has direct contact with food
Food Safety and Standards (Packaging Regulations, 2018), India	• The packaging materials that come in direct contact with food packaging shall be food-grade quality • Ensure the packaging material is suitable for the type of food, the storage conditions and also transportation conditions • The packaging materials must be able to withstand the mechanical, chemical and thermal stresses during the transportation conditions • The packaging materials used as food packaging shall not alter the quality and safety of packaging materials that might cause the food to become unsafe for consumers

microorganisms and be unsafe for the consumers, or cause food wastage. Because of this, the packaging materials must possess good mechanical properties to withstand external impacts. The leaching of chemicals from the packaging materials to the contacted food is also a serious concern in the food industry. The leaching of chemicals into the contacted food could be prevented by reducing the use of some harmful additives in polymer packaging materials, such as antioxidants, colourants, etc. Basically, the food packaging requirements of most countries are quite similar in terms of prohibition of chemical substance migration, suitable mechanical and thermal stress properties, and other stipulations. According to Table 5.3, the basic requirements of food packaging in most countries focus on preventing the packaging from contaminating the food products with harmful substances, having good packaging mechanical strength and thermal resistance, and prohibiting recycled plastic materials as food packaging materials. In most countries, the additives used in food packaging must be harmless or food-grade packaging materials. In addition, packaging with additives will be tested for migration effects, to identify any possibility of the chemical substances migrating from the packages to the contained food. The mechanical and thermal properties of the food packaging must also fulfil certain requirements, since the food products will likely experience vibrations or temperature changes during distribution, transportation and storage. This could further cause the packaging to be damaged and could contaminate the food in the packages.

Active packaging is an efficient technology that can be used to protect packaged food from recontamination by microbials. In other words, active packaging incorporates additives known as active components to absorb or release chemical substances from the prepacked food or from the environment. The active component additives are selected based on the requirements of the uses: for example, antioxidants, antimicrobial agents, oxygen scavengers, etc. Basically, active packaging has been improved, exhibiting three main stages: initially only having antimicrobial activity or in situ antioxidants, then the presence of release behaviour and, finally, active packaging with the behaviour of a controlled release effect. This is the protective action of active packaging that can be achieved through the release of antimicrobial substances either in the form of volatiles or nonvolatiles from the packaging materials to the contacted food. The efficiency and effectiveness of active packaging is mainly dependent on the release rate of the active components used and the oxidation kinetics of the food. It is important for active packaging to achieve a good balance between these two factors, because if the antioxidant release rate is too slow or too fast, the available quantity of the released antioxidant to inhibit the oxidative effect cannot be maintained on the surface of the food.

As mentioned earlier, the balance effects between the food oxidation kinetic effect and the release rate of active compounds are the main factors to decide the effectiveness of the active packaging in applications. The release rate of the antioxidant requires good control at a suitable rate, so that it can control and maintain the optimum quantity of the released substance in inhibiting the oxidative activity (Kuai et al., 2021). The release rate of the antioxidant from the packaging needs to be precisely determined to obtain an effective active packaging. Generally, the active packaging is divided into two systems: the packaging-food system and packaging-headspace system. The active packaging system generally consists of three main sections, which are the active substances, packaging material and food. The packaging-food system basically refers to the direct contact between the food and the packaging materials. The antioxidants will diffuse in the polymer matrix through the structure of the polymer, and then some of them will transfer from the polymer matrix to the food surface and others will distribute between interfaces. The equilibrium distribution of the antioxidant between the polymer matrix, food and interfaces is the major factor that affects the migration phenomenon. The mechanism of antioxidant release basically operates in accordance with Fick's Law. The release of antioxidants is controlled by the equilibrium of antioxidants at the interphases by adjustment of the concentration ratio at the interphases.

5.3 Labelling requirements

In the food industry, it is the responsibility of the manufacturer or importer to provide full required labelling information on the food packaging. All the products must follow all the requirements set by the government to secure and prove the quality of the products to consumers. The labelling on food packages must also state clearly the ingredients of the food to alert those people with allergy responses to certain foods and keep them from ingesting them accidentally. In Malaysia, the food manufacturer or importer should display and provide all the requirements in food labelling as proposed by the Food Regulations (1985), which were approved by the Ministry of Health. The purpose of labelling is also to provide clear information on product ingredients, required storage methods and also relevant certification to customers for reference purposes. In comparing the food packages labelling regulations of many countries, including the United States, United Kingdom, Europe, Malaysia, Singapore and others, all the labelling requirements for food packaging are quite similar with each other, as summarized in Table 5.4. As seen in this table, the product's name, brand name, food quantity, ingredients, manufacturer/importer/distributor, nutrition facts, expiration date, etc. are compulsory for most countries. The product labelling provides clear information to the consumers before they purchase the foods, especially the expiration date and ingredients list. The expiration date provides clear information about the time duration period for the packed food to still be in good condition. In addition, the ingredients of the foods are compulsory to be included on the food packaging labelling requirements for the United States, Malaysia, Singapore, United Kingdom, etc. All the ingredients must be stated clearly, especially the presence of allergens such as nuts, seafood, etc. on the label. This information could prevent someone with an allergy from experiencing an allergic response after mistakenly ingesting the food.

Table 5.4 The summarized general labelling requirements for countries of United States, United Kingdom, Malaysia, Europe, Singapore, etc.

No.	Labelling requirements	Descriptions
1	Illustrations	Provide a recommended consuming method, product picture
2	Logo	Approved brand logo, and include other relevant logos such as relevant certification, etc.
3	Product information	The brand name and product name are compulsory. Other information such as food nutrition, calories of the food
4	Quantity of food	The quantity of food labelling is mandatory. The food quantity can be in weight, volume number, etc.
5	Ingredients	All the content in the food needs to be clearly stated. Foods that contain ingredients that may cause hypersensitivity must be stated on the label. Foods containing allergens such as nuts, gluten, fish, prawns, lactose, eggs, etc. need to be stated clearly
6	Nutrition facts	Food packaging has a total surface area less than 100 cm^2. According to food regulations, the terms of 'free' and 'low' can be used. As an example: alcohol-free, low fat, fat free, etc.
7	Manufacturer/ importer	The name and address of the manufacturer/distributor/importer of the manufacturing right must be included
8	Expiration date	The expiry date is mandatory on all food packaging. Need to provide suitable food storage time frame during which the food may retain its quality
9	Bar code	Bar code is required to ensure the business can proceed more efficiently
10	Descriptions	It is noncompulsory for the brand to include a short description
11	Instruction	Instructions such as the storage conditions like temperature, etc., and also instructions related to food storage after opening the package

Information collected and compiled from: Food and Drug Administration, US Department Agriculture. (2020). Food standards: Labelling and composition. Department of Environment, Food, & Rural, Affair. Updated on 2020, Food Standards. (2020). UK Act 1999. Updated on 2020. Food Standards Agency. https://www.legislation.gov.uk/ukpga/1999/28/contents, Food Regulations Malaysia. (1985). Malaysian Ministry of Health, Singapore Food Agency Act 2019 (Act 11 of 2019). Regulation under Sale of Food Act, Roche, K. A. (2016). Food labelling: Applications. Encyclopedia of food and health.

5.4 Consumer behaviour and tips to identify environmentally friendly packaging

Recently, rapid improvements in information technology have eased the access to consumer information regarding the impact of plastic waste on environmental pollution issues. Consumers are being better alerted to the environmental impacts of plastic products, causing them to be more conscious of the selection of food packaging materials. The contribution of plastic waste to environmental pollution has further caused consumers to prefer products with alternative sustainable packaging materials. In 2018, various research studies were conducted to investigate the willingness of consumers to spend more money for goods with environmental packaging in Germany, Sweden, Finland and France (Brovensiepen et al., 2018; Otto, Strenger, Maier-Noth, & Schmid, 2021). According to Table 5.5, 77% of consumers in Germany refuse to pay more for environmentally friendly packaging materials. Basically, the consumers are more concerned about the use of packaging materials in fulfilling several factors, such as the food protection effects of the packaging, handling purposes, labelling of the product and packaging design and materials used, as shown in Fig. 5.4; this is due to the fact that effective food protection can improve the food sustainability. In this study, many consumers also were primarily concerned about the use of packaging in handling the food properly and for storage purposes. Packaging materials that can help in proper food handling, due to good packaging design and suitable materials, are strongly preferred by consumers. However, some researchers also investigated the preferences of consumers in Sweden, Finland and France with regard to being willing to pay more for environmentally friendly packaging materials in 2018. They found that a majority of the consumers in these countries were willing to pay more for using sustainable packaging materials. In comparing these studies, the perspectives of consumers on the use of environmentally friendly

Table 5.5 Investigations of the willingness of consumers to pay more for sustainable packaging materials in 2018.

Country	Willingness to pay more for sustainable packaging materials
Germany (Brovensiepen et al., 2018)	77% of consumers **will not** pay more for sustainable packaging. Reason: • Consumers are more concerned about the food protection and handling purposes of packaging materials • From this study, most consumers consider the sustainable packaging aspect to be their last concern in food packaging issues
Sweden (Lindh et al., 2016)	80% of the consumers can accept paying more for sustainable packaging
Finland and France (Orset et al., 2017)	Majority of people are willing to pay more for the use of sustainable packaging

Important concerns of consumers on the usage of food packaging materials:

1. Food protection (protect the food from contamination, preservation effect)
2. Food Handling and clear labelling of the ingredients (proper handling during transportation process; labelling must able to provide clear information)
3. Packaging design
4. Sustainable packaging materials

Fig. 5.4 Importance concerns of consumers on the selection of packaging materials.

packaging are strongly influenced by the structure of populations. This is attributed to the fact that different countries have different implementations of sustainable packaging in the food industry and the availability of sustainable packaging materials in a country will affect the selection. Another investigation was conducted after 2 years by Popovic et al. (2020) on the consumers in 11 countries. They observed that almost 73% of the consumers from these countries were willing to pay more for goods with more environmentally friendly packaging materials.

The expectations of customers have encouraged the relevant industries to develop more possible sustainable packaging alternatives to replace the conventional plastic packaging materials. In the perspective of most consumers, the nonbiodegradable behaviour of plastic packaging waste alerted them to the use of nonbiodegradable plastic in plastic packaging materials. This is because the generation of huge amounts of plastic packaging material in these consumers' daily lives caused them to have negative judgements on the use of nonbiodegradable polymers in packaging materials. The huge amounts of nondegradable plastic packaging material waste caused available landfills to reach saturation. The largest drawback of plastic disposal in landfills is the poor ability of plastic material to degrade in a suitable time period. Thus this has further caused the rapid accumulation of plastic waste in landfills. In order to reduce the use of nonbiodegradable plastic as packaging materials, some researchers and companies have focused on the replacement of plastic packaging materials with other materials, such as paper, for packaging purposes. Although the selection of paper as a packaging material could help to reduce the nonbiodegradable polymer waste, the use of paper also has a negative impact on energy sustainability. Paper materials in general offer poor protection for products, especially those in the electronics, medical, and food industries. This poor protective effect of paper packaging materials severely increases the negative impact on sustainability of the products. In addition, the sustainability of paper use is also questionable because of paper packaging waste with special coated treatments, which cannot be recycled for other applications. Eventually, this paper packaging waste will end up in landfills for disposal.

In actual practice, the use of plastic packaging materials is crucial in most industries, especially in the food, medical consumable and electronics component industries, because plastic packaging materials have a positive impact on the sustainability of valuable products by providing better protection. The protection provided by plastic packaging to valuable products or goods is summarized in the following list:

(1) Plastic materials offer a good barrier effect compared to other materials such as paper. Plastic material packaging can effectively prevent direct contamination of the products/goods.
(2) The excellent gas permeability resistance of plastic packaging materials can prolong and extend the shelf life of the products and goods.
(3) Secondary functions of plastic packaging, such as communication of information and the promotion of articles.

The poor protective effects of paper packaging materials have limited their application in food packaging materials. Various research studies have been conducted to develop new green plastic materials and new methods for recycling the plastic waste as packaging materials in food packaging. The protective effect of packaging materials on prepacked products is crucial, especially in preserving the quality and safety of food during the transporting and distribution processes and during food storage Otto et al., 2021. Protective packaging materials can effectively reduce food waste and improve the sustainability of the food industry. Due to this consideration, the use of new packaging innovations like biodegradable plastics, packaging materials from recycled plastic and packaging materials with reduced plastic volume are being selected by industry to replace the conventional plastic packaging materials. This is attributed to the use of conventional plastic packaging in the food industry having provided a better environmental impact from the perspective of food products. This is because the conventional plastic packaging can effectively preserve and protect the food product from contamination and other external damage during the shipping process and furthermore it can extend the preservation time of the food product, thus reducing the amount of food loss and waste. As a result, the food industry prefers to replace the conventional plastic packaging materials with other newly developed biodegradable plastic or new plastic packaging concepts, such as recycled plastic packaging or multilayer packaging methods.

The perspective of consumers, however, is quite different from that of the food industry; they are focused on the contribution of plastic packaging waste to environmental pollution and tend to use other packaging materials that are more environmentally friendly than nonbiodegradable plastic packaging. In comparison to plastic packaging materials, the environmental impact of traditional packaging materials is observed to be different from plastic packaging materials in terms of total greenhouse gas emissions, biodegradability effects, the amount of food waste and the lifetime of the packaging materials (Otto et al., 2021). Sometimes, customers prefer to buy unpackaged food to reduce the use of polymer packaging materials. In addition, customers are also more willing to purchase food with traditional packaging such as glass, biodegradable polymers, paper, recycled plastic packaging, and others when compared to food in virgin plastic packaging. To fulfil the requirements of preserving the food and meeting the needs of the customers, many researchers have been working on the development of new biodegradable polymers and new packaging methods. Various research has been carried out to produce sustainable plastic packaging materials with good biodegradability, multilayer packaging material made from recycled plastic and other traditional packaging materials such as glass, paper bags, and others.

As already stated, the use of plastic packaging is important, especially in the food industry, due to its good preservation and protective effects on food products. The largest drawback of plastic waste disposal in landfills is the fact that plastic materials do not degrade in a reasonable time period. The poor and very slow degradability of nondegradable plastic materials has led to the rapid accumulation of plastic waste in landfills. In addition, the production of fossil-based plastic materials has been found to have a high global warming potential (GWP) value, which indicates the release of greenhouse emissions to the environment, as summarized in Table 5.6. In order to reduce the negative impact of food packaging materials on the environment, poly(lactic acid) (PLA) is commonly used as a replacement alternative to other

Table 5.6 Common plastic materials for beverage bottles and food packaging.

Plastic materials	Description
Poly(lactic acid) (PLA)	• Commonly used biodegradable polymer in food packaging and beverage bottle industries • PLA is basically used to replace the use of polyethylene terephthalate (PET) in beverage bottles and food packaging • The biodegradation of PLA occurs at temperature of 50–60°C • The production of PLA causes an average GWP of 3.05 kg CO_2 eq/kg
Polyethylene terephthalate (PET)	• Currently, PET is commonly used in beverage bottles and food packaging applications • The production of PET was found to cause an average GWP of 2.95 kg CO_2 eq/kg • PET bottles are reusable and can be refilled up to 25 times
Polypropylene (PP)	• One of the most commonly used polymer materials in the application of beverage bottles and food packaging • Nonbiodegradable polymer produced from fossil fuel • The production of a PP container caused an average of 3.94 kg CO_2 eq/kg

nonbiodegradable polymers such as polypropylene (PP), polyethylene terephthalate (PET), etc. Currently, PLA is the most commonly used biodegradable plastic material in food packaging. Due to the environmentally friendly goal, PLA has been selected by various packaging industries to replace nonbiodegradable polymers such as PET, PP, and others in beverage bottles and other food packaging (Otto et al., 2021).

Currently, PLA has been selected by some food industries to replace nonbiodegradable plastics such as PET, PP, etc. as beverage bottles and other food packaging materials due to its biodegradable behaviour. However, some researchers have found that the replacement of nonbiodegradable plastic packaging with PLA is unable to reduce the saturation of landfill disposal. This is because the degradation of PLA is highly dependent on the temperature and moisture of the landfill. Suitable temperature and moisture can promote a hydrolysis reaction of PLA chains and thus further accelerate the biodegradation process. According to some researchers (Castro-Aguirre, Iniguez-Franco, Samsudin, Fang, & Auras, 2016; Tokiwa & Calabia, 2006), PLA will only experience biodegradation at a temperature range of 50–60°C. It would be difficult for landfill surroundings to reach this high temperature range. Due to this, the PLA waste degrades slowly in soil, as the surrounding temperature is lower than 40°C. On the other hand, the production of PLA, PET and PP was observed to have a GWP of 3.05 kg CO_2 eq/kg, 2.95 kg CO_2 eq/kg and 3.94 kg CO_2 eq/kg, respectively. According to Table 5.2, the GWP of PET was found to be slightly lower than the GWP of the biodegradable polymer PLA. This is attributed to the fact that degradation of PLA in normal natural environmental temperatures is very slow and the rate of the decomposition of PLA is similar to that of PET, which is very slow in environmental biodegradation (Otto et al., 2021). The advantages of PLA to the environment are mainly attributed to its origin from renewable resources, which causes it to be preferred to the use of PET. However, when compared with PP, both PET and PLA have lower GWP. In other words, PLA and PET have a better sustainability effect than PP.

Currently, many consumers do not really understand the important factors of the actual environmental impacts. They are mainly focused on recycling knowledge, environmental pollution problems, greenhouse emissions issues and biodegradability effects of packaging materials. This has caused them to ignore the main sustainable issues in the other factors, such as generation of more food waste due to poor preservation properties of the packaging used, design of the packaging and high efficiency in handling during transportation. This also indicates that most consumers lack knowledge of the importance and influence of packaging on the environment. The generation of food waste due to improper handling of packaging can lead to the release of greenhouse emissions to the environment. Due to consumers' low awareness and understanding of sustainable packaging, it is important to emphasize that the following are the most important criteria in selecting packaging materials that are more environmentally friendly:

- Packaging materials that are recyclable.
- The use of packaging such that no product waste is generated.
- Packaging materials that are biodegradable.
- Packaging materials that can be recovered as energy.
- Packaging waste that consists of as little packaging materials as possible.

Currently, many consumers resist the use of plastic due to the poor perceptions of and feelings about the slow degradability of nonbiodegradable plastic. However, as stated, the use of plastic packaging has a positive effect by providing good protection to products, further reducing the probability of product waste from damage or spoiling. Basically, most consumers prefer to select plastic packing material that can be recycled. In Europe, most people will sort out the plastic bottles and plastics packaging waste for recycling purposes. Most Europeans wish for nonrecyclable plastic to be replaced with the production of plastic materials that are recyclable (Lindh et al., 2016; Otto et al., 2021). A study conducted on Swedish consumers found that 62% of the consumers had the most negative impression of the effect of plastic waste on the environment, as shown in Fig. 5.5. In the opinions of most consumers, the nonbiodegradability of plastic materials has caused some severe negative environmental impacts, especially regarding the disposal of plastic waste in landfills and the ocean, as the disposal of plastic materials into the ocean has had a severely harmful impact on aquatic life. In addition, the disposal of plastic materials also requires more land to be used as landfills to dispose of these huge amounts of plastic wastes. On the other hand, most consumers believe that the use of paper packaging in certain applications has the least negative impact on the environment. A study reported by Otto et al. (2021) found that 80% of consumers agreed that paper had the least negative environmental impact among all types of packaging materials (Fig. 5.6). Then, glass was seen as having the second least negative environmental impact among packaging materials, as shown in Fig. 5.6. Most consumers prefer to use paper as a packaging material in food packaging applications over other materials, because most consumers have positive attitudes and emotions about paper, such as it being a biologically natural and easily recyclable material, due to the familiarity of many consumers with sorting out paper packaging wastes for recycling purposes.

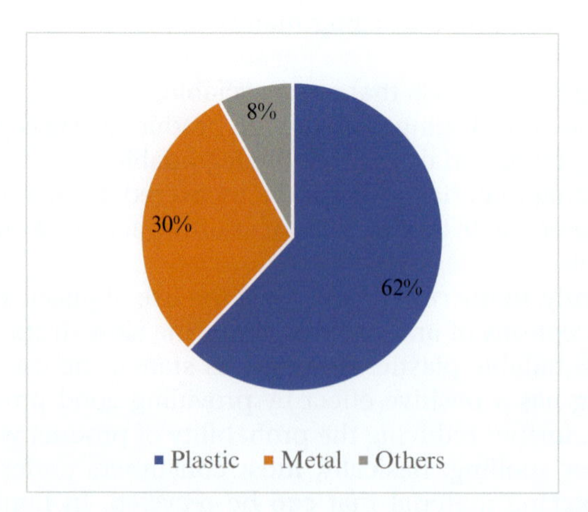

Fig. 5.5 The packaging material with the most negative environmental impact in consumers' view regarding food packaging use.

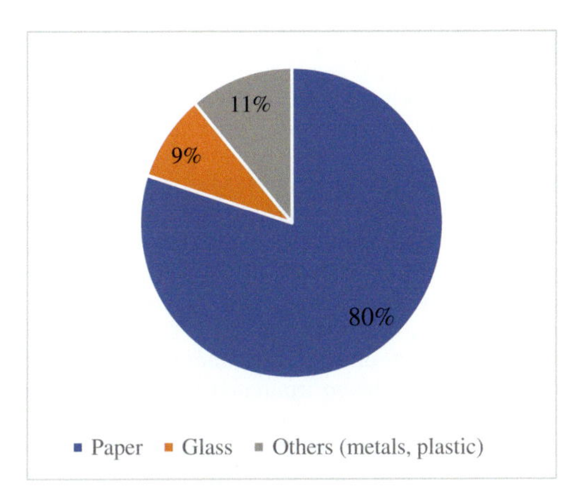

Fig. 5.6 The packaging material with the least negative environmental impact in consumers' view regarding food packaging use.

5.5 Conclusion

Packaging is important in various industries, especially in the food industry. Currently, the use of multilayer packaging in the food industry has drawn increasing attention due to its excellent gas permeability resistance, provision of a good water vapour barrier and considerable mechanical properties. The application of multilayer packaging basically can reduce the volume usage of polymer materials while providing optimum properties and thus can reduce food packaging wastes in landfills. Furthermore, the use of multilayers of polymer materials can optimize the properties of multilayer packaging by the use of different polymer materials with different benefits. In addition, the combination of active packaging and multilayer packaging has been found to efficiently improve resistance to antimicrobial activity. When the food packaging has direct contact with food, the polymer materials used in the packaging are strictly controlled in many countries. The food regulations in many countries state that the polymer materials that release harmful chemical substances, such as polymers with some hazardous additives, are prohibited from being used as food primary packaging that directly contacts food. A labelling requirement on the food packaging is necessary to fulfil the food packaging rules and regulations of different countries. Product information such as ingredients and expiration date are important, since information such as allergen content could avoid unnecessary accidents. Labelling on food packaging is thus necessary to avoid any misleading or insufficient information.

References

Albaar, N., Budiastra, W., & Hariyadi, Y. (2016). Influence of secondary packaging on quality of carrots during transportation. *Agriculture and Agricultural Science Procedia, 9*, 348–352.

Asgher, M., Qamar, S. A., Bilal, M., & Iqbal, H. M. N. (2020). Bio-based active food packaging materials: Sustainable alternative to convectional petrochemical-based packaging materials. *Food Research International, 137*, 109025–109036.

Aulin, C., Gallstedt, M., & Lindstrom, T. (2010). Oxygen and oil barrier properties of microfibrillated cellulose films and coatings. *Cellulose, 17*(3), 559–574.

Bee, S. T., Hassan, A., Ratnam, C. T., Tee, T. T., Sin, L. T., & Hui, D. (2014). Dispersion and roles of montmorillonite on structural, flammability, thermal and mechanical behaviours of electron beam irradiated flame retarded nanocomposite. *Composites Part B: Engineering, 161*, 41–48.

Bharadwaj, R. K. (2001). Modeling the barrier properties of polymer-layered silicate nanocomposites. *Macromolecules, 34*(26), 9189–9192.

Brovensiepen, G., et al. (2018). Verpackungen im Fokus - Die Rolle von Circular Economy auf dem Weg zu mehr Nachhaltigkeit. *Retrieved from Frankfurt am Main, Germany.*

Castro-Aguirre, E., Iniguez-Franco, F., Samsudin, H., Fang, X., Auras, R., et al. (2016). Poly(lactic acid)-mass production, processing, industrial applications and end of life. *Advanced Drug Delivery Reviews., 107*, 333–366.

Chung, S. H., Ma, H. L., & Chan, H. K. (2018). Maximizing recyclability and reuse of tertiary packaging in production and distribution network. *Resources, Concervation and Recycling, 128*, 259–266.

Chytiri, S., Goulas, A. E., Riganakos, K. A., & Kontominas, M. G. (2006). Thermal, mechanical and permeation properties of gamma-irradiated multilayer food packaging films containing a buried layer of recycled low-density polyethylene. *Radiation Physics and Chemistry, 75*, 416–423.

Ferrara, C., Zigarelli, V., & Feo, G. D. (2020). Attitudes of a sample of consumers towards more sustainable wine packaging alternatives. *Journal of Cleaner Production, 271*, 122581–122592.

Food Regulations Malaysia. (1985). *Malaysian Ministry of Health.*

Food Standards. (2020). UK Act 1999. Updated on 2020. Food Standards Agency. https://www.legislation.gov.uk/ukpga/1999/28/contents.

Food Standards Australia New Zealand. (2021). Monitored by Ministry for Primary Industries and Public Health Units (New Zealand) and The Australian Government Department of Agriculture, Water and Environment (Australia). *Food Safety Regulations.*

Gherardi, R., Becerril, R., Nerin, C., & Bosetti, O. (2016). Development of a multilayer antimicrobial packaging material for tomato puree using an innovative technology. *LWT-Food Science and Technology, 72*(2), 361–367.

Hagstrum, D. W., & Sbbramanyam, B. (2009). *Stored-product insect resource.* Saint Paul, MN: AACC International.

Kaiser, K., Schmid, M., & Schlummer, M. (2018). Recycling of polymer-based multilayer packaging: A review. *Recycling, 3*, 1–26.

Kuai, L., Liu, F., Chiou, B. S., Avena-Bustillos, R. J., McHugh, T. H., & Zhong, F. (2021). Controlled release of antioxidants from active food packaging: A review. *Food Hydrocolloids, 120*, 106992.

Lindh, H., Olsson, A., & Williams, H. (2016). Consumer perceptions of food packaging: Contributing to or counteracting environmentally sustainable development? *Packaging Technology and Science, 29*(1), 3–23.

Mahmoudi, M., & Parviziomran, I. (2020). Reusable packaging in supply chains: A review of environmental and economic impacts, logistics system designs, and operations management. *International Journal of Production Economics, 228,* 107730–107744.

Mieth, A., Hoekstra, E., & Simoneau, C. (2016). Guidance for the Identification of Polymers in Multilayer Films Used in Food Contact Materials. *>uropean Commission JRC Technical Reports.*

Nguyen, A. T., Parker, L., Brennan, L., & Lockrey, S. (2020). A consumer definition of eco-friendly packaging. *Journal of Cleaner Production, 252*(119792), 1–11.

Orset, C., Barret, N., & Lemaire, A. (2017). How consumers of plastic water bottles are responding to environmental policies? *Waste Management, 61,* 13–27.

Otto, S., Strenger, M., Maier-Noth, A., & Schmid, M. (2021). Food packaging and sustainability—Consumer perception vs. correlated scientific facts: A review. *Journal of Cleaner Production, 298,* 126733.

Pandey, V. K., Upadhyay, S. N., Niranjan, K., & Misha, P. K. (2020). Antimicrobial biodegradable chitosan-based composite nano-layers for food packaging. *International Journal of Biological Macromolecules, 157,* 212–219.

Pauer, E., Tacker, M., Gabriel, V., & Krauter, V. (2020). Sustainability of flexible multilayer packaging: Environmental impacts and recyclability of packaging for bacon in block. *Cleaner Environmental Systems, 1,* 100001–100012.

Popovic, I., Bossink, B. A. G., van der Sijde, P. C., & Fong, C. Y. M. (2020). Why are consumers willing to pay more for liquid foods in environmental friendly packaging? A dual attitudes perspective. *Sustainability, 11*(22).

Ragaert, K., Delva, L., & Geem, K. V. (2017). Mechanical and chemical recycling of solid plastic waste. *Waste Management, 69,* 24–58.

Rudmann, L., Langenmair, M., Hahn, B., Ordonez, J. S., & Stieglitz, T. (2020). Novel desiccant-based very low humidity indicator for condition monitoring in miniaturized hermetic packages of active implants. *Sensors and Actuators B: Chemical, 322,* 128555–128567.

Shamini, G., & Yusoh, K. (2014). Gas permeability properties of thermoplastic polyurethane modified clay nanocomposites. *International Journal of Chemical Engineering and Application, 5*(1), 64.

Sharanyakanth, P. S., & Radhakrishnan, M. (2020). Synthesis of metal-organic frameworks (MOFs) and its application in food packaging: A critical review. *Trends in Food Science & Technology, 104,* 102–116.

Soares, C. T. D. M., Ek, M., Ostmark, E., & Gallstedt, M. (2022). Recycling of multimaterial multilayer plastic packaging: Current trends and future scenarious. *Resources, Conservation and Recycling, 176,* 105905.

Stejskal, V., Bostlova, M., Nesvorna, M., Volek, V., Dolezal, V., & Hubert, J. (2017). Comparison of the resistance of mono- and multilayer packaging films to stored-product insects in laboratory test. *Food Control, 73,* 566–573.

Thellen, C., Schirmer, S., Ratto, J. A., Finnigan, B., & Schmidt, D. (2009). Co-extrusion of multilayer poly(m-xylylene adipimide) nanocomposite films for high oxygen barrier packaging applications. *Journal of Membrane Science, 340,* 45–51.

Tokiwa, Y., & Calabia, B. P. (2006). Biodegradability and biodegradation of poly(lactide). *Applied Microbiology and Biotachnology, 72*(2), 244–251.

Visanko, M., Liimatainen, H., Sirvio, J. A., Mikkonen, K. S., Tenkanen, M., Sliz, R., et al. (2015). Butylamino-functionalized cellulose nanocrystals films: Barrier properties and mechanical strength. *RSC Advances, 5*(20), 15140–15146.

Wan, Y. J., Li, G., Yao, Y. M., Zeng, X. L., Zhu, P. L., & Sun, R. (2020). Recent advances in polymer-based electronic packaging materials. *Composites Communications, 19,* 154–167.

Wang, L., Chen, C., Wang, J., Gardner, D. J., & Tajvidi, M. (2020). Cellulose nanofibrils versus cellulose nanocrystals: Comparison of performance in flexible multilayer films for packaging applications. *Food Packaging and Shelf Life, 23,* 100464–100474.

Wang, X. D., Yao, C. H., Wang, F., & Li, Z. D. (2017). Cellulose-based nanomaterials for energy applications. *Small, 34*(26), 9189–9192.

Zhang, S., Luo, K., Zhang, L., Mei, X., Cao, S., & Wang, B. (2015). Interfacial separation and characterization of Al–PE composites during delamination of post-consumer Tetra Pak materials. *Journal of Chemical Technology and Biotechnology, 90*(6), 1152–1159.

6

Recycling and circular economy of plastics

6.1 Introduction

Plastics recycling is a commonly heard term in modern life. There are two main reasons why plastic recycling is important. First, plastics are produced from nonrenewable petroleum and excessive use of petroleum can lead to depletion in the near future. Second, disposal of plastics is an environmental burden because petroleum-derived plastic wastes degrade very slowly, and the plastics can transform into smaller fragments, causing another problem called 'microplastics'. The small fragments/particles of half-degraded plastic wastes can be accidentally consumed by lifeforms, causing accumulation in the food chain and subsequently endangering the stability of the ecosystem. Thus the rationale of recycling is to reduce the amount of plastic disposed of in landfills, as well as introduce a second life to plastics, to minimize fossil source consumption. In this topic, recycling refers to plastic wastes that are being collected, segregated, cleaned and that undergo a physical process such as injection moulding, blow moulding, extrusion, blown film and other technology to rework the recycled plastic resin into 'second-time' plastic products. Such recycled plastic resin can be either used directly to produce second-grade plastic products, or mixed with virgin resins (with the intention to cut the cost of materials) to produce plastic products with a second life. Furthermore, most of the recycled plastics are thermoplastic, which means the plastic wastes can be crushed into small particles, undergo a melting process and be moulded into new shapes. However, these recycled plastic products usually can only be recycled once and then they need to be discarded, because the quality of these products is usually somewhat compromised and undergoing a further recycling process tends to cause the

Plastics and Sustainability. https://doi.org/10.1016/B978-0-12-824489-0.00005-2
Copyright © 2023 Elsevier Inc. All rights reserved.

quality to drop drastically, with the plastic becoming brittle, yellowish, fragmented or easily broken, leading to safety concerns. Hence, the waste plastics need to be discarded in landfills, which is the traditional waste management method.

Nowadays, scientists have addressed concerns about the limitations of plastic recycling (i.e. that plastics still need to be thrown away eventually). A more practical approach to manage these plastic wastes is through chemical recycling or incineration methods. In short, chemical recycling is a pyrolysis process to transform the waste plastic to simple hydrocarbon molecules including CH_4, C_2H_6, C_2H_4, C_3H_8 and others; such simple hydrocarbon molecules are the basic chemicals for petrochemical industry processes. The pyrolysis oil, consisting of long chains of carbon molecules, can be used as fuel oil for steam generation and electricity production. On the other hand, in the incineration process the waste plastics are combusted, and the thermal energy generated is used to operate turbines for electricity generation. Both technologies are actually discussed in detail in Chapter 3. Thus this chapter mainly focuses on the plastic recycling products on the market so that readers can gain a better understanding of the production process, or perhaps be inspired to venture into a business in this area.

6.2 General concept of plastic circular economy

In general, the plastic circular economy is about the recycling process. It is about how to restore and regenerate the materials to be used in a closed loop system, rather than being designed for a single use and discarded without maximizing the value of plastics for the benefit of mankind and the economy. From the diagram shown in Fig. 6.1, the linear economy is a one-way process in which people utilize natural resources to make plastics, and after one use the product is discarded in a landfill. However, in the circular economy concept, the plastic materials are recycled to extract the maximum value, and at the final stage, the plastics should either undergo a chemical recycling process to convert the plastic back to a monomer for re-polymerization to produce virgin plastic, or the plastic waste should be sent to incineration to generate electricity, contributing to the economy loop. In other words, this is a waste-to-wealth approach. From a science and technology perspective, according to the Second Law of Thermodynamics, none of the process is reversible or generates zero waste throughout the process. For instance, the processing of plastic still requires electricity sources from fossil fuel. Anyone

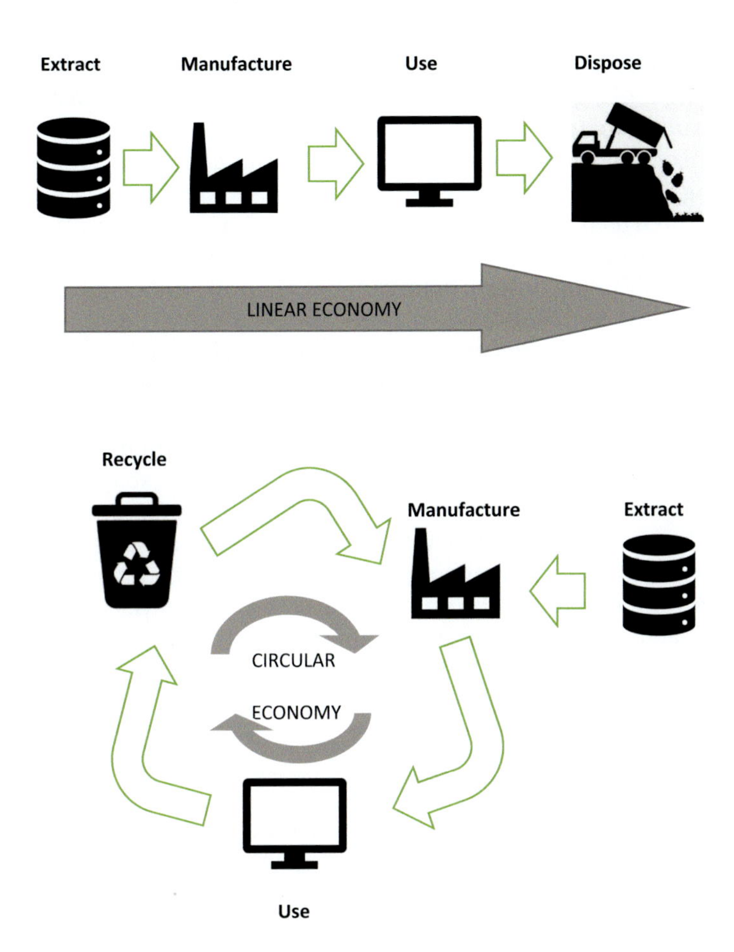

Fig. 6.1 Simplified diagram of linear economy and circular economy.

may argue that waste-to-energy is the solution, but the energy generated from the waste is insufficient to operate the plastic machinery. Hence, emissions and a carbon footprint are unavoidable and the most appropriate approach is to minimize consumption by using all the natural resources responsibly. Reduce, reuse and recycle are always the best approaches to minimize the environmental impacts.

6.2.1 Recycled plastic products

Recycled plastic products refers to plastic products that have been used by the consumer the first time, then collected, remanufactured and sold a second time for other uses. This kind of recycled plastic product is usually from a postconsumer source. It is usually found in nonhygiene-required products like storage containers, stools, chairs, pails, clothes hangers, flowerpots, gardening tools, garbage bags and others. This is because most post-consumer materials can have hygiene concerns and problems due to mixed plastic types from printing, labelling or multilayer packaging. Thus only products with compromised quality can be manufactured using recycled plastic resin from domestic postconsumer collection point sources. On the other hand, some products such as clothes, drinking bottles, detergent bottles, and food containers are claimed to be made from recycled materials, but these types of recycled plastic items are actually from known sources of preconsumer materials. For instance, thermoformed plastic trays can be made from rejected plastic bottles from a bottle manufacturing plant. These plastic bottles are rejected due to aesthetic issues like shrink marks, 'fish eyes', distortion or irregular shapes. These rejected bottles are then sold to another factory for manufacturing of other products, and the hygiene issue does not arise because the source and quality are guaranteed. As a savvy consumer, it is very important to identify the appropriate sources of recycled products to safeguard our lives and the environment, while avoiding misleading marketing gimmicks that affect our efforts to save the environment.

6.2.2 Textile products

Commonly found recycled textiles on the market are recycled polyester T-shirts originating from polyethylene terephthalate (PET) bottles. PET is commonly used for bottles manufacturing due to its transparency, inertness and gas barrier resistance. One of the commonly found applications is bottling of carbonated drinks. With PET, pressurized carbon dioxide cannot easily penetrate through the wall of PET bottles. Meanwhile, cooking oils, fruit cordials, sauces, shower soaps, cosmetics, etc. are also packed in PET because of its outstanding gas barrier. PET can restrict oxygen penetration into the bottle contents and the shelf life of the products can thus be prolonged. Since there are thousands of applications of PET bottles, much of the domestic plastics waste is made up of huge amounts of bottles, in addition to plastic bags. Often improper disposal of PET bottles causes blockage of drainage to rivers and the ocean, with bottles ending up as plastic pollution during flooding in the rainy season.

PET bottles can be recycled to produce textiles for apparel. In fact, PET is the most widely used fibre in the textile industry, with about 52% of fibres produced globally being sourced from PET. A nonprofit organization called Textile Exchange is standing up to initiate the 2025 Recycled Polyester Challenge, together with the United Nations Framework Convention on Climate Change's Fashion Industry Charter for Climate Action, by urging the current apparel industry to increase their usage of recycled polyester to 45% at 17.1 million metric tons by 2025 (Textile Exchange, 2021). Inline with the initiative, Textile Exchange has produced voluntary standards called Recycled Claim Standard (RCS) and Global Recycled Standard (RGS) to be followed by the textile industry players in order to claim sustainability of their products. There are certain criteria that need to be met in order to claim the textiles are made of recycled or reclaimed fibres, as shown in Table 6.1, for general classification. As seen in the table, certain products cannot be called recycled and reclaimed textiles, to avoid confusing the public. For instance, fabric and fabric waste collected from postconsumer or preconsumer garments sewn into new garments, bags, curtains, carpet, etc. can only be classified as reused and not recycled according to the Textile Exchange standards.

Basically, the process to recycle PET bottles into textiles involves a physical recycling process. The process of recycling, shown in Fig. 6.2, involves four main steps. First, the postconsumer PET bottles are collected, crushed to reduce the size and washed. Since PET is very moisture sensitive and this tends to affect the quality of the products (causes brittleness and low

Table 6.1 General consideration as reclaimed or recycled postconsumer and preconsumer textiles (Textile Exchange, 2021).

Material source	Accepted as reclaimed		Not accepted as reclaimed	
	Postconsumer	*Preconsumer*	*Postconsumer*	*Preconsumer*
Sources	Any material generated from the end users and that no longer can be used as its intended purpose, including material from distribution chains	Any material diverted from the waste stream during the manufacturing process Not applicable for reuse of materials such as rework, regrind, or scrap generation in process which can be reused on the spot	Any materials which can be reused yet were discarded during manufacturing Any materials sold in the second market without any changes	Any materials which can be reused yet were discarded during the manufacturing Any materials sold in the second market without any changes

Continued

Table 6.1 General consideration as reclaimed or recycled postconsumer and preconsumer textiles (Textile Exchange, 2021)—cont'd

Material source	Accepted as reclaimed		Not accepted as reclaimed	
	Postconsumer	*Preconsumer*	*Postconsumer*	*Preconsumer*
Brand/retailer	Old garments collection activities	Consumer or warranty returns due to damage Third party distribution chain returns due to inferior quality	Secondhand or second quality	Leftover stock Unsold goods and materials Aged goods with damage: more than 1 year Aged goods without damage at process categories: more than 1 year Sampling goods at any stage
Customer/end user	Old garments	–	–	–
Collectors either government owned, commercial, nonprofit organization, etc.	PET bottles, fishing nets, plastics, tires, etc.	–	Garment/finished products from reused, renting, trading, swapping, borrowing, inheriting, etc.	–

mechanical strength), vacuum drying is the most suitable method to remove the moisture content. This is followed by drawing and spinning into filaments, and finally it is turned into fibres.

Many famous fashion industry players nowadays are putting forth substantial efforts to make apparel from recycled polyester. For instance, Uniqlo markets a Polo shirt made of up to 75% of recycled polyester. Other companies also accepted the 2025 Recycled Polyester Challenge and the number keeps growing, as shown in Table 6.2, which represents a move in a very positive and healthy direction.

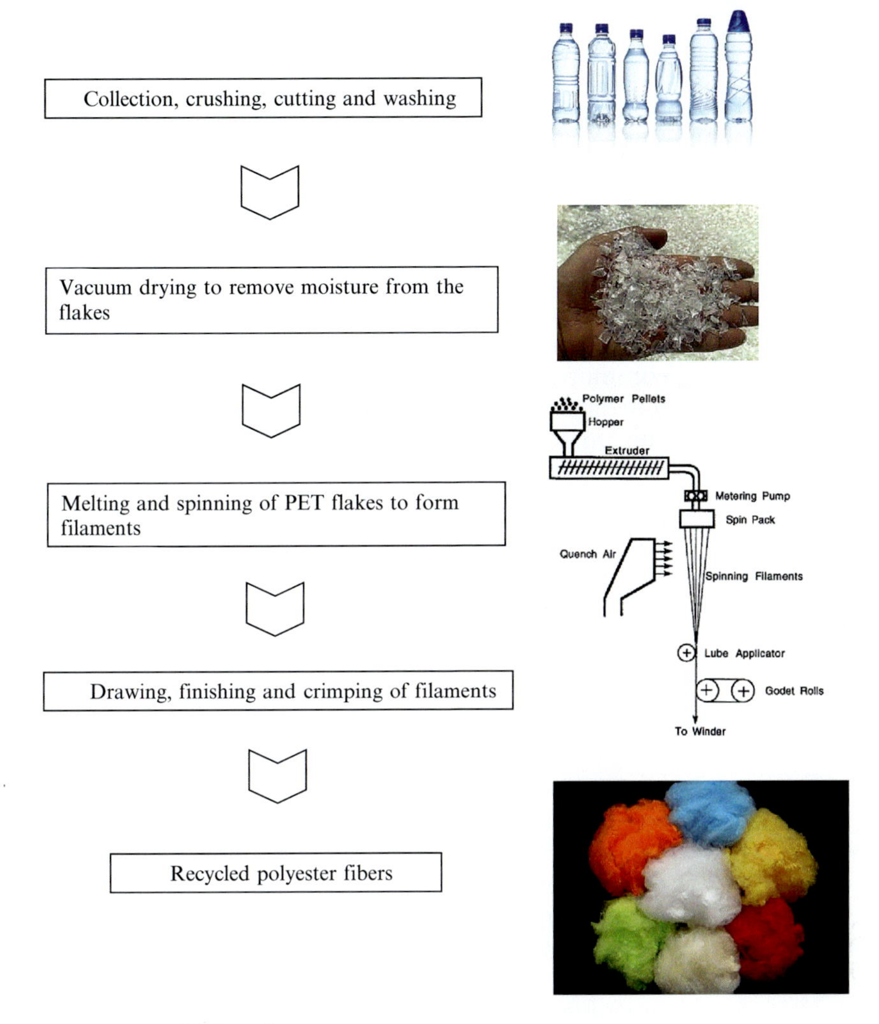

Fig. 6.2 Recycling process of PET to fibre.

Table 6.2 Companies committed to 2025 recycled polyester challenge.

Adidas	Adore Me	Alphine Group	AMUR	Armstrong
ATHLETA	Babyshop Group	Banana Republic	DARE2B	DEDICATED
Elevate Textiles	Ereks	FERRE	FOREVER NEW	GAP
Ginatricot	G-STAR RAW	GYMSHARK	H&M Group	HOUSE OF BAUKJEN
HUGO BOSS	INDITEX	J.CREW	JURITEX	KNICKEY
LULUMELON	MANGO	Mantis World	MARA HOFFMAN	MARIA McMANUS
M&S	MUSTO	NEW BALANCE	OCEAN BALANCE	OUTERKNOWN
Pacsafe	Pact	Piping Hot	Puma	Reformation
Remedy	ROYA ROBBINS	Sapphire	Roya Robbins	Uniqlo

Besides recycled PET, other plastic types can be recycled to textiles, including nylon (also known as polyamide). A trade name for recycled polyamide is ECONYL produced by Italian manufacturer Aquafil. These recycled nylons are sourced from discarded fishing nets, carpet fluff, tulle, etc. The products made from ECONYL yarn are PARISE products, Gucci handbags, swimwear, Breitling watch straps, carpets/rugs by Delos, etc. Recycled polypropylene is also used to produce carpets/rugs, woven and nonwoven fabrics. The common uses for the woven and nonwoven fabrics are to make tote bags, aprons, napkins and other items. Recycled polypropylene possesses good weathering resistance and is suitable for prolonged use.

6.2.3 Furniture

Furniture manufacturing using recycled plastics is innovative and many large furniture producers have seen this as a corporate social responsibility. Indeed, the corporations should play their roles responsibly, to handle products sold to consumers such that when the products reach their end of life, the disposal of postconsumer furniture can be done in a responsible manner. Ikea, which is one of the largest furniture chains, provided such a buy-back scheme in 2021, pledging to move towards a circular model of consumption by looking into alternatives for reuse and recycle, rather than simply disposing of furniture in the landfill. Many furniture manufacturers found that during the COVID-19 pandemic period, social lockdown and work-from-home arrangements actually boosted the demand to replace aging furniture. Many consumers wanted to refurnish their homes to better their working environment and to be more comfortable at home with family members. Hence, the handling of furniture disposal has become an overwhelming issue in recent years.

First, the most commonly found recycled plastic used in furniture is the upholstery or cushions. Instead of filling the cushion with virgin polyester from PET fibre, recycled PET fibre from PET bottles is a good replacement. For instance, Ikea claims that the two-seater sofa Glostad uses a minimum 90% recycled polyester for the back cushion and 100% recycled polyester fibre for the backrest cushion, as shown in Fig. 6.3. In addition, many waste thermoplastics can be easily melted and reprocessed into a variety of furniture items, like stools, cabinets, coffee tables, floor tiles, etc. Thermoplastic types of waste plastics can be easily formed into the desired shape by the compression moulding method, as shown in Fig. 6.4. The compression moulding does not require great skill of operation and the quality is tolerable. For instance,

Fig. 6.3 Ikea two-seater sofa Glostad used a minimum of 90% recycled polyester for the back and 100% recycled polyester fibre backrest cushion.

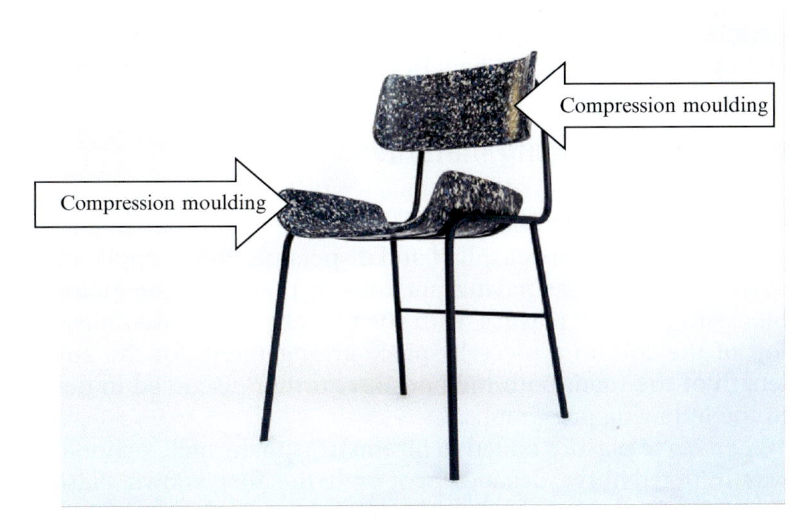

Fig. 6.4 (Top) A chair in which the back support and seat are produced from a compression moulding process, as shown in arrows; (Bottom) A typical type of plastic compression moulding machine. Chair picture from Precious Plastic (2022) available at https://preciousplastic.com/.

mixing a small amount of polypropylene (PP) into high-density polyethylene (HDPE) does not significantly affect the properties of the products. Furthermore, the presence of additives in the waste plastics is also unlikely to cause large differences in processability. However, the drawback of compression moulding is the fact that the rate of productivity is slower compared to injection moulding. Although injection moulding has high productivity and the ability to produce more interesting shapes, as shown in Fig. 6.5, it requires high skills of operation and the setting parameters can be crucial for certain types of plastics. For instance, injection moulding is more suitable for a high melt flow index (lower in melt viscosity) in order to successfully fill up the mould cavity. Injection moulding is undeniable higher in precision, and the cost of setting up the mould is several times higher than a compression mould. As such, consistency of quality of recycled plastics in injection moulding can be achieved, enough to meet the expectations of manufacturers as well as consumers. In short, recycled furniture is a good approach to create value chains of the circular economy for the benefit of environmental protection.

6.2.4 Roadbuilding materials

Recently, waste plastics have been widely used as roadbuilding materials. There are two main approaches: adding waste plastic particles in bitumen or asphalt and dispersing, before application as the binder for road paving; and making prefabrication modular blocks from waste plastics, with such blocks being assembled on top of the soil, in a piece-by-piece arrangement, for the entire length of the road. Both methods are further discussed in detail in the following paragraphs.

For waste plastics added to bitumen/asphalt, such technology was initiated three decades ago, with the first known plastic-asphalt road project in Canada pioneering in the early 1980s. Fisher-Price toys were ground up to repave the City of Edmonton. With positive feedback from road users, such effort was further expanded to larger cities, including Highway 401 around Toronto, by paving with recycled milk jugs. Continual improvements of recycled plastic paving have enabled it to be used for the Vancouver International Airport runway, South Fraser Perimeter Road, the expressway along the south side of the Fraser River in a collaboration with Goodwood Plastic Products Ltd., of Shortts Lake, NS. Other related projects using recycled plastic-added bituminous roads are summarized in Table 6.3.

In addition, a company called PlasticRoad in The Netherlands has introduced a prefabricated slab using recycled plastics, shown

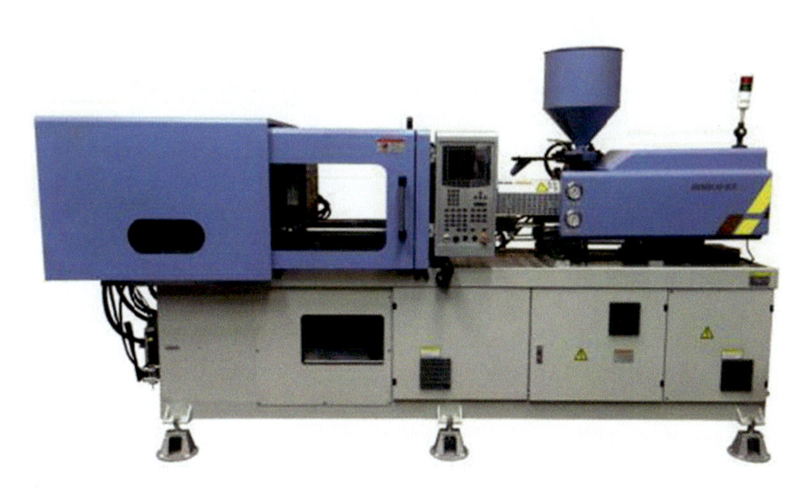

Fig. 6.5 (Top) An injection moulding chair made from recycled plastic; (Bottom) A typical injection moulding machine. Chair picture from Gracious Living. (2021). Available at https://blog.erema.com/products-made-from-recycled-plastic.

in Fig. 6.6, to substitute for bituminous pavements. The main applications are footpaths, bike paths, light vehicle parking spaces, platforms, schoolyards, playing fields and sports infrastructure. There are two designs, CCL200 and CCL300. CCL300 is designed with additional space underneath (shown with cover

Table 6.3 Road and highway built using recycled plastic bituminous/asphalt.

Country	Status and development
The Philippines[a]	• San Miguel initiated a project involving the first road combining plastic scraps with asphalt in November 2019. The surface material, developed with Dow Chemical Co., used 900 kg (1984 lbs) of plastic to pave a 1500 m^2 (16,145 ft^2) test site near the capital of The Philippines
Malaysia[b]	• In October 2020, Setia Bintang laid a pilot stretch of 117 m with PBE MacRebur using 810 kg of plastic waste, located at the KM 21 layby on the Kuala Lumpur—Karak Highway in the Gombak area
Australia[c]	• Australia, New South Wales (NSW) Environment Protection Authority has approved the use of Downer's Reconophalt road surfacing material. Reconophalt can contain high recycled content from materials such as soft plastics, toner, glass and reclaimed road • Reconophalt was used to construct a local road in Craigieburn, Victoria and the usage has expanded to Australia Capital Territory, New South Wales, South Australia, Tasmania and Western Australia
Thailand[d]	• A recycled plastic asphalt has been developed between the collaboration of SCG and Dow Chemical and first used to build the recycled plastic road in the RIL Industrial Estate in Rayong The project began with a plastic waste sorting program at Chemicals Business, SCG and communities in Rayong, in which plastic and thermal bags were shredded and mixed into the asphalt used to pave the road • With the size of 220 m in length, 3 m in width, and 6 cm in thickness, the prototype road creates value for plastics that are not generally reused or recycled appropriately and brings sustainability to communities, society, and the business
Ghana[e]	• A road running through Accra, Ghana's capital, looks like any other blacktop. Yet what most drivers don't realize is that the asphalt under them contains a slurry of used plastics—shredded and melted bags, bottles, and snack wraps—that otherwise were destined for a landfill • The impetus for many similar road projects underway in Ghana was an ambitious plan announced by President Akufo-Addo in 2018
China[f]	• Dow and Shiny Meadow collaborate to turn milk bottles into polymer-modified asphalt using Dow's Elvaloy RET asphalt modification technology
Canada[g]	• The road of City of Edmonton was used waste plastic pave way. Such effort is further expanded to larger cities road of Highway 401 around Toronto by paving with waste milk jugs. Continual improvement of recycled plastics paving has enabled it to be used for Vancouver International Airport runway, South Fraser Perimeter Road, expressway along the south side of the Fraser River in the collaboration of Goodwood Plastic Products Ltd.
India[g]	• Since 2015, the Indian government has laid 2500 km of waste plastic bituminous road with 703 km of national highway constructed in India using 6%–8% of plastics mixed with 92%–94% bitumen • A road from New Delhi to Meerut used plastic waste to replace 10% of the bitumen • Dow Chemical Co. recently worked with government officials in India and waste collectors in Bangalore and Pune to pave 40 km of roads using self-developed Elvaloy polymer-modified asphalt technology by taking off 25 million flexible pouches from landfills

Table 6.3 Road and highway built using recycled plastic bituminous/asphalt—cont'd

Country	Status and development
California, United States[g]	• In California, a paving company was responsible for a pilot project in July 2020 that repaved three lanes on a 1000-ft section of highway in the town of Oroville using recycled asphalt pavement and plastic sourced from single-use plastic bottles—the first time the California Department of Transportation has paved a road using 100% recycled materials • As developed by Redding, California-based TechniSoil Industrial, the process employs a recycling train of equipment that grinds up the top 3 in. of pavement and mixes the grindings with a liquid plastic polymer binder, which comes from a large amount of recycled, single-use bottles
United Kingdom[g]	• Lockerbie, Scotland-based asphalt specialist MacRebur produce three different types of pellets that vary in durability and application, and which are sold internationally. MacRebur claims that each kilometre of road laid with its pellet mix uses up to the equivalent weight of 740,541 single-use plastic bags, or 1 tonne of MacRebur mix contains the equivalent of 80,000 plastic bottles. MacRebur has paved roads throughout the United Kingdom—including a project in June 2020 that involved repaving a major road in Carlisle, in the North-west of England

[a]https://atlasofthefuture.org/project/smc-recycled-plastic-road/.
[b]https://focusmalaysia.my/msian-roads-to-be-paved-with-recycled-plastics/.
[c]https://www.theconstructionindex.co.uk/news/view/plastic-roads-move-forward-in-australia-and-europe.
[d]https://www.allaroundplastics.com/en/article/innovation-en/2189.
[e]https://www.goodnewsnetwork.org/paving-with-plastic-dent-global-waste-problem-yale/.
[f]https://www.dairyreporter.com/Article/2021/04/22/Dow-and-Shiny-Meadow-making-roads-from-milk-bottles-in-China.
[g]https://www.canplastics.com/features/the-plastic-road-ahead/.

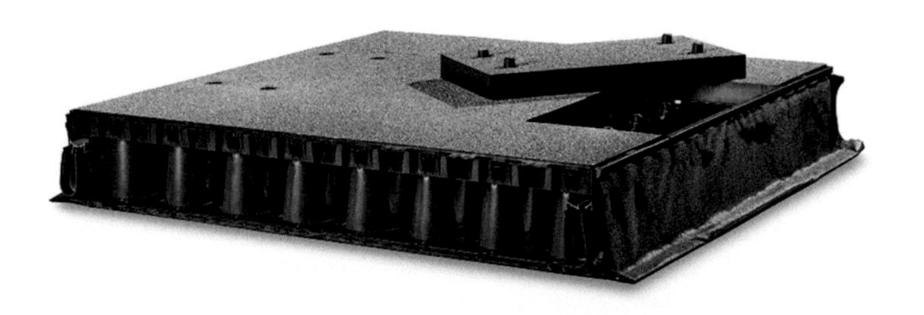

Fig. 6.6 (Top) Modular system of CCL200; (Bottom) Modular system of CCL300 with cover to store sensors. Both are produced by PlasticRoad using 100% recycled plastic materials.

Fig. 6.7 Bicycle path built using PlasticRoad system in Zwolle.

in Fig. 6.6 bottom) to store sensors to monitor the road condition and water drainage system. Both designs have the ability to collect excess water to protect local subsoil. An example of an application can be found in Fig. 6.7 of building a bicycle path in Zwolle in The Netherlands. In conclusion, adding waste plastic into a bituminous mixture or transforming it into a modular system for road-building are well accepted globally to massively reduce the impacts of waste plastics on the environment. This approach remains a good way to embrace a circular economy for sustainable future development.

6.2.5 Multilayer packaging

It is very common for food packaging to use multilayer materials, particularly for perishable food such as milk, oils and fats, cordials, carbonated drinks, crispy foods, read-to-eat food, etc. There are two main reasons to use multilayer packaging instead of single layer. The first reason is to improve the shelf life by preventing germ and atmospheric gas penetration through the package walls from spoiling the contents. The second reason is mainly for artistic and advertisement purposes, to attract consumers to buy the products. However, over the years, multilayer packaging and excessive packaging have become major issues to environmentalists. Multilayer packaging undeniably has recycling problems due to nonsegregation and the fact that mixing different types of plastics together can cause incompatible blending problems. Subsequently, the low quality of the recycled plastics makes them unlikely to be sold at good prices. On the other hand, industrial players have modified multilayer packaging assembly so that recycled material can be added in the interlayers of the plastic packaging. This allows replacing the internal layers, which are not in contact with food, with recycled plastic materials. This can minimize the use of virgin plastics content in the plastic packaging, while lowering the packaging material cost due to the cheaper recycled interlayers. The details of the multilayer recycled packaging film are discussed in the next paragraph.

Producing plastic film requires a plastic blown film machine. The common single-layer blown film process is shown in Fig. 6.8 (top), where the plastic pellets are fed to the extruder to turn the solid state of plastics into a molten state, followed by passing through the die and expanded using pressurized air. The thickness of the film can be adjusted through the mould, as well as the expansion by the pressurized air Fig. 6.8 (bottom). In the final step, the film is collapsed and collected in the rolling machine. Most of the time, before reaching the rolling collector, a corona or infrared treatment is applied on the surface of the film. This is to improve the printing ability; otherwise the plastic film without treatment has a smoother and paraffin-rich surface that can reduce the printing quality, such that the ink can be easily peeled off, even by gently rubbing on the surface. Fig. 6.9 shows the blown film of a coextrusion multilayer film with multiple die head (refer to Figs 6.10 and 6.11) simultaneously in operation. In the multilayer film, the outer layers are usually made from virgin materials, while certain internal layers can be made from recycled materials. For instance, as shown in Fig. 6.11, a three-layer film with a middle high barrier layer can be made from recycled plastics, which can reduce the cost and in the meantime does not compromise safety due to being sandwiched by two virgin plastic layers of PET. A similar concept is also applicable for a thermoforming process with multiple-layer sheets produced by coextrusion. Commonly, in single-use packaging like food trays two layered sheets are produced, where the layer in contact with food is made of virgin polymer while another layer, not in contact with food, is a recycled plastic material. Such thermoformed plastic trays are usually made of similar types of plastic material like polypropylene or PET. Since similar types of materials are used in the product, it can be easily recycled as is and does not require

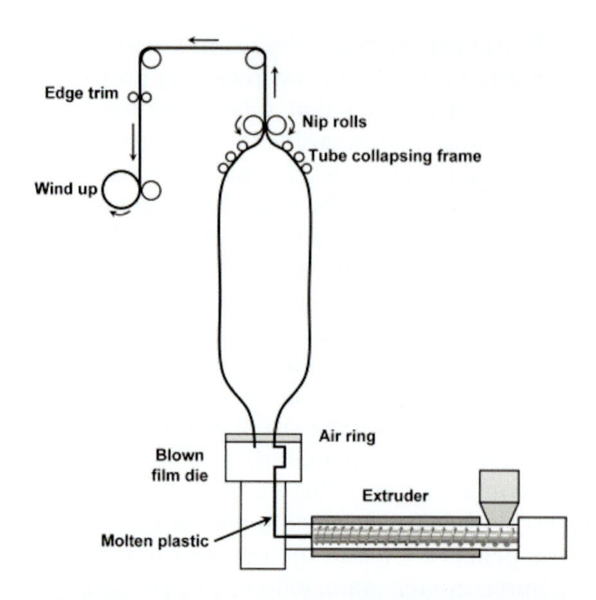

Fig. 6.8 (Top) schematic of blown film process; (Bottom) expansion process of blown film. From McKeen, L. W. (2017). *Permeability properties of plastics and elastomers* (4th ed.). Elsevier, with permission of Elsevier.

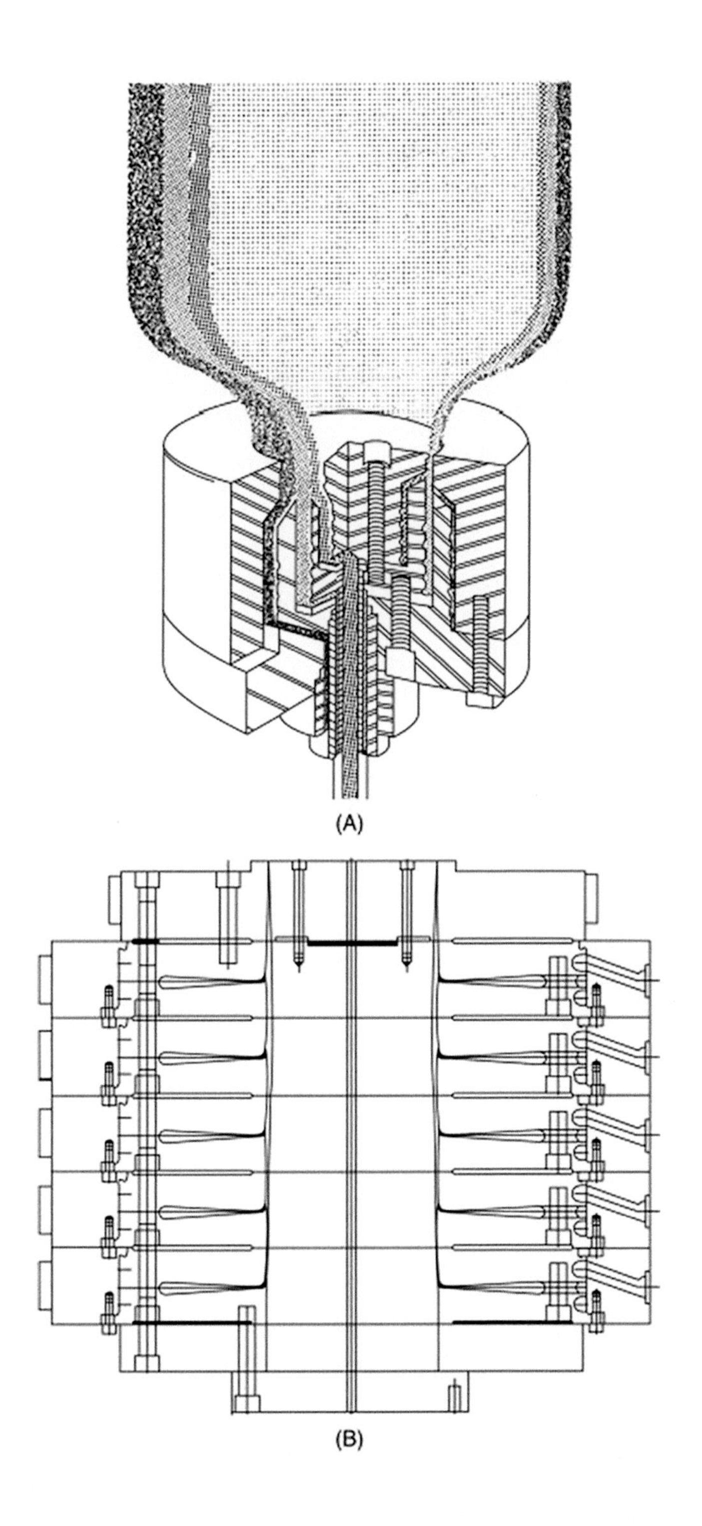

Fig. 6.9 A typical drawing of blown film dies: (A) three layers die and (B) five layers die. From McKeen, L. W. (2017). *Permeability properties of plastics and elastomers* (4th ed.). Elsevier, with permission of Elsevier.

Fig. 6.10 Nine-layer blown film die and bubble equipment. From McKeen, L. W. (2017). *Permeability properties of plastics and elastomers* (4th ed.). Elsevier, with permission of Elsevier.

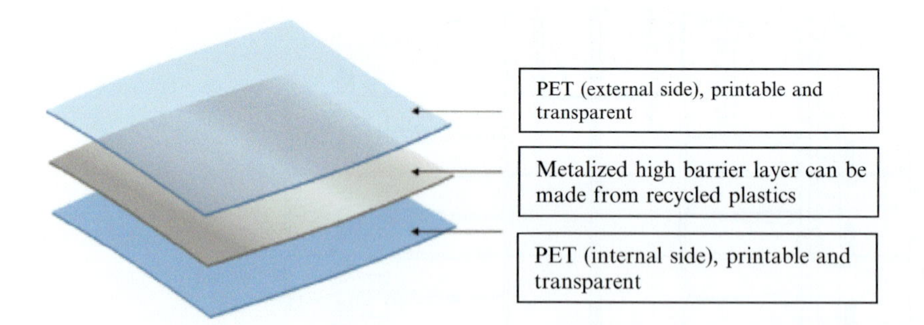

Fig. 6.11 Three-layer coextrusion of blown film with possible substitute of the middle layers with recycled plastic material.

separation in the early stage. This is unlike blown film, where the middle layers can be made of ethylene vinyl alcohol (EVOH), nylon, etc., and the recycling of such plastic packaging can be more troublesome compared to a single type. One of the common problems is that the mixing of different types of material cannot be easily recycled and the result of recycling can be low quality. These details of recycling are covered thoroughly in Chapter 3.

6.2.6 Automotive

The use of recycled plastics in the automotive industry has increased since the enforcement of Directive 2000/53/EC from 1 January 2015, where the automotive manufacturers are obliged to reuse 95% of end-of-life vehicles, consisting of 85% of the materials they are made of and the remaining 10% to produce energy. The use of recycled plastic materials is mainly for nonengineering or nonloaded parts, due to the fact that recycled plastic materials have questionable reliability and performance. As such, automotive makers choose to use recycled plastics for insulation and upholstery. Honda-marketed Acura cars used 2800 tonnes of textile waste, such as blue jeans and towels, for acoustic insulation and recycled plastic for car seat fabric (Waste360, 2018). The list of automakers that utilized recycled plastics in their products is seen in Table 6.4. Major automakers show committed efforts to make their vehicles eco-friendlier for the betterment of users and the environment.

6.3 Waste collection, segregation and methods of recycling

For any recycling effort to be successful, the crucial factor is the segregation process. The ability to segregate properly can be strengthened by continual community education. The community needs to be taught about the benefits of participating in the recycling of plastics, not only because the segregation and collection of domestic plastic waste according to recycler

Table 6.4 List of major automakers committed to utilize recycled plastics in their products.

Automaker	Parts and components from recycled plastics
Honda	Acoustic insulation, car seat fabric, splash guards and mud guards
Skoda	Engine cover, underneath engine cover, dashboard and car boot compartment
General Motors	Sound deadening covers, radiator covers
Ford	Cushion
Chrysler	Wheel liners, seat cushion from polyurethane foam
Nissan	Sound deadening materials for car dashboard
Volkswagen	Spare wheel compartment covers, wheel arch inserts, floor covering
Volvo	Tunnel console, floor carpet, sound-absorbing material, seats from PET bottles

expectations can yield monetary rewards, but also because the recycling efforts should be treated as efforts to reduce the impact of plastics pollution on the environment for the benefit of current and future generations. More accurately, reductions of plastic waste impacts are the responsibility of everyone. Thus everyone should learn the right way to perform plastic recycling.

Prior to 1st Step

It is important to note that the most important approach to minimize the impacts of plastic waste is to reduce, or whenever possible avoid, unnecessary use of plastics. For instance, plastic carrier bags provided free of charge by retailers can be replaced by self-brought reusable bags. Reduce usage of single-use food containers and cutlery with take-away food. Also, reuse food containers and beverage bottles; for example, PET mineral water bottles can be refilled several times before disposal. Reduce and reuse are the two important approaches to be taken prior to recycling in order to significantly minimize plastic disposal in the environment.

1st Step: Collection and segregation

The collection and segregation from the consumer end can be simple when the manufacturers have properly labelled and categorized according to the International Resin Identification Code, as shown in Fig. 6.12. As an example of domestic plastic waste recycling, a householder with empty plastic containers and bottles needs to look for the label appearing on the containers and bottles, and then segregate according to the international Resin Identification Code. The segregation is not limited to the major portion of the container only, but also includes the components as well. As shown in Fig. 6.13, a common bottle filled with cooking oil is made up of bottle body, cap, pourer and handle. There are three types of plastics in use, which are polyethylene terephthalate (PET) in the main bottle, cap and pourer are low-density polyethylene (LDPE) and the handle is polypropylene (PP). A shrink film or paper

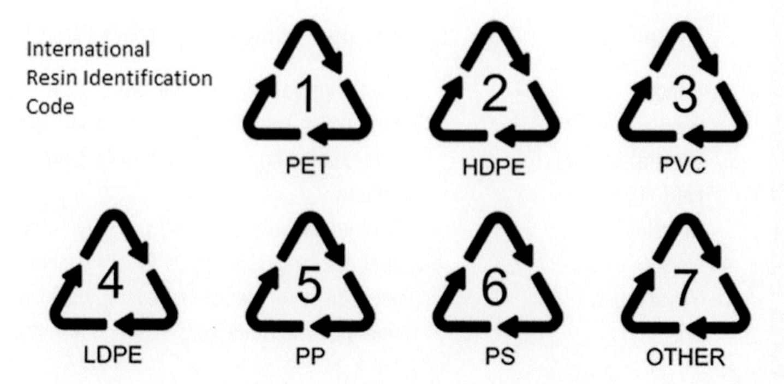

Fig. 6.12 International Resin Identification Code.

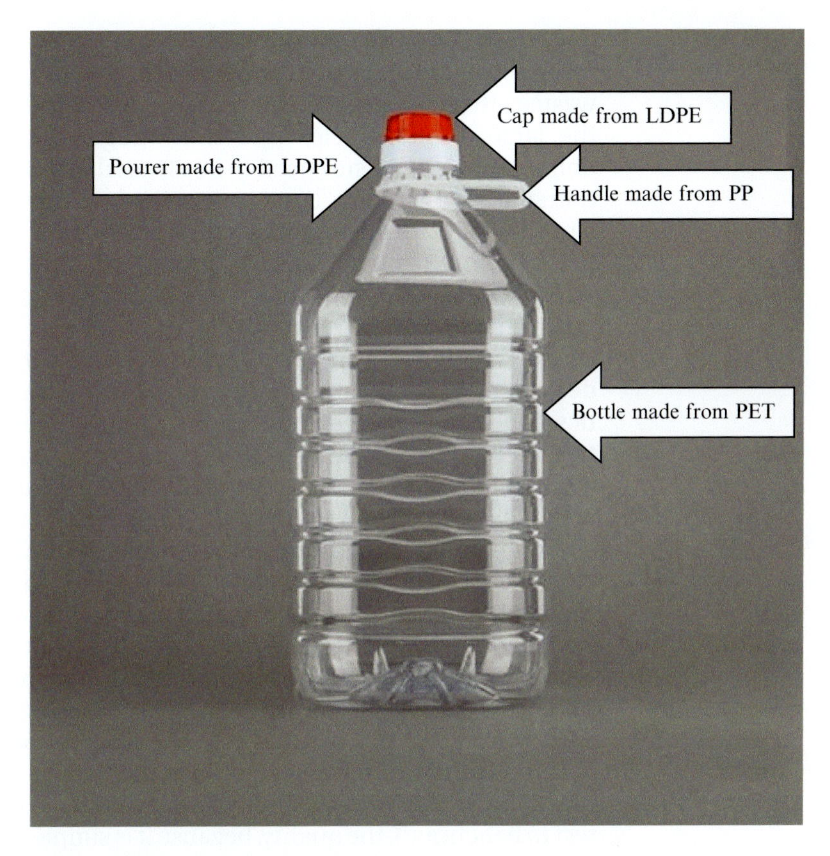

Fig. 6.13 The cooking oil bottles commonly found in the market.

label attached to the bottles also needs to be removed. All these plastics should be separated accordingly before sending to the recyclers. Nonetheless, many household consumers do not practise such detailed segregation, which means the segregation process must be further carried out by the factory workers, causing a less efficient factory workflow. On the other hand, factory-to-factory or in-house factory recycling can be more straightforward, where rejected HDPE beverage bottles at one factory can be taken by another factory to produce HDPE bottles for engine lubricant. Such a recycling arrangement is easier because the type of plastic is well known and is not contaminated, so it can be directly used for another plastic processing process.

In conclusion, for this step, household plastic wastes require more detailed separation followed by segregation according to the type of plastics. Otherwise, the recycling process is unable to proceed effectively.

2nd Step: Cleaning and grinding

The cleaning process is important for domestic plastic wastes after segregation and is followed by grinding to reduce the large plastic articles into small fragments. Indeed, the domestic plastic wastes need to be cleaned before the segregation step by consumers. For instance, food containers need to be cleaned with soap to remove oil and grease prior to sending to the recycling centre. When those sorted plastic wastes have arrived at the recycling factory, the cleaning process is carried out again to ensure meeting the requirements of industrial users. This cleaning step requires a huge amount of water and detergent. As a result, this substantial washing of recycled plastics can lead to water pollution issues. A proper wastewater treatment system is needed by the factory operators to avoid contamination of the water source. After the cleaning process, all the items need to undergo a drying process to remove water content. Finally, the grinding process reduces the size so that it can be used in mixtures together with virgin resin. To maintain the quality control of the recycled plastic resin, an important parameter to be tested is the melt flow index, which is a parameter that measures weight/10 min of the hot extruded resin under the force from a piston at a specific weight, using the common standard of testing ASTM D1238, 'Standard Test Method for Melt Flow Rates of Thermoplastics by Extrusion Plastometer'. Many small and medium industrial users use this test to benchmark the quality, because it is simple to use and the results are quite robust to exhibit the quality and processability of the recycled plastic resin directly. Other quality control can be crucial for specific plastics like PET, which require an intrinsic viscosity examination because the recycling of moisture-sensitive PET can cause significant reduction in molecular weight from the reversal reaction of condensation polymerization. Such a condition also occurs in recycling of nylon.

3rd Step: Melt blending and quality control

Recycled plastic resin is very unlikely to be used alone due to the fluctuation in quality when sourced from domestic plastic waste. Recycled plastic resin is commonly mixed with virgin plastic resin so that the variation impact can be minimized. For instance, the melt viscosity of the molten plastics can be kept at a lower range for the extrusion process of the composition of recycled plastic resin 10 wt.%–60 wt.%. Nonetheless, when the source of recycled plastic resin is factory-to-factory, the composition of recycled plastic resin can be as high as 90 wt.%. There are two approaches to utilizing recycled plastic resin. The first approach is physically mixing the recycled plastic resin with virgin resin and feeding it into the machinery, such as injection moulding, to produce plastic items. The second approach is melt blending recycled plastic resin and virgin resin in a twin screws extruder to produce a plastic compound. This compound can be customized at the recycler site, with additives, so that the users can use them without knowing how to operate the compounder as well as formulation development at this melt blending stage. The quality control can vary, including melt flow index, colour, moisture content, mechanical strength and others, according to the user demands.

4th Step: Resin and end-products

The recycled plastic resin is now ready to be used in various processing methods like injection moulding, extrusion, blow moulding, rotational moulding, thermoforming, etc. In fact, usually recycled plastic resin has identical characteristics to virgin resin, except it is lower in viscosity due to degradation during the first-time processing and use, degradation during crushing, thermal effects and degradation caused by contamination. Thus the perception that plastic can be recycled endlessly is wrong. A visible phenomenon is that the recycled resin has a darker colour, such as yellowish or brownish, compared to virgin plastic resin, which is why most of the recycled plastic resin has colourant added to minimize the visible discolouration.

Finally, the manufacturers who use recycled plastics are urged to label clearly what types of recycled plastics have been used, based on the Resin Identification Code, so that future recycling processes can still be carried on for both physical recycling and chemical recycling (also known as pyrolysis). At the end of life, plastic wastes that are very contaminated and degraded are no longer suitable for further recycling. At this point, incineration is the most promising approach to convert waste to energy, while landfilling is an unfavourable solution which can lead to microplastic pollution and long-term impacts on the environment.

6.3.1 Equipment requirements, capital expenditure and operation expenditure

As mentioned earlier, the most common recycling method of plastics is the physical recycling method as compared to the chemical recycling method. A comparison of physical and chemical recycling of waste plastics can be found in Table 6.5. In general, physical recycling is the most popular technology, widely adopted by small and medium industrial players due to the affordable capital expenditure and the fact that there is no great need for expertise. In the view of these benefits, many of the physical recycling technology providers establish a turnkey project and training for prospective or existing industrial clients on operating this technology. However, the main issue for physical recycling technology remains the difficulty of waste plastic segregation and pollution problems (resulting from the wastewater generated during cleaning and unrecyclable plastic illegally disposed of after sorting). Thus the technology operator is urged to follow strict environmental regulations, while enabling plastic recycling to proceed for the benefit of the community and future generations.

Table 6.5 Comparison of physical and chemical recycling of plastics.

Recycling method	Advantages	Disadvantages
Physical	– Well-established technology with acceptable cost of capital (starting cost can be as low as USD $100,000 with a single screw extruder) – The process can be easily adopted in any existing plastic manufacturing factory to recycle from self-produced reject products – Able to handle small quantities of wastes, i.e. <200 kg/day – Moderate knowledge is needed through practitioner experience on plastic materials – Acceptable quality of recycled plastics depending on the quality of segregated plastics	– Limited to recycling well-known types of plastics only. Difficult to recycle plastics with different composition of additives – Preferable to recycle virgin plastics – Plastic segregation is needed according to types of plastics. Poor quality of segregation can cause low quality of produced recycled plastics – Operators can have lack of awareness about appropriate operation of factories which can lead to soil, sound, air and water pollution – The prices of recycled plastics are low and the recycled plastics usually can only be used to blend with similar type of plastics
Chemical	– Most of the plastics are able to undergo a pyrolysis process to break down at high temperature into monomer or pyrolysis oil – Acceptable quality of pyrolysis oil depending on the quality of segregated plastics – Pyrolysis products, i.e. low molecular weight chemicals can be used for wider industries to produce new/virgin materials – Better operation unlikely to cause substantial pollution	– Requires high technical knowledge to manage the pyrolysis process – Very complicated and costly setup process with estimated capital cost > USD $1 million – Required consistency of raw material quality – Process/technology is set up for limited types of input – Requires more attention on the emission handling process to avoid air pollution

A common lineup for a waste plastic recycling line should consist of crusher/shedder, feeder and extruder and finally a pelletizer to convert the pellet form, while a specific apparatus can be added for cleaning, drying, controllability and productivity of the recycled plastics. Tables 6.6–6.9 summarize the main recycling technology of the major industry players. In general, the full set of recycling machinery can cost starting from Euro €100,000 to 3,000,000, depending on the technology requirement. For instance, recycling of in-house edge trim film at a scale of 50 kg/h that does not require a washing and drying process can cost €150,000, whereas postconsumer PET bottles and required food-grade quality of recycled pellets can cost up to €3,000,000 with quality control equipment included, such as inline measurement of intrinsic viscosity.

Table 6.6 Next Generation Recyclingmaschinen GmBH (NGR) (Additional information at www.ngr-world.com).

Technology	Functions
E:GRAN	Special design to recycle plastic film. The plastic films are cut in the feeding section followed by conveying area and extruder along a single operation shaft. Such an assembly makes optimal use of heat from the chopping process and operation in one step chopper-feeder-extruder combination system. Two sizes are available: E:GRAN 50-12 maximum output 50 kg/h and E:GRAN 75-16 maximum output 120 kg/h. Suitable for HDPE, LDPE and PP
A:GRAN	This model uses a heavy-duty shredder to operate at low speed to reduce the size gently by using a pneumatic ram to push the material into the shredder drum, followed by feeding directly to the extruder. It is designed to process large, thick or bulky waste without preshredding into pieces so the size can pass through the hopper opening. Two sizes are available A:Gran 65-40 maximum output 100 kg/h and A:Grant 70-40 maximum output 120 kg/h. Suitable for PET, HDPE, LDPE, PP, PS/EPS, etc.
S:GRAN	This model is designed to be versatile to tackle a wide range of materials and shapes including fibres and thick-walled scrap without a preshredding step. The shredder and extruder operate independently based on load. There are eight sizes of this model with the output range from 220 to 800 kg/h. Suitable for PET, HDPE, LDPE, PP, PS/EPS, etc.
X:GRAN	A larger and comprehensive design to cater to all types of plastic wastes, including bulky materials such as large bales and carpet pads. There are six sizes of this model with the output ranging from 900 to 2700 kg/h. Suitable for PET, HDPE, LDPE, PP, PS/EPS, etc.
P:REACT	A specially made system for PET recycling using an advanced liquid state polycondensation process to achieve better intrinsic viscosity. When the molten PET feeds into P:REACT, liquid state polycondensation occurs and the intrinsic viscosity increment can be manipulated by the residence time and vacuum pressure. P:REACT also possesses a good decontamination function and the recycled PET has good safety standards

Continued

Table 6.6 Next Generation Recyclingmaschinen GmBH (NGR) (Additional information at wwwngr-world.com)—cont'd

Technology	Functions
F-GRAN	Designed for processing of film flakes to overcome the flakes from bridging and ensure flowability. The feeding screw acts like a feeder between the transition area between flake conveyor and extruder input so that the flakes can be fed steadily into extruder. Available in seven sizes with range of output 750–2300 kg/h. Suitable for PET, HDPE, LDPE, PP, PS/EPS, etc.
C:GRAN	This model is specially designed to cater to flakes and films with residue moisture up to 25%. This model includes a rotating blade to shred the plastic scrap, and it is compressed, heated and fed into the extruder. There are six sizes of this model with maximum output range of 450–2500 kg/h. Suitable for PET, HDPE, LDPE, PP, PS/EPS, etc.

PET, polyethylene terephthalate; *HDPE*, high-density polyethylene; *LDPE*, low-density polyethylene; *PS*, polystyrene; *EPS*, expanded polystyrene.

Table 6.7 EREMA Engineering Recycling Maschinen und Anlagen Ges.m.b.H. (Additional information at www.erema.com).

Technology	Functions
INTAREMA	INTAREMA series of recycling machines have high throughput for recycling a wide range of waste plastics. It uses patented Counter Current technology by changing the direction of rotation inside the cutter/compactor. When the materials pass from the cutter/compactor to the extruder, the extruder acts as a sharp edge, cutting up the waste plastic efficiently. The output for polyethylene ranges from 50 to 2800 kg/h, biaxial oriented polypropylene 50 to 3000 kg/h and biaxial oriented polyethylene terephthalate 80 to 2100 kg/h. Meanwhile, INTAREMA TVEplus is designed for high moisture waste plastics with addition of a degassing process. For recycling of trim edge, INTAREMA K is designed for this purpose
VACUREMA	This model is specially designed for recycling of PET and the product is suitable for food contact with good intrinsic viscosity. During the process, the decontamination process is carried out in the early stage in PET flake form, followed by a drying process in order to make the intrinsic viscosity value consistent. Finally, an extrusion process forms the recycled PET into pellet form
VACUNITE	This model is designed for bottle-to-bottle recycling, combining VACUREMA technology and vacuum assisted V-LeaN Solid State Polycondensation (SSP). The products are recycled PET with good colour values, intrinsic viscosity stability and low energy consumption
VACUNITE	A larger and more comprehensive design to cater to all types of plastic wastes, including bulky materials such as large bales and carpet pads. There are six sizes of this model with the output ranging from 900 to 2700 kg/h. Suitable for PET, HDPE, LDPE, PP, PS/EPS, etc.
COREMA	This system is designed by combining recycling and compounding in one step. For instance, nonwoven polypropylene, polyethylene edge trims, etc. are processed by INTAREMA and the melt is directly fed into a corotating twin-screw compounder with the additives, filler and reinforcing filler added to produce resin directly

Table 6.8 Starlinger Technology (Additional information at https //www.starlinger.com/en/recycling/).

Technology	Functions
recoSTAR direct	This system is designed to handle large quantities of materials such as polyethylene, polypropylene, polystyrene, nylon and polycarbonate. The recoSTAR direct has an output range of 130–1200 kg/h. The complete set of machines can be equipped with a degassing, high vacuum, backflushing filter pelletizing system
recoSTAR dynamic	A specially designed system for recycling waste plastic for hygroscopic, wet and/or preground and foamed materials at low energy consumption. This model covers six size ranges from 150 to 2600 kg/h
recoSTAR universal	A more versatile design to process films, fibres, filaments, nonwovens, tapes, fabrics, solid lumps, start-up scrap, and injection moulding of waste plastics. Precutting process plastic wastes do not required whereas the materials can be fed into shredder and cut in between rotary and stationary knives before passing to extruder. Five machine sizes available for output 150–1300 kg/h
recoSTAR PET: Nonfood	The recoSTAR PET process line is recycling technology designed for recycling of PET flake source in-house process or postconsumer either from bottles, preforms, sheets or strapping bands. The flakes undergo a washing process and drying followed by melt-filtrated to produce uniform granulates that are suitable for a wide range of applications (containers, sheets, fibres, and nonwovens and strapping). In order to ensure good energy savings and high crystallinity, an inline crystallization after underwater pelletizing can be installed. This system is also versatile enough to include a solid state polymerization process to improve demand for intrinsic viscosity. The recoSTAR PET and recoSTAR PET HC (High Capacity) series has five different machine sizes of 150–3.600 kg/h
recoSTAR PET: Food	A more stringent PET recycling process with decontamination capability to meet food contact requirements with stringent washing requirements. Four models available to choose from: (a) recoSTAR PET FG: Solution for sheet application with good intrinsic viscosity keeping (b) recoSTAR PET FG+: Increased intrinsic viscosity and suitable noncarbonated soft drink preforms (c) recoSTAR (HC) iV+: Good solution for bottle-to-bottle to meet food-contact requirements with adjustable intrinsic viscosity with high capacity production (d) recoSTAR PET iV+ Superior: Solution for postconsumer waste PET material with capability of decontamination level being highest to reduce organic compounds

Table 6.9 Polystar Machinery Co. Ltd. (Additional information from www.polystarco.com).

Technology	Functions
Repro-Flex	Repro-Flex is made for a one-step process to recycle thermoplastic polyethylene and polypropylene wastes including flexible packaging. It is also possible to process foam material as well. In general, Repro-Flex is an integrated machine without preshredding process. There are seven sizes for this model for rate of output range 80–1200 kg/h
Repro-Flex Plus	Repro-Flex Plus is a two-stage recycling machine to cater to the recycling of heavily printed material (up to 95% printed surface) which requires filtration before degassing. Triple degassing is developed with double degassing in the first extruder follow by another venting area in connection between the first and second extruder to provide subsequent removal of ink and extra humidity level from the material
Repro-Air	Repro-Air is a light machine system developed for blown film and cast film producers with air-cooled system. It can cater to the recycling of polyethylene films scraps, start-up or changeover film rolls, edge-trim waste, lightly printed film. Good machine for in-house recycling purposes. Since it is air cooled and the pellet is fully dried, the recycled plastic pellet can be directly fed to reprocessing immediately
Repro-Direct	Repro-Direct is a versatile recycling system to handle large quantities of precrushed, heavy rigid regrind scraps such as bottles, pipes, containers and lumps in the form of granules
Repro-One	The one-step recycling machine Repro-One has an integration of a heavy-duty single-shaft shredder that is directly connected to the extruder pelletizer. This is suitable for PP raffia/PP nonwoven that can be easily redone when the in-house process generates a reject of raffia waste. The recycled PP material can still possess good mechanical properties which can mix at high percentages with virgin resin
Sequeezer Drying Machine	The machine is made to easily process a high level of water moisture in wasted plastic after the prewashing line. The wasted plastics can be postconsumer PE film, PP woven, PP jumbo bags and agricultural film, in the form of flakes after washing with 10%–35% moisture to reduce 1%–5%. This squeezer drying machine method is preferable to conventional drying methods such as centrifugal drying or blow drying using hot air. It is coupled with squeezing and drying process requiring less thermal energy, thus less material degradation for better quality and reusability of recycled pellets

6.4 Conclusion

In conclusion, consumers should take the initiative to participate in plastics recycling while embracing the circular economy. Understanding the recycling process, industry requirements, responsibility and benefits of plastic recycling can ensure the effectiveness of plastic recycling efforts in the communities. Thus the impacts of plastic pollution can be mitigated to achieve a better living environment and the sustainable growth of future generations.

References

Textile Exchange. (2021). *Guide to recycled inputs GRS-202-V1.0-2021.09.22 RCS-202-V1.0-2021/09/22*. Available at www.textileexchange.org.
Waste360. (2018). Available at https://www.waste360.com/business-operations/honda-s-history-using-recycled-materials-continues.
Precious Plastic. Available at: https://preciousplastic.com/. 2022.

7

Biopolymers and challenges

7.1 Introduction

Biopolymers have been used by humans for thousands of years. In the early days, biopolymers were natural product extracts from plants, such as gums, which are types of adhesive substances with a vegetable origin. According to Patten et al. (2010), plant gums are derived from complex carbohydrate polymers such as hydroxyproline-rich proteins, resins, or other components. These gums can be soluble, partly soluble or insoluble. Gums were used as waterproofing adhesives and coatings for boats during ancient times. Nevertheless, the applications of natural polymers were very limited at that time because humans had limited knowledge of polymer synthesis, and did not know how to modify polymers or natural gums to improve the applications.

About 200 years ago, scientists found a method to improve rubber latex harvested from the rubber tree *Hevea brasiliensis*. Charles Goodyear in 1844 discovered how to utilize sulphur to vulcanize rubber. Before that, natural rubber latex was merely obtained in a sticky lump and had an elastic characteristic after drying. Leo Hendrik Baekeland in 1907 synthesized the first synthetic polymer, commonly known as Bakelite, or chemically called phenol-formaldehyde resin. Nowadays, synthetic polymers such as polyethylene, polypropylene, polystyrene, poly(vinyl chloride) (PVC), polyethylene terephthalate (PET) and others are widely used in daily life. Petroleum-based polymers are used in just about every aspect of life. However, the drawbacks of petroleum-based polymers are due to their mass production at cheap cost, so that their uses are exploited to meet the demands for single-use products, such as disposable packaging, food containers, food utensils, electronic gadgets, and others. The lack of awareness of the global community of plastic pollution due to the convenience of petroleum-based polymers has led to nondegradable petroleum-based polymer pollution becoming a very serious

Plastics and Sustainability. https://doi.org/10.1016/B978-0-12-824489-0.00008-8
Copyright © 2023 Elsevier Inc. All rights reserved.

global issue. These nondegradable polymers usually end up in a landfill, with less than 10% being recycled (Pellegrio et al., 2019). These types of polymers take hundreds of years to fully degrade into harmless components. Thus scientists are recommending biopolymers as potential candidates to replace petroleum-based and nondegradable polymers. As implied by their name, biopolymers are biodegradable and it is preferred that they come from renewable sources, especially plant-based inputs. As suggested by Sin and Tueen (2019), the current domestic polymer development follows the trends shown in Fig. 7.1.

Based on the number of publications in biopolymer-related fields over the last few decades, it can be observed that the amount of research on biopolymers is overwhelming. Fig. 7.2 shows the trends of publications for about 50 years (1970–2019). Biopolymer research has been such a popular research field because biopolymers are expected to replace nonrenewable petroleum resources, which are currently major components of plastics production, to fulfil global market demand.

On the other hand, there are several understandings of the term biopolymer. From the general context of sustainable biopolymers, this can include biodegradable polymers and biobased

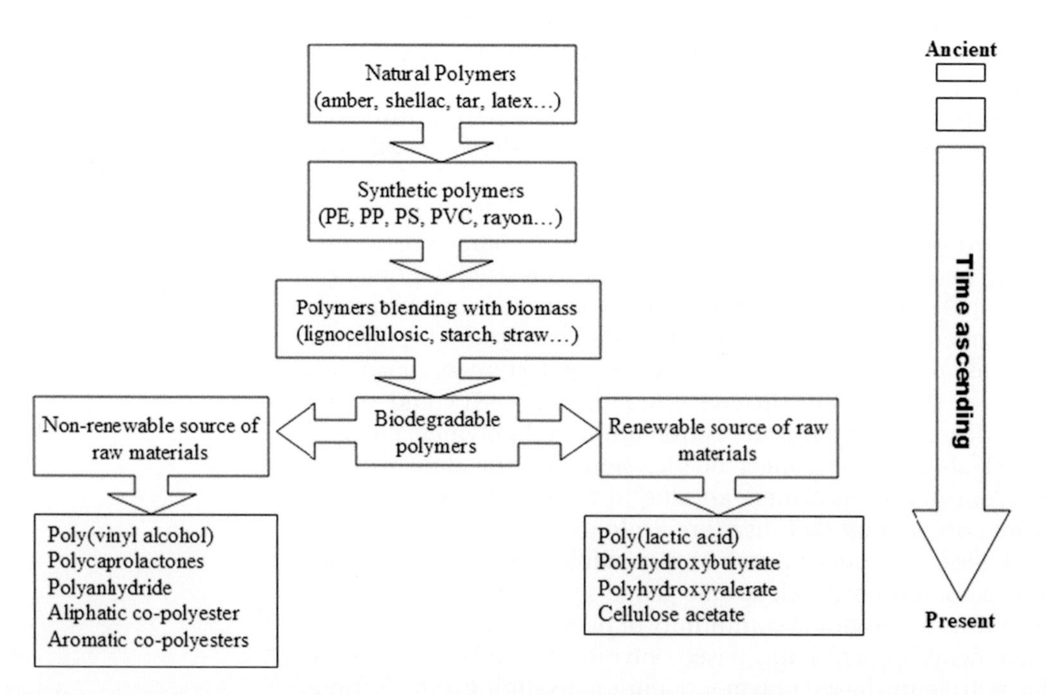

Fig. 7.1 Trends of polymer development. From Sin, L. T., & Tueen, B.-T. (2019). Polylactic acid. A practical guide for the processing, manufacturing, and application of PLA. United States: Elsevier with permission of Elsevier.

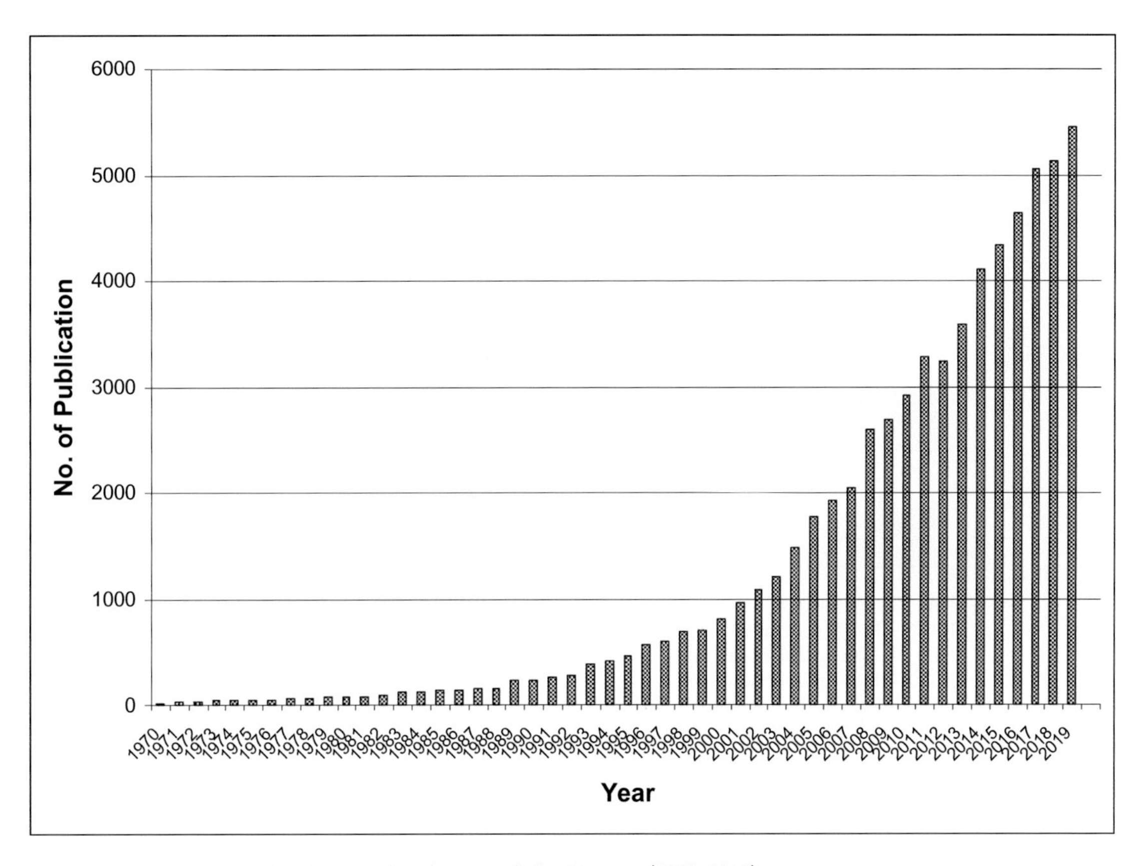

Fig. 7.2 Publications on biopolymer-related research for 50 years (1970–2019). Data from Scopus.

polymers (i.e. polymers derived from plant sources to replace petroleum sources). Sin and Tueen (2019) summarized biodegradable polyester derived from both renewable and nonrenewable sources, shown in Fig. 7.3. Note that the biodegradable polymers are mainly produced from the polyester family, which contains oxygen elements in the chemical chains. The presence of oxygen is essential, because oxygen is needed to initiate a chain scissioning reaction after contact with moisture, and the chains are further broken down into smaller fragments, ready for micro-oganism consumption. In addition, there is another type of polymer product called the oxo-degradable polymer. Commonly, oxo-degradable polymers are used to produce plastic bags that undergo disintegration after a short period of time. Oxo-degradable polymers typically are based on pro-degradant additives in fossil-based polymers, such as polyethylene, polypropylene, polystyrene, etc., to initiate chain scissioning. There are

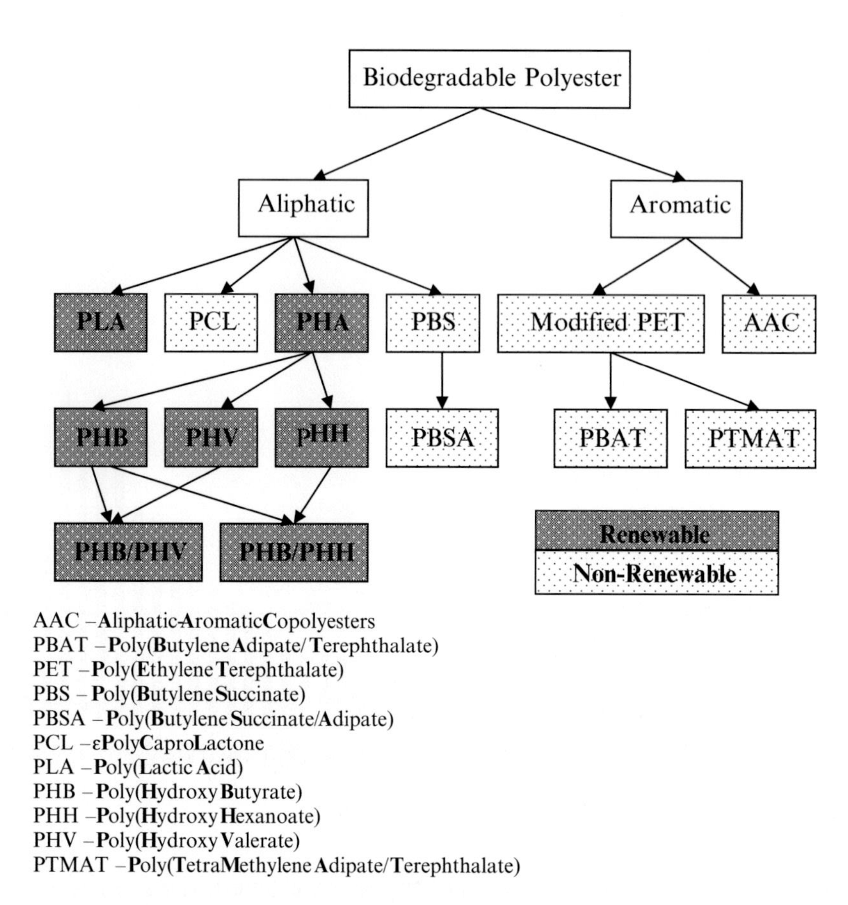

Fig. 7.3 Biodegradable polyester family. From Sin, L. T., & Tueen, B.-T. (2019). Polylactic acid. A practical guide for the processing, manufacturing, and application of PLA. United States: Elsevier, with permission from Elsevier.

notable pros and cons of oxo-degradable polymers. The pros are that only small quantities of pro-degradants are added, which does not affect the production parameters in the factory, while the cost of materials is relatively insignificant. However, the drawbacks are that pro-degradant additives consist of transition metal compounds, which pose a risk of causing environmental pollution, in addition to the fact that disintegration of plastic chains can generate microplastics in the environment, also leading to pollution.

There are many major producers of biopolymers in the global market, as listed in Table 7.1. In general, the demands for biopolymers have been steadily growing for over a decade, because consumers are aware of the importance of biodegradable polymers in reducing plastic waste pollution. Nonetheless, the price of biodegradable polymers remains the main constraint to substantial

Table 7.1 Players in biodegradable polymer/packaging industry.

Company	Country
BASF	Germany
Biomatera	Canada
Biome Bioplastics Ltd.	United Kingdom
Biomer	Germany
BIOP Biopolymer Technologies Ag	Netherlands
Biotec-Biologische Naturverpackungen GmbH & Co.	Germany
Cereplast	Italy
Corbion	France
Danimer Scientific	United States
FKuR Plastics Corp.	United States
Futerro	Belgium
Galactic SA	Belgium
Huhtamaki Group	Finland
Mitsui Chemicals	Japan
Natureworks LLC	United States
Novamont S.p.A.	Italy
Plantic Technologies Ltd	Australia
Rodenburg Biopolymers B.V.	Netherlands
Synbra Technology B.V.	Netherlands
Teijin Limited	Japan
Teknor Apex	Singapore
Tianan Biologic Material Co. Ltd.	China
Tianjin Guoyun Biological Materials Co. Ltd.	China
Toray Industries, Inc.	Japan
Toyobo Co. Ltd.	Japan
Zhejiang Hisun Biomaterials Co. Ltd.	China

acceptance by a wider spectrum of consumers, because the cost of biodegradable polymers can be at least two times that of petroleum-based nondegradable polymers. Hence, many of the developing and underdeveloped countries are not keen to use biodegradable polymers. As shown in Fig. 7.4, the biodegradable industry players are only found in developed and high gross domestic product (GDP) regions. This indicates that biodegradable polymers still lack penetration into middle and low GDP regions. Ironically, those developing and underdeveloped countries are facing the most serious plastic pollution, particularly landfills overflowing with nondegradable plastic waste as well

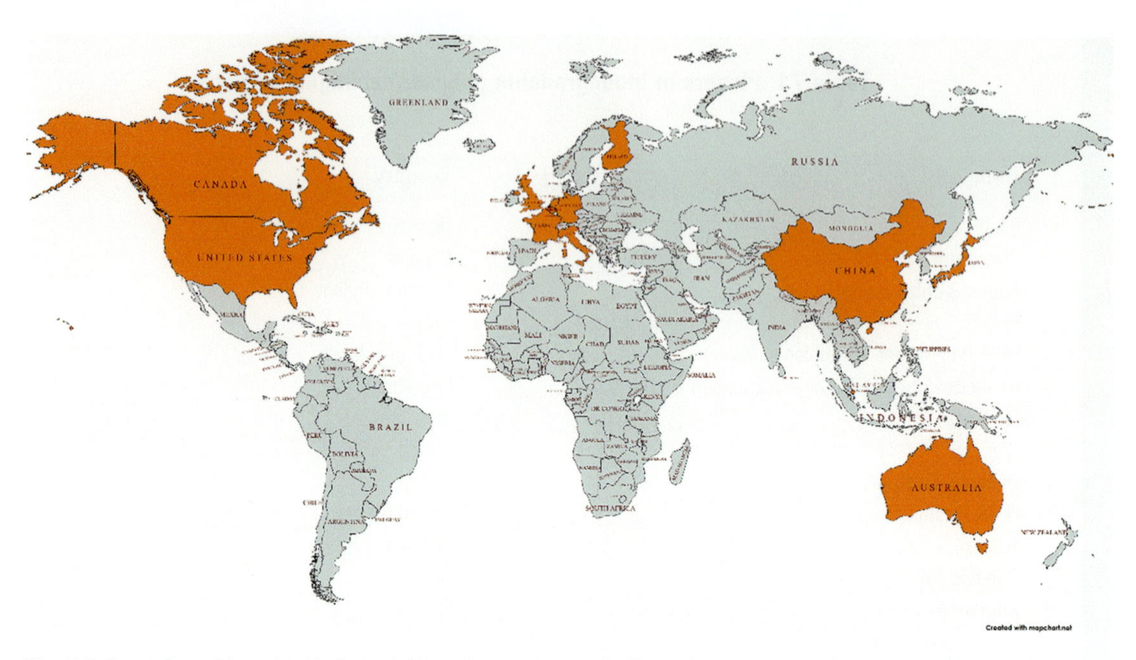

Fig. 7.4 Countries with major biodegradable polymer players (indicated in *dark* colour) through 2020.

as the fact that the rivers have tons of plastic garbage floating on their surface. When the use of biodegradable plastics is impractical due to high prices, the next-most appropriate approach is to educate the community about handling of plastic wastes in order to reduce their impacts over the long run.

7.2 Market demand of biopolymers

The market for biopolymers has shown tremendous growth over the last few years, particularly in the European region. This is driven by the awareness of European community support for sustainable products. Moreover, many countries have started to impose regulations and bans against single-use plastics such as food packaging and straws, which has further improved the market demand for alternatives to nondegradable polymeric materials. According to IHS Markit's *Chemical economics handbook: Biodegradable polymer report*, the market value of biodegradable plastics reached $1.1 billion in 2019 and it is expected to increase to $1.7 billion by 2023, while the consumption of biodegradable polymers reached 484.7 kilotons in 2019 and possibly will reach 984.8 kilotons in 2022, with a compound annual growth rate of 15.2% during 2017–22 (BCC Research). Although the outbreak of the COVID-19 pandemic has tumbled the global economy, leading to lower demand for consumer products, the plastics products have remained resilient due to high demand for hygienic packaging for food service, e.g. most of the airline operators have refined their standard operating procedure for in-flight food service with individually packaged foods for passengers. In addition, many event organizers of conferences, meetings, celebrations, parties and wedding ceremonies turned their food service into prepackaged food in order to reduce contact, as compared to serving buffet style as previously. All these 'new normal' practises have enhanced demand for biodegradable polymers in order to reduce the impact of single-use plastics.

As shown in Fig. 7.5, currently the main consumer of biodegradable polymers is Western Europe, followed by Asia and Oceania and North America, including both the United States and Canada. This is mainly because these governments have spent a great deal of effort in educating their communities to be responsible in handling and disposal of single-use plastics. The consumer acceptance in these regions is higher compared to the rest of the world (see Fig. 7.6). Moreover, these countries also have developed environmental protection frameworks so that consumers can experience the benefits of choosing biodegradable plastics as packaging. The people are well informed that biodegradable and compostable plastics can also be transformed into energy through the incineration process, as well as composting instead of ending up in landfills. For instance, Japan has encouraged food producers to use biodegradable plastics while also providing sufficient collection spots to segregate biodegradable polymers from other commodity plastics like polyethylene, polypropylene or polystyrene, so that biodegradable plastics can be properly managed to be transformed into value-added products. As strong initiatives from governments have driven biodegradable polymer demand in these countries, subsequently the biopolymer

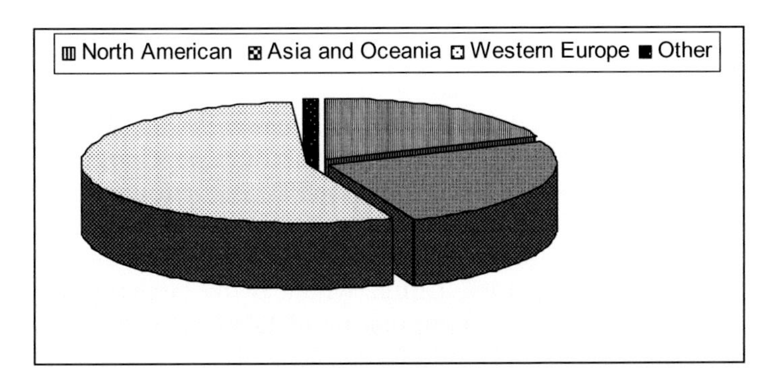

Fig. 7.5 Distribution of world consumption of biodegradable polymers in 2018.

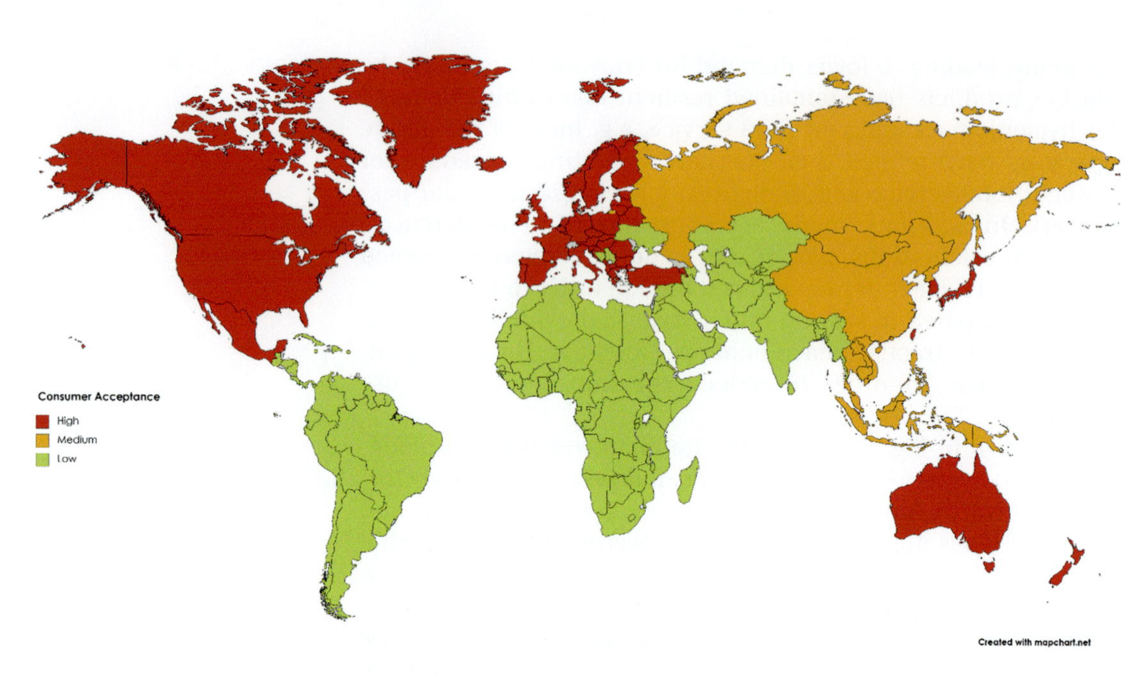

Fig. 7.6 Consumer acceptance of biopolymers based on region.

industry has flourished and become a worldwide pioneer in this clean technology.

As stated earlier, biodegradable polymers can be produced from both petroleum and renewable sources. Both types of biodegradable polymers have gained attention in the industry. Particularly, the renewable biodegradable polymers are not just biodegradable, but are also sourced from plants with environmental credits. Meanwhile, petroleum-based biodegradable polymers may also help to overcome the accumulation of nondegradable plastic waste, according to the data published by nova-Institut GmBH (2019), as shown in Fig. 7.7. First, it can be seen that the largest share of biopolymers is consumer goods, followed by building and construction industry goods. In the 5 years from 2018 to 2023, the market share of both consumer goods and building and construction industry goods will further increase 1% and 2%, respectively. This is because of government policies and successful education that have pushed consumers to avoid fossil-sourced plastic materials while replacing them with biopolymers. In fact, many countries have implemented regulations to reduce or ban the use of nondegradable plastics for environmental protection. This is the main spark that has led to the demand for biopolymers. For instance, China, which is the largest polymer-consuming country with a population of 1.44 billion, has banned

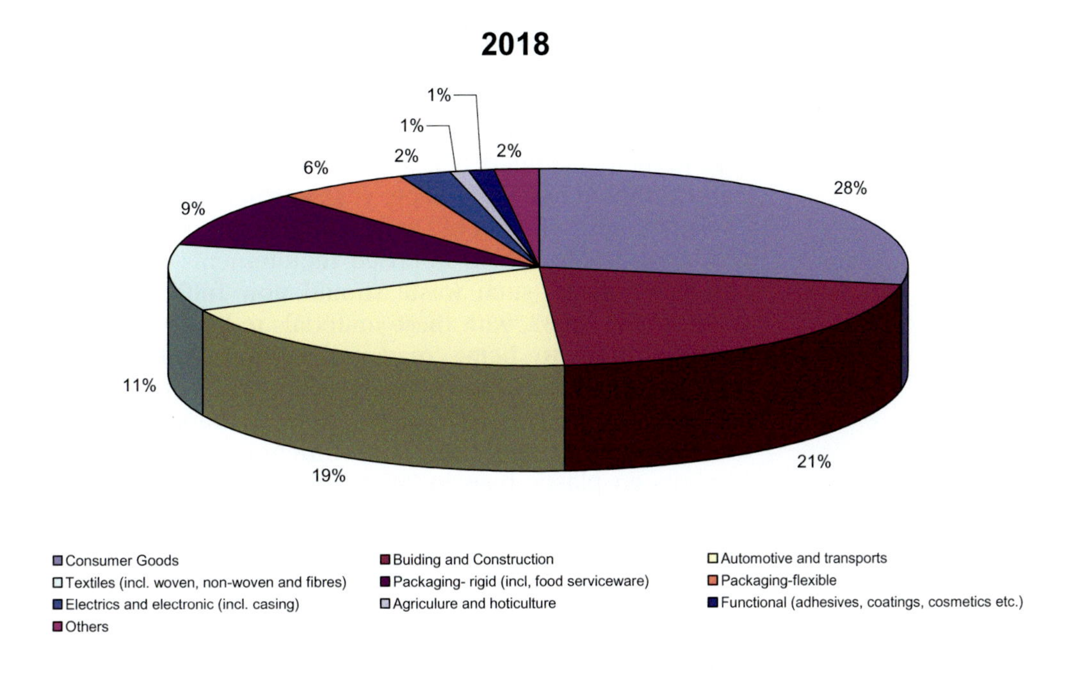

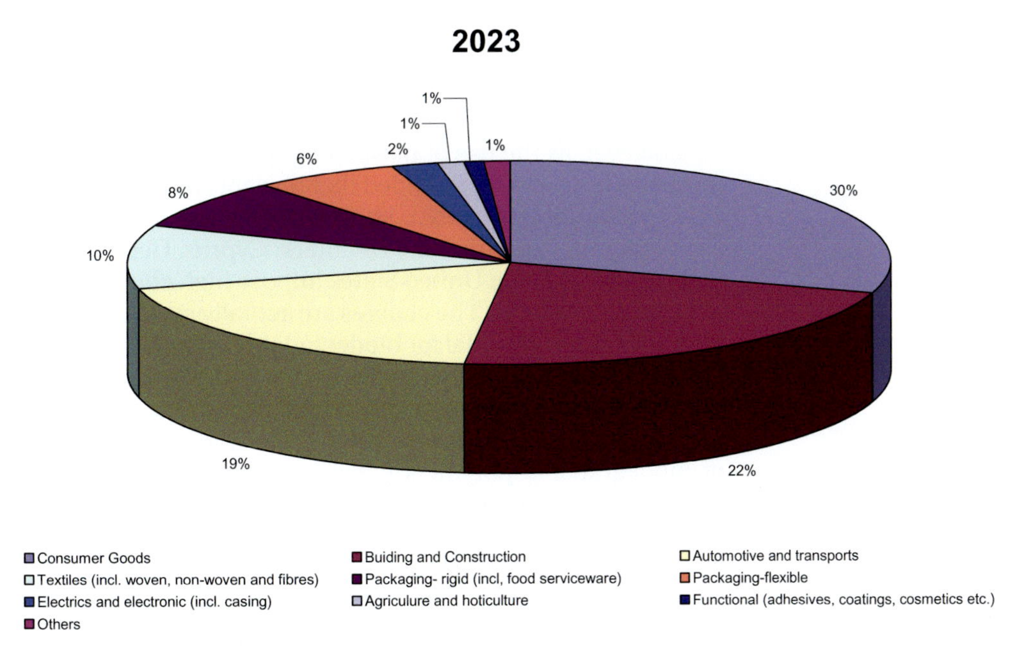

Fig. 7.7 Percentage of market share of biopolymers in 2018 and 2023.

the use of plastic bags, particularly in the restaurant industry. Both production and sale of plastic bags less than 0.025 mm thick as well as straws are banned, whilst no single-use plastic items will be offered by 2025. Major supermarkets do not provide free plastic bags to their customers. These regulations have helped to save at least 60 million barrels of oil per year. Furthermore in Europe, following the introduction of Directive 94/62/EC on Packaging and Packaging Waste, which imposed requirements for plastic and packaging waste, such waste should now fulfil the European standard EN 13432, with these materials to be declared as compostable prior to being marketed to the public (European Bioplastics, 2009). For instance, Ireland was one of the first countries to introduce a plastic bag levy. Ireland's Department of Environment, Heritage and Local Government introduced a charge of 15 cents on plastic bags in 2002. This move had an immediate effect, reducing the use of plastic bags from 328 to 21 bags per capita. After this encouraging outcome, the Irish Government increased the levy to 22 cents, further reducing the use of plastic bags. On the other hand, in the United Kingdom, consumers were initially required to pay 5 penny for all single-use plastic carrier bags. The amount to be paid for these bags was doubled to 10 penny after the end of 2018. From this requirement, the estimated consumption of plastic bags in the United Kingdom dropped from 140 bags per person to about 28 bags per person. Currently, there are about 55 countries around the world that have imposed complete bans on single-use plastic carrier bags.

However, as the use of plastic bags is not entirely avoidable in modern life, it is recommended that reusable plastic bags be made of a compostable material, so that disposal will not burden the environment. Countries like Denmark, Cyprus, Germany, Poland, Italy, France and the United States impose high charges on single-use plastic products. The charges are actually several times higher than the cost of material for biodegradable polymers. Such a move has forced consumers to choose either to bring their own carrier bags when shopping, or to pay a high cost to buy biodegradable bags from supermarket operators. Subsequently, the production of biodegradable polymers has remained profitable and is well accepted by consumers. For Southeast Asian countries like Malaysia and Singapore, the supermarket operators are charging 20 cents per carrier bags as per the country's currency. The local authorities of several states of Malaysia have initiated a soft landing strategy by initially imposing charging for plastic carrier bags on Saturday on the first 2 years of implementation, followed by fully paid plastic bag carriers everyday after 2 years of getting a positive response from the consumers.

As people become more aware of using biopolymer types of packaging, many companies have increased their product range to include biopolymer materials. Consequently, varieties of 'eco-packaging' are available on the market, yet their biodegradability and compostability remain questionable. Subsequently, such eco-plastic products need to demonstrate proof according to standards to verify their biodegradability and compostability. In the European Union, compostable packaging must fulfil the requirements of EN 13432, while other countries have their own standards to be met in order to allow the use of a compostable logo (see Table 7.2). There are also labels for biomaterials which fulfil

Table 7.2 Certification of compostable plastic for respective countries.

Certification body	Standard of reference	Logo
Australia Bioplastics Association (Australia) www.bioplastics.org.au	EN 13432: 2000	
Association for Organics Recycling (United Kingdom) www.organics-recycling.org.uk	EN 13432: 2000	
Polish Packaging Research and Development Centre (Poland) www.cobro.org.pl/en	EN 13432: 2000	
DIN Certco (Germany) http://www.dincertco.de/en/	EN 13432: 2000	
Keurmerkinstituut (Netherlands) www.keurmerk.nl	EN 13432: 2000	
Vincotte (Belgium) www.okcompost.be	EN 13432: 2000	

Continued

Table 7.2 Certification of compostable plastic for respective countries—cont'd

Certification body	Standard of reference	Logo
Jätelaito-syhdistys (Finland) www.jly.fi	EN 13432: 2000	
Certiquality/CIC (Italy) www.compostabile.com	EN 13432: 2000	
Biodegradable Products Institute (USA) www.bpiworld.org	ASTM D 6400-04	
Bureau de normalisation du Québec (Canada) www.bnq.qc.ca	BNQ 9011-911/2007	
Japan BioPlastics Association (Japan) www.jbpaweb.net	Green Plastic Certification system	
Biodegradable Products Institute (North America) www.bpiworld.org	ASTM D6400 or ASTMD6868	

Table 7.2 Certification of compostable plastic for respective countries—cont'd

Certification body	Standard of reference	Logo
Nordic Ecolabeling http://www.nordic-ecolabel.org/	EN 16640:21 and EN 16785-1:2015	
DIN-Geprüft https://www.dincertco.de/	AS 5810 and NF T51-800	

the criteria of environmentally friendly products, as listed in Table 7.3. These are the labels that can be used for biobased products, for instance biobased polyethylene, polypropylene, etc. which have inputs from biobased resources (i.e. renewable sources), yet they are neither biodegradable nor compostable materials.

7.3 Types of biopolymers available in the market and current applications

The common understanding of the word 'biopolymer' is broad; that is, biopolymers can be refined into biodegradable polymers (regardless of whether petroleum based or renewable based) and biobased polymers (regardless of their biodegradability). In this section, the common examples of biopolymers are discussed, as shown in Table 7.4.

7.3.1 Polylactic acid

Polylactic acid (PLA) is currently the most established biopolymer. PLA is produced from lactic acid, which possesses both L and D isomers, as shown in Fig. 7.8. In general, there are two routes to produce PLA: the direct polycondensation route and the ring opening polymerization route, as shown in Fig. 7.9. PLA was

Table 7.3 Certification for sustainable ecoplastic or related products.

Certification body	Label
EU Ecolabel	
Roundtable on Sustainable Biomaterials	
International Sustainability and Carbon Certification	
REDcert	
Blue Angel	

Table 7.4 Classification of common biopolymers.

Biodegradable and biobased	Biodegradable and petroleum-based	Biobased and nonbiodegradable
Polylactic acid	Polyvinyl alcohol	Polyethylene terephthalate
Polyhydroxyalkanoates	Polycaprolactone	Polyethylene
	Poly(butylene succinate)	Polypropylene
	Aliphatic-Aromatic Copolyesters	
	Modified polyethylene terephthalate	

L-lactic acid D-lactic acid

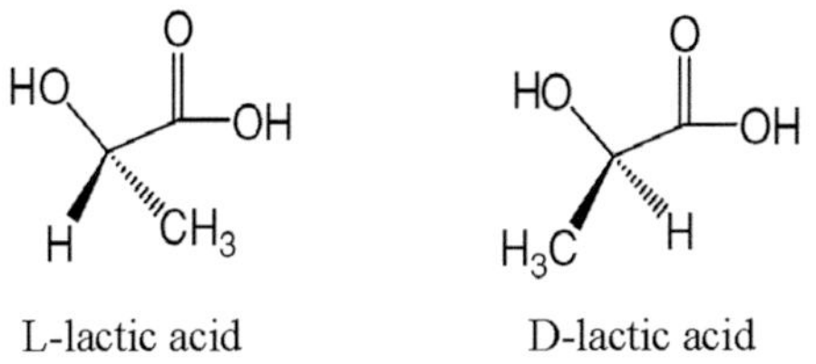

Fig. 7.8 Chemical structures of L- and D-lactic acid with melting point 16.8°C.

successfully synthesized as early as the 1800s, but the development of PLA has taken a long time to reach production viability for domestic usage. During the early stages of production, PLA was limited to use in biomedical devices, because the cost of synthesis was high and PLA was only produced at lab scale or pilot scale. Mostly at this time the direct polycondensation method was employed, which required critical parameter control in order to achieve a high-molecular-weight PLA. In the 1990s, the market for PLA started to expand, with the first pilot plant set up in 1992 by Cargill, using the indirect ring opening polymerization of lactides (see Fig. 7.10) to enhance the production yield of PLA. In 1997, a Cargill and Dow Chemical joint venture founded the company NatureWorks, with their preliminary commercial products coming to market under the name Ingeo. Currently, the production capacity of NatureWorks is 150,000 metric tonnes per year.

On the other hand, Corbion (formerly known as Purac), currently the world's largest lactic acid producer, operates a lactic acid plant in Thailand with an annual output of 120,000 MT in 2016. Corbion's lactic acid technology does not generate by-products of gypsum. This plant has planned to increase to a capacity of 205,000 MT annually in the future. Currently, Corbion

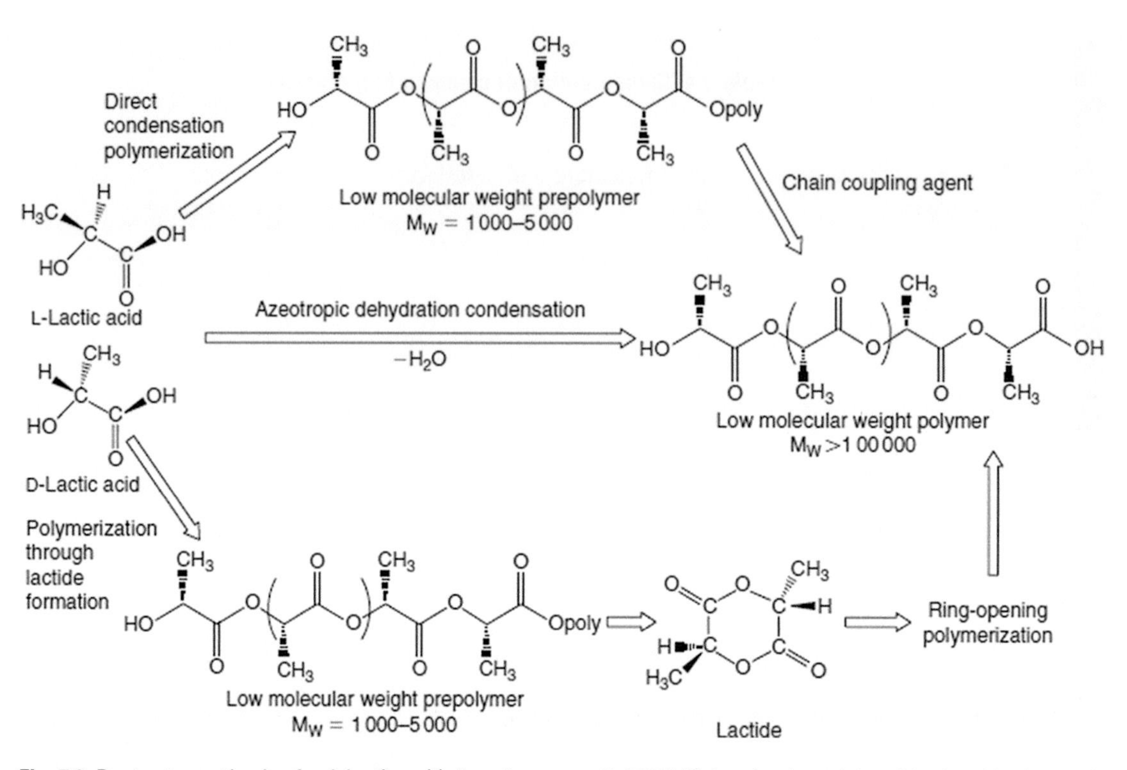

Fig. 7.9 Routes to synthesis of polylactic acid. From Hartmanna, H. (1998). High molecular weight polylactic acid polymers. In D. L. Kalan (Ed.), *Biopolymer from renewable resources* (pp. 367–411). Berlin: Springer-Verlag, with Springer Nature permission.

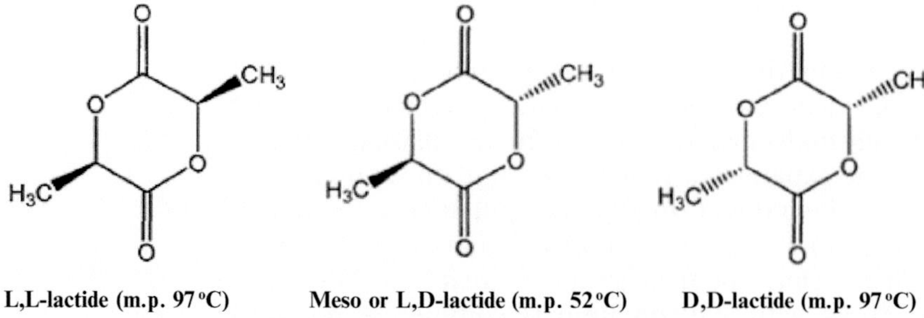

L,L-lactide (m.p. 97°C) **Meso or L,D-lactide (m.p. 52°C)** **D,D-lactide (m.p. 97°C)**

Fig. 7.10 Chemical structures of L,L-, meso- and D,D-lactides (m.p. is melting point).

supplies over 60% of lactic acid globally from its operation facilities located in the Netherlands, Spain, Brazil and the United States. For decades, Corbion-Purac has been manufacturing PLA and PLA copolymers for biomedical applications such as sutures, pins, screws and tissue scaffolding materials. With the current promising market of PLA for domestic applications, Corbion has a further business expansion strategy to utilize its lactic acid production expertise for PLA production. Based on the existing high-volume production of lactic acid, Corbion has the opportunity to convert lactic acid into L-lactide and D-lactide under the brand name PURALACT. Corbion has invested EUR45 million to produce 75,000 MT of PLA at its lactide plant in Thailand.

Purac in the Netherlands and Sulzer Chemtech AG in Switzerland have joined forces to produce PLA foam. Synbra, a company in Etten-Leur, The Netherlands, has been engaged to set up PLA foam technology for Purac-Sulzer, expanding their product range, which includes a green polymer foam called BioFoam. Synbra has been in the styrofoam manufacturing line for more than 70 years. The expandable PLA of Synbra utilizes lactide produced by Purac's lactide facility in Spain. In addition, Purac is also collaborating with Toyobo, a Japanese film, fibre and biotechnology firm, to make an amorphous and biodegradable PLA product for the European market under the brand name Vyloecol. Unlike the production technology used by Purac-Sulzer, Vyloecol developed by Purac-Toyobo is a patented amorphous PLA for application in coatings or adhesives for packaging films and materials.

Purac is also active in PLA production in the European Union, with Galactic and Total Petrochemicals. They established a 50/50 joint venture—Futerro—in September 2007 to develop PLA technology. The preliminary project was to construct a demonstration plant with a 1500 MT PLA production capacity; this pilot unit cost $15 million. The Galactic production site is located at Escanaffles, Belgium, and the monomer, lactide, is obtained from fermenting sugar beet.

Uhde Inventa-Fischer as part of the Thyssenkrupp Industrial Solutions AG produces PLA with good elongation at break by polymerizing and blending with crystalline PLA. Uhde Inventa-Fischer developed the PLAneo technology and set up a plant in China owned by COFCO Biochemical, with 10,800 tonnes per year PLA.

In Asia many companies have been established to explore PLA technology. Japan was the first country to be involved in the research and development of PLA. China then followed, as the market for PLA started to grow. Although Japan was involved in PLA technology earlier than other Asian countries, some of the large, ambitious companies halted production due to high production costs, lack of availability of raw materials and an immature market to accept premium plastics with a higher price. Shizmadu initially operated a pilot plant to produce small commercial quantities of PLA. Since then, production has ceased and the technology was sold to the Toyota Motor Corporation. Toyota increased production to 1000 MT per year, mainly for automotive applications. In 2008, the plant was sold to Teijin, and now Teijin is expanding production for its BIOFRONT products. The company increased the productivity of BIOFRONT to 5000 MT per year in 2011.

Unitika Ltd., a 120-year-old textile company, has marketed PLA products under the Teramac brand. Teramac resin can be processed using a wide range of plastic technologies, including injection, extrusion, blow, foam and emulsion. The Korean company Toray has launched a full-scale commercialization of Ecodear PLA films and sheets. Ecodear exhibits heat and impact resistance as well as flexibility and high transparency equivalent to petroleum-based plastic films.

Since 2007 many projects have been announced in China. However, many of these have seen a lack of further development. Zhejiang Hisun Biomaterial was the first company in China to produce PLA on a commercial scale, with an annual production of 5000 MT per year. Other companies had smaller plants at the time: Shanghai Tongjieliang BioMaterial had a pilot plant producing a maximum 1000 MT per year PLA, and Nantong Jiuding Biological Engineering had a larger facility that could produce up to 1000 MT per year. At the end of 2009, Nantong Jiuding Biological Engineering secured funding of US$1.4 million from the National Development Reform Commission to expand its PLA project (CCM International Limited, 2010). This was followed by an expansion project, involving a total investment of US$19 million, to boost production to 20,000 MT per year. Henan Piaoan Group, a medical equipment and supplies manufacturer, has purchased the patented PLA technology of Japan's Hitachi Plant Technologies Ltd. The Henan Piaoan plant is expected to produce 10,000 MT of PLA annually. Most of the PLA produced in China is for export rather than internal use, because the biodegradable market in China is still in its infancy and there is a lack of local regulation of biodegradable polymer use for environmental protection.

A list of PLA resin producers worldwide is given in Table 7.5.

7.3.2 Polyhydroxyalkanoates

Polyhydroxyalkanoate (PHA) refers to a group of natural biodegradable polymers produced from microorganism activity. Unlike PLA, which is produced from the lactic acid obtained from fermentation of bacteria followed by a polymerization process, PHAs are mainly produced directly from microorganisms and the chemical composition of PHAs depends on the strains of the

Table 7.5 Polylactic acid resin major producers.

Producer	Capacity (MT/year)	Location
NatureWorks	150,000	Nebraska, United States
Total Corbion	75,000	Rayong, Thailand
Galactic-Futerro	1500	Belgium
Zhejiang Hisun Biomaterial	5000	Zhejiang, China
Unitika-Terramac	5000	Japan
Tyssenkrupp-UIF Polycondensation Technologies	500	China

microorganisms. PHAs are a form of energy storage within the cells of microorganisms, as part of the normal survival process that occurs within the microorganism. According to Li, Yang, and Loh (2016), PHAs are found to be short-molecular-chain PHAs (C3–C5), which have 3–5 carbon monomers, and medium-molecular-chain PHAs (C6–C14), which have 6–14 carbon monomers in the 3-hydroxyalkanoate units. For instance, poly(3-hydroxybutyrate) (PHB), poly(3-hydroxyvalerate) (PHV) and their copolymer poly(3-hydroxybutyrate-co-3-hydroxyvalerate) (PHBV) are typical examples of short-molecular-chain PHAs. Meanwhile, poly(3-hydroxyoctanoate) (PHO) and poly(3-hydroxynonanoate) (PHN), which are synthesized as copolymers with 3-hydroxyhexanoate (HHx), 3-hydroxyheptanoate (HH) and/or 3-hydroxydecanoate (HD), are typical examples of medium-molecular-chain PHAs.

So far, researchers have found >150 different PHA monomers, which means PHAs are the largest group of natural polyesters. All these are summarized in Fig. 7.11. Although PHAs have a wide range of chemical compositions, most of them possess relatively poor mechanical properties compared to common fossil-type polymers used for packaging applications. In general, the typical weaknesses of PHAs include:

(a) The melting temperature of PHAs is too close to the thermal decomposition/degradation temperature, causing PHAs to be unable to be processed by conventional plastic processing methods such as injection moulding, extrusion, blow moulding, etc.

(b) Poor mechanical properties due to a smaller chain length limit biomedical applications such as vascular and controlled drug delivery applications.

(c) Low molecular weight of PHAs causes viscous and tacky characteristics.

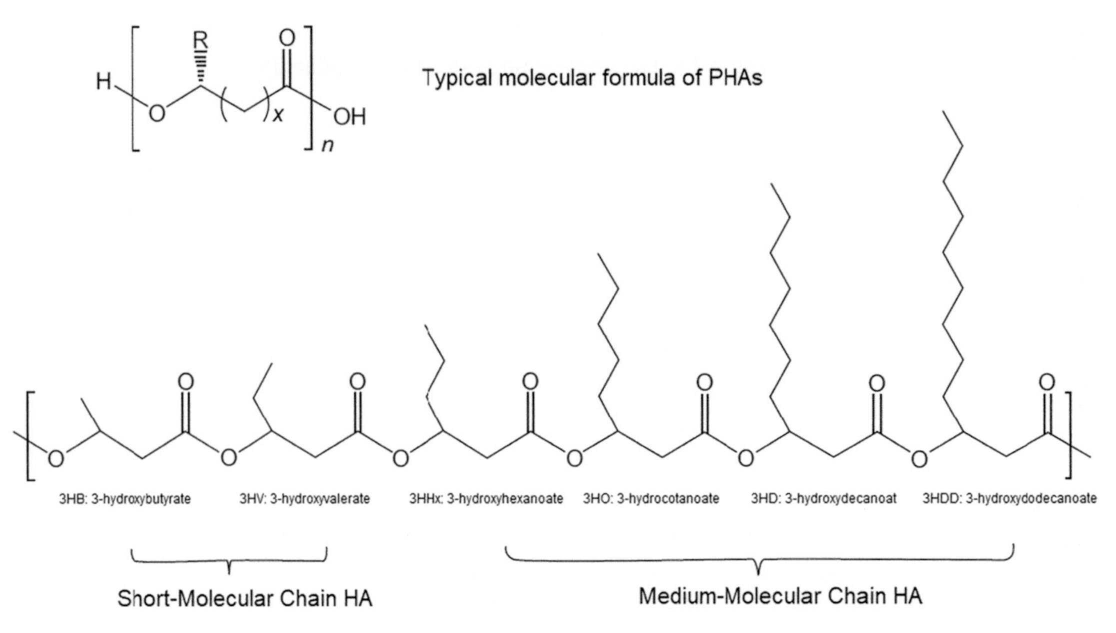

Fig. 7.11 Chemical formula and structure of PHA group of polyesters.

(d) Hydrophobicity of PHAs is unfavourable for advanced biomedical applications with low rates of biodegradation.

Hence, PHAs require modifications through blending with other compounds such as PLA, polycaprolactone (PCL), starch or chemical grafting to improve the characteristics. A review by Sudesh et al. (2000) compared the properties with those of commodity polymers. It was found that the properties of PHAs are equally as good as polypropylene and low-density polyethylene (refer to Table 7.6). This indicates that PHAs have potential for use equally good to substitute commodity polymers. However, as mentioned earlier, there are still limitations, such as the thermal degradation temperature being too close to the melt temperature, which cause the moderation of processing parameters is needed before applying in the existing polymer processing technology. While the cost and availability of PHAs remains a main concern. For instances, managing complicated biological research, gene clone, catalytic mechanism, substrate specificities of enzymes and tailor-made biosynthesis of PHAs properties are the main factors that affect the quality and quantity of PHAs to be introduced to the market.

There are several companies involved in the production of PHAs (refer to Table 7.6). One of the most established producers is Bio-On in Bologna, Italy, which has been the largest PHA producer in the world since 2018, with an output from 5000 to

Table 7.6 Comparison of properties of PHA polymers and commodity polymers (Sudesh et al., 2000).

Type of Polymer	Melting temperature T_m (°C)	Glass-transition temperature T_g (°C)	Young's modulus (GPa)	Tensile strength (MPa)	Elongation at break (%)
P(3HB)	180	4	3.5	40	5
P(3HB-co-20 mol% 3HV)	145	−1	0.8	20	50
P(3HB-co-6 mol% 3HA)	133	−8	0.2	17	680
Polypropylene	176	−10	1.7	38	400
Low-density polyethylene	130	−30	0.2	10	620

10,000 tons per years. Bio-On markets their product under the tradename of MINERV PHA and it is able to withstand production temperatures up to 180°C. MINERV is suitable for applications ranging from cosmetics and toys, to packaging, furniture and pollution treatment. Another producer, Danimer Scientific, which was a codeveloper of NatureWorks PLA in earlier years, bought the medium-molecular PHA chain from Procter & Gamble and launched their product called Nodax. Danimer Scientific's PHA plant is located in Winchester, Kentucky, United States. Nodax has been widely used for producing beverage bottles with the ability to fully biodegrade in 18 months. Another producer, Metabolix, now known as Yield10 BioScience, has filed a patent to produce PHB (a type of PHA), which can be found in the seeds of the plant known as *Camelina sativa*. PHB is mainly used in water treatments, where it acts to disrupt the growth of bacteria by denitrifying them in the water system. According to Yield10 BioScience, the PHB pellets are highly effective biomaterial that is sustainable, cost-effective, has zero waste and is maintenance free to reduce nitrate in water systems.

In China, Tianan Biologic produces PHBV at a production facility at Beilun Port, Ningbo City, China at 2000 MT per year. One of their products, called ENMATTM thermoplastics resin Y1000P, can be processed using the conventional methods of injection moulding, thermoforming, blown films and extrusion. This PHBV has a tensile strength of 39 MPa, elongation at break 2% and melting point of 170–176°C. In addition, PolyFerm in Canada produce a medium molecular chain PHA from sugar and vegetable oil derived fatty acid using microorganisms in a bioreactor. VersaMer has a thermal degradation of 240°C, glass transition temperature of −45°C to −35°C, melting points at 45–60°C and average molecular weight of 100,000–150,000 with polydispersity index of 1.75–1.83. The typical tensile strength is 7–16 MPa and Young's modulus is 2–12 MPa. The details of the products from PolyFerm can be found in Table 7.6. In short, the family of PHAs remains a good choice of biodegradable polymer for domestic applications in the near future (Tables 7.7 and 7.8).

7.3.3 Polycaprolactone

Polycaprolactone (PCL) is one of the most important biodegradable polymers (Hartmanna, 1998). PCL is as popular as PLA because it is a biodegradable and bioabsorbable polymer mainly for biomedical applications, particularly for drug delivery with a controlled release function. The application of PCL in consumer packaging remains limited due to its high price, making it unlikely

Table 7.7 List of worldwide producers of PHAs.

Producer	Location	Trade name	Applications: Client
Bio-On	Castel San Pietro, Bologna, Italy	MINERV	Consumer packaging: Zeropack Cosmestic: MyKAI Furniture: Kartell
Danimer Scientific	Winchester, Kentucky, United States	Nodax	Biodegradable straw: WinCup Food Packaging: Genpak Water bottle: Nestle Spirits Bottle: Bacardi
Yield10 Bioscience	Saskatoon, Canada	–	Wastewater treatment to remove nitrate
Tianan Biologic Materials Co. Ltd.	Beilun Port, Ningbo, China	ENMAT	Thermoplastics Fibre and Nonwoven Denitrification: Water treatment
Kaneka	Takasago, Japan	Kaneka Biodegradable Polymer PHBH	Packaging and cutlery
PolyFerm	Canada	VersaMer	Packaging, medical such as sutures, surgical meshes, etc., adhesives
Telles-Metabolix and Archer Daniels Midland Company	Clinton, Iowa	Mirel	Film, packaging, injection moulding, extrusion and thermoforming

Table 7.8 Detailed information of Polyferm PHAs.

Product	Chemical name	Composition (mol%)	Molecular weight	Polydispersity	Melting temperature	Decomposition temperature
VersaMer-10-65	Poly(3hydroxydecanoate-co-3hydroxyoctanoate)	HD (65), HO (32), HHx (3),	No data	No data	No data	No data
VersaMer-9-70	Poly(3hydroxynonanoate-co-3hyroxyheptanoate)	HN (70), HHp (29), HV (1)	104,000	1.75	46°C	240°C
VersaMer-9-90	Poly(3hydroxynonanoate-co-3hydroxyheptanoate)	HN (90), HHP (10)	110,000	1.75	59°C	245°C
VersaMer-8-90	Poly(3hydroxyoctanoate-co-3hydroxyhexanoate)	HO (93), HHx (7)	130,000	1.83	4°C	240°C
VersaMer-9-U18	Poly(3hydroxynonanoate-co3hydroxyheptonoate-co-3hydroxynonenoate-co3hydroxyundecenoate)	HN (51), HHp (28), HN (9), HUD (8), HV (1.4), HHp (1.2), other HA (1.4)	110,000	1.90	44°C	280°C
VersaMer-9-U50	Poly(3hydroxynonanoate-co-3hydroxyheptanoate-co-3hydroxynonenoate-co-3hydroxyundecenoate)	HN and HHP combined (50); HN and HUD combined (50)	No data	No data	No data	No data
VersaMer-9-U90	Poly(3hydroxynonanoate-co-3hydroxyaheptanoate-co-3hydroxynonenoate-co-3hydroxyundecenoate)	HN and HHp combined (10); HN and HUD combined (90)	No data	No data	No data	No data
BIOPOL	Poly(3hydroxybutyrate-co-3hydroxyvalerate)	HB (81), HV (19)	No data	No data	136°C	No data
PHBHHx	Poly(3-hydroxbutyrate-co-3hydroxyhexanoate)	HB (95), HHx (5)	No data	No data	155°C	No data

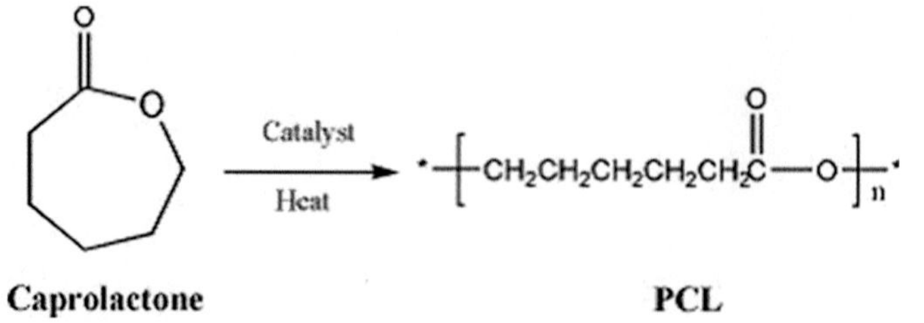

Fig. 7.12 Reaction to produce PCL.

for general usage (Woodruff & Hutmaacher, 2010). PCL is prepared by either ring opening polymerization of ε-caprolactone using a variety of anionic, cationic and coordination catalysts or via a free-radical indirect method called ring opening polymerization of 2-methylene-1-3-dioxepane. PCL is a semicrystalline polymer with a glass transition temperature of −60°C and a melting point at 59∼64°C. With these thermal characteristics, PCL again has limited applications in packaging, especially for hot content. This is because when PCL is subjected to high temperatures, it is very likely to collapse due to the loss of its physical strength from thermal softening. Fig. 7.12 shows the reaction to produce PCL.

DURECT is one of the renowned marketed Lactel polymer series of absorbable polymers consisting of PCL as well as the copolymer of poly (DL-lactide-co-ε-caprolactone), with the information as listed in Table 7.9. In addition, PURAC marketed their biomedical grade of PCL under the trade name of Purasorb. Most of the time, PCL used in sutures and implants takes more than 12 months to biodegrade and is not harmful to the body. Sun et al. (2006) traced PCL in the body of rats for 3 years, with the

Table 7.9 DURECT LACTEL PCL and its copolymer specifications.

Polymer	Tensile strength (psi)	Elongation (%)	Modulus (psi)	Crystalline melt transition (°C)	Glass transition temperature (°C)
PCL	3000–5000	300–500	3×10^4 – 5×10^4	58–63	−65 to −60
25:75 PDLA-CL	∼1100	∼600	5×10^4 – 2×10^5	Amorphous	−50 to −40
80:20 PDLA-CL	∼750	∼150	6×10^4 – 2×10^5	Amorphous	17–23

PCL, polycaprolactone; *PDLA-CL*, poly (DL-lactide-co-caprolactone).

distribution, absorption and excretion of PCL traced in the rats by radioactive labelling. The results showed that PCL capsules with initial molecular weight of 66,000 remained intact in shape during the 2-year implantation. It decomposed into low molecular weight, 8000, fragments at the end of 30 months.

PCL can also be used when high hydrolytic stability and good light fastness are needed. For instance, polyurethane elastomers added to PCL are good for dispersion to make textiles and leather. PCL can also be used as a coating, with excellent weatherability and chemical resistance. Some of the major producers of PCL for such applications are BASF-Capromer, PERSTORP-Capa and DAICEL-PLACCEL (Perstorp Trade Brochure, n.d.). Furthermore, PERSTORP also produces a thermoplastic type of PCL marketed under the tradename of CAPA thermoplastic. The details of the melt flow index and melting range of CAPA are listed in Table 7.10. Although PCL is considered to be high cost, Novamont is the first company to produce a PCL/starch compound under the trademark of Mater-Bi. According to Novamont, the PCL/starch can be applied as consumer packaging, carrier bags and food service items, and at the end of its usage life, it can be sent to recycling in compositing plants to become fertile compost. It can also be used as mulch film for agricultural uses, and is able to biodegrade in soil and reduce horticulture workload upon degrading and becoming part of the soil. In short, application of PCL for domestic use remains limited and the main focus is on biomedical applications.

7.3.4 Poly(butylene succinate)

Poly(butylene succinate) (PBS) is one of the petroleum-based yet biodegradable polymers. It is also a type of aliphatic polyester with the advantages of good processibility by most of the existing plastic technology, such as extrusion, injection moulding, melt blow, filament and yarns. Showa Denko was one of the pioneers of PBS producing, with the tradename Bionolle. Nevertheless, in a news release, Showa Denko (2016) stated that they had decided to terminate production of Bionolle with the sale of the resin ending on December 2019 due to the harsh market environment. From a report by Fujimaki (1998), Showa Denko has marketed four grades of PBS including 1000 series (PBS), 2000 and 3000 series (poly(butylene succinate adipate), PBSA), and 6000 series (poly(ethylene succinate)), as given in Table 7.11.

On the other hand, there are several producers still in operation, such as AnQing Hexing Chemical established in 1998, which

Table 7.10 Melt flow index and melting range of Perstorp PCL CAPA.

Grade	Approximation of molecular weight	Appearance	Melt flow index[a]	Melting range (°C)
Capa 6200	20,000	Solid	15	58∼60
Capa 6250	25,000	Granules	9	58∼60
Capa 6400	37,000	Solid	40	58∼60
Capa 6430	43,000	Granules	13	58∼60
Capa 6500	50,000	Granules	7	58∼60
Capa 6500C	50,000	Granules	7	58∼60
Capa 6506	50,000	Powder	7	58∼60
Capa 6800	80,000	Granules	3[b]	58∼60

[a]Melt flow index measure with 1″ PVC die, 2.16kg weight, g/10min at indicated temperature; for 6200 series at 80°C and 6500 series at 160°C.
[b]Measure at 5kg at 190°C.

is one of the largest PBS producers in China with annual production of 10,000 tonnes. The PBS of AnQing Hexing has a melt flow index of 14g/min (190°C@2.16kg), heat deflection temperature 89°C, melting point 115°C, strength at break 30MPa, elongation at break 540%, tensile strength 30MPa, flexural modulus

Table 7.11 Specification of Showa Denko Bionolle PBS.

Property	PBSU #1000	PBSU #2000	PBSU #3000	PESU #6000
Melt flow index@190°C (g/10min)	1.5	4.0	28	3.5
Density (g/cm^3)	1.26	1.25	1.23	1.32
Melting point (°C)	114	104	96	104
Glass transition temperature (°C)	−32	−39	−45	−10
Yield strength (kg/cm^2)	336	270	192	209
Elongation (%)	560	710	807	200
Stiffness 10^3 (kg/cm^3)	5.6	4.2	3.3	5.9
Izod impact strength (kg-cm/cm) 20°C	30	36	40	10
Combustion heat (cal/g)	5550	5640	5720	4490

440 MPa and impact strength 9 kJ/m^2. The resin is suitable for injection moulding and extrusion to produce disposable items such as combs, food trays, food services, etc. Another China producer of PBS is Zhejiang HangZhou Xinfu with the production capacity of 20,000 tonnes for injection moulding applications. PTT MCC Biochem, a joint venture of PTT Public Company and Mitsubishi Chemical, has set up a plant in Thailand and has produced PBS called BioPBS since 2017(PTT MCC, n.d). BioPBS is able to decompose at composting facilities at 30°C and is processable using the existing polymer technology. The typical specifications of BioPBS are summarized in Table 7.12. In short, PBS is a promising biodegradable polymer with favourable thermoplastic properties with the potential to directly substitute for commodity polymers such as polyethylene and polystyrene. PBS is known to be suitable for cups, paper cup coating and flexible packaging, yet the main drawback of PBS is that it is produced from petrochemical sources.

Table 7.12 Specifications of BioPBS.

Properties	Unit	BioPBS FZ71 (Standard grade)	BioPBS FZ91 (Standard grade)	BioPBS FD92 (Flexible grade)
Density	g/cm^3	1.26	1.26	1.24
Melt flow rate (190°C, 2.16 kg)	g/10 min	22	5	4
Melting point	°C	115	115	84
Yield stress	MPa	40	40	17
Stress at break	MPa	30	36	24
Strain at break	%	170	210	380
Flexural modulus	MPa	630	650	250
Flexural strength	MPa	40	40	18
Izod impact strength (23°C)	kJ/m^2	7	7	47
Heat deflection temperature (0.45 MPa)	°C	95	95	63
Rockwell hardness	R Scale	107	107	56

PTT MCC Biochemm BioPBS.

7.3.5 Aliphatic-aromatic copolyesters

The aliphatic-aromatic copolyesters (AACs) are also one of the biodegradable yet fossil-type polymers. The advantage of AAC is that its properties are as competitive as those of commodity polymers such as polyethyelene and polypropylene. AAC is able to be easily processed using existing polymer processing technology to replace commodity polymers, such as blown film, injection moulding and extrusion. AAC is also suitable to be blended with other biopolymers to improve processability, such as PLA and PHB. One of the renowned producers of AAC is BASF who has marketed the product Ecoflex since 1998. Ecoflex is synthesized from 1,4-butanediol, adipic acid and terephthalic acid to produce poly(butylenes adipate-co-butylene terephthalate) (PBAT). It was invented by taking advantage of the good flexibility and biodegradability of aliphatic polyesters, while the strength and toughness are contributed from the copolymerization with aromatic structure, as shown in Fig. 7.13.

The production of Ecoflex by BASF was increased from the initial 14,000–74,000 metric tonnes per year, at their plant located in Ludwigshafen, Germany. Since Ecoflex is a fossil-type biodegradable polymer, BASF has taken the initiative to reduce the fossil content by substitution with PLA, subsequently marketed under the tradename Ecovio, a product suitable for wider applications like mulch film, compostable carrier bags, fruit and vegetable bags, as well as a sealing layer for multilayer structures. Most of the time, Ecoflex is used as a performance enabler, blended with other ingredients such as PLA or starch in order to meet the performance requirement to be as good as commodity polymers like polyethylene and others. From the studies performed by Siegenthaler et al. (2011), Tables 7.13 and 7.14 compare the properties of low-density polyethylene and high-density polyethylene, respectively, with Ecoflex and their compounds. In short, Ecoflex and Ecovio are both promising biodegradable polymers, not only

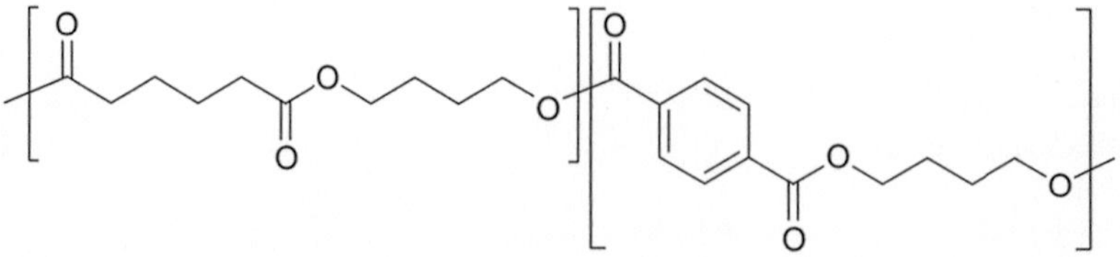

Fig. 7.13 Chemical structure of aliphatic-aromatic polyester Ecoflex. From Siegenthaler, K. O., Künkel A., Skupin, G., Tamamoto, M. (2011). Ecoflex® and Ecovio®: Biodegradable, performance-enabling plastics. *Advances in Polymer Science, 245*, 91–135, with permission Springer-Verlag, Berlin, Heidelberg.

Table 7.13 Comparison of low-density polyethylene, Ecoflex and its compounds for blown films (Siegenthaler et al., 2011).

Test	Low-density polyethylene carrier bags	Ecoflex	Ecoflex with granular starch compound	Ecoflex with thermoplastics starch compound
Transparency	Opaque	Translucent	Opaque	Opaque
Printability	Eight colour flexoprint	Eight colour flexoprint	Poor printability due to hydrophilicity of starch content	Eight colour flexoprint with promising quality
Modulus of elasticity (MPa)	~330	~110	~150	~270
Tensile strength (MPa)	~32	~35	~23	~21
Elongation (%)	~460	~640	~390	~21
Puncture resistance (J/mm)	17	26	9	19
Oxygen permeability (mL/m^2 d bar)	4800	2000	–	–
Water vapour permeability (g/m^2 d)	3	240	–	–
Food contact	No limitation	No limitation	Dry food	Dry food
Biodegradability	No	Yes	Yes	Yes

suitable for immediate application but also possessing the ability to improve other biopolymers through melt blending methods.

7.3.6 Polyvinyl alcohol

Polyvinyl alcohol (PVOH) is a water-soluble polymer and its solubility greatly depends on the hydrolyzed percentage, i.e., percentage of hydrolysis of polyvinyl acetate. PVOH is fully biodegradable. PVOH is produced by no less than 20 companies worldwide; the renowned producers are Kuraray, Denka, Sekisui, Synthomer and many more. The water-soluble characteristic of PVOH makes it extensively used as a binder, emulsifier and thickening agent for a wide range of products such as paper,

Table 7.14 Comparison of high-density polyethylene (HDPE), Ecoflex and its compounds for blown films (Siegenthaler et al., 2011).

Test	High-density polyethylene blown films	55% Ecoflex + 45% PLA dry mixed before blown	55% Ecoflex + 45% PLA compound (melt blended)	55% Ecoflex + 45% PLA compound (melt blended)
Transparency	Opaque	Opaque	Opaque	Translucent
Printability	Eight colour flexoprint	Eight colour flexoprint	Eight colour flexoprint	Eight colour flexoprint
Modulus of elasticity (MPa)	~650	~1180	~1020	~1560
Tensile strength (MPa)	~45	~39	~50	~47
Elongation (%)	~640	~360	~430	~160
Puncture resistance (J/mm)	42	19	31	31
Oxygen permeability (mL/m^2 d bar)	2000	–	1400	–
Water vapour permeability (g/m^2 d)	1.3	–	160	–
Food contact	No limitation	No limitation	No limitation	No limitation
Biodegradability	No	Yes	Yes	Yes

remoistenable adhesives, textile finish, paper surface treatment as well as cosmetics, food and pharmaceutical products. For packaging purposes, PVOH is suitable for food, drugs and cosmetics, since it has outstanding barrier properties to gases such as oxygen, nitrogen and carbon dioxide.

7.3.7 Biobased nonbiodegradable polymers

The biobased polyethylene, polycarbonate and bio-polyamide are all from nonbiodegradable polymers. However, compared to the fossil polymers, biobased nonbiodegradable polymers are derived from renewable feedstock.

7.3.7.1 Biobased polyethylene and polypropylene

Dow Chemical partnered with UPM Biofuels to produce biobased polyethylene in Terneuzen, The Netherlands in 2019. The process produced biobased polyethylene from the renewable naphtha invented by UPM BioVerno in Lappenranta, Finland from crude tall oil, which is the residue of the paper pulp process. Since paper pulp is obtained from the feedstock of sustainable managed forests, the biobased polyethylene process provides added value to the supply chain of pulp production. Elopak has been using biobased low-density polyethylene produced by Dow to coat liquid carton containers as well as production carton caps.

In addition, LyondellBasell and Neste, the world's largest producer of renewable diesel from waste and residue oils, reached an agreement to make renewable biobased polyethylene and polypropylene marketed by LyondellBasell under the tradenames Circulen and Circulen Plus. The renewable input to produce biobased polyethylene and polypropylene is based on bionaphtha and biogas derived from organic waste and vegetable oils. The specifications of Circulen and Circulen Plus are listed in Table 7.15. Furthermore, Mitsui Chemicals also joined the trend by announcing collaboration with Kaisen Inc. to commercialize biobased polypropylene starting in 2024. Mitsui-Kaisen is adopting a process to produce isopropanol from fermentation of nonedible plants, followed by dehydrating to obtain propylene. The industry can expect that more companies will join the trend to produce polyethylene and polypropylene from biosources in the very near future (Table 7.16).

7.3.7.2 Biobased polycarbonates

Mitsubishi Chemical produces DURABIO, a biobased polycarbonate obtained from the plant source isosorbide. DURABIO is one of the safest types of polycarbonate because it is different from conventional polycarbonates made from Biophenol A (commonly known as BPA). Some research studies have reported that BPA can migrate into food and beverages when it comes into contact with them. Excessive exposure to BPA is a concern because of health effects on brain and prostate gland of foetuses, infants and children. DURABIO is derived from plant glucose and transformed into sorbitol followed by the isosorbide monomer. Subsequently, isosorbide is polymerized to become a biobased polycarbonate following the scheme shown in Fig. 7.14. Although the biobased polycarbonate is derived from a plant source, DURABIO is not biodegradable and it possesses outstanding optical properties, transparency and scratch resistance to

Table 7.15 Biobased polyethylene and polypropylene from LyondellBasell's Circulen and Circulen Plus.

Property	Circulen EP540P	Circulen HP456J	Circulen HP483R	Circulen HP500N	Circulen HP501H	Circulen 2420D Plus	Circulen 2420K Plus	Circulen EP448T Plus	Circulen 2320F Plus	Circulen EP310M HP	Circulen 2426F Plus
Type	Polypropylene, impact copolymer	Polypropylene	Polypropylene	Polypropylene	Polypropylene	Low-density polyethylene	Low-density polyethylene	Polypropylene	Low-density polyethylene	Polypropylene, impact copolymer	Low-density polyethylene
Melt flow index (230°C/2.16 kg), g/10 min	15	3.4	27	12	2.1	0.25	4	48	0.75	7.5	0.875
Tensile modulus (MPa)	1300	–	1300	1400	1450	260	260	1250	260	1050	260
Tensile stress at yield (MPa)	28	34	32	35	33	10	11	27	11	21	11
Tensile strain at yield (%)	6	11	9	10	9	–	–	5	–	6	–
Charpy Impact strength, notched (23°C, Type 1, Edgewise, Notch A) (kJ/m^2)	7	–	3	4	8	–	–	5	–	45	–
Tensile strength machine direction (MPa)	–	–	–	–	–	27	22	–	26		24
Heat deflection temperature B (0.45 MPa, unannealed)	90	90	80	95	90	–	–	90	–	80	–
Applications	Caps, closures, housewares, luggage, opaque containers	Intermediate bulk containers, raffia/tapes	Caps, closures, furniture, housewares, opaque containers	Furniture, housewares	Caps, closures, housewares	Agriculture film, bags, shrink film, stretch hood	Coating, food packaging film, hygiene film, shrink film	Sports, leisure, toys	Agricultural film bags, pouches, hygiene film, liner film	Impact modifier, lamination film	Hygiene film, shrink film, food packaging film

Table 7.16 Biobased nonbiodegradable polymers.

Type of biobased polymers	Plant source	Feedstock	Process	Applications
Polyethylene terephthalate	Sugarcane	Sugar	Starting with ethanol produced from fermentation followed by conversion to monoethylene glycol from bio-ethanol and combined with purified terephthalic acid	PlantBottle for CocaCola and Dasani 
Polyethylene	Sugarcane	Sugar	Starting with ethanol produced from fermentation followed by dehydration to ethylene and additional polymerization	Carrier bags and tubes 

Continued

Table 7.16 Biobased nonbiodegradable polymers—cont'd

Type of biobased polymers	Plant source	Feedstock	Process	Applications
Polycarbonate	Corn	Isosorbide	Produced sorbitol from hydrogenation of glucose followed by dehydration of sorbitol to produce isosorbide	DURABIO marketed by Mitsubishi Chemical for automotive sunglasses, LED lighting, optical film, etc. 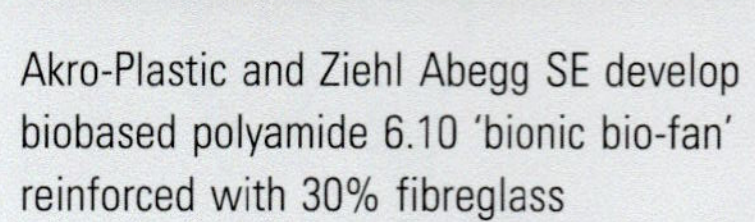
Polyamide (PA 4,10/PA 6,10)	Castor oil	Sebacic acid	Sebacic acid as the monomer to produce polyamide is produced from castor oil	Akro-Plastic and Ziehl Abegg SE develop biobased polyamide 6.10 'bionic bio-fan' reinforced with 30% fibreglass 

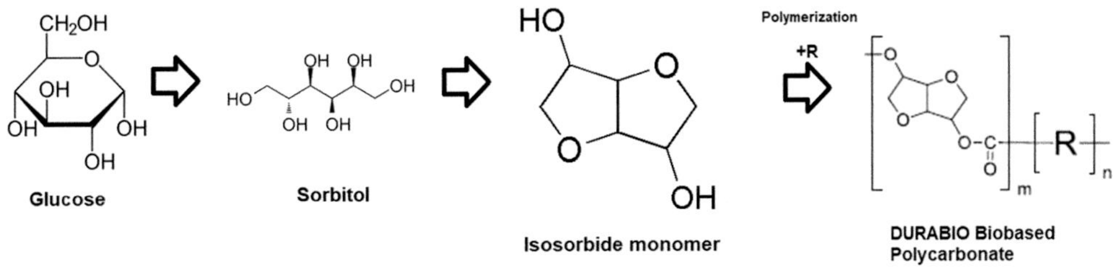

Glucose
Sorbitol
Isosorbide monomer
Polymerization
DURABIO Biobased Polycarbonate

Fig. 7.14 Process to produce DURABIO.

Table 7.17 Specifications of biobased polycarbonate DURABIO.

Property	Unit@condition	D7340R injection	D6350R injection	D5360R injection	D5380R extrusion
Density	g/cm^3	1.6	1.34	1.31	1.31
Water absorption 23°C, 24h	wt.%@23°C, 24h	0.3	0.3	0.2	0.2
Melt flow rate 230°C, 2.16kg	g/10min@23°C, 2.16kg	10	12	13	5
Tensile modulus	MPa	2700	2200	2300	2300
Tensile strength	MPa	79	70	64	64
Elongation at break	%	72	110	130	120
Flexural modulus	MPa	2700	2500	2100	2100
Flexural strength	MPa	116	95	94	94
Charpy impact strength	kJ/m^2@with notch	7	8	9	10
Total light transmittance	%@3mmt	92	92	92	92
Heat deflection temperature	°C@1.80MPa	102	90	82	82

meet the crucial requirements of electronic equipment, automotive housings and front panels for smartphones. The typical grades of DURABIO are listed in Table 7.17.

7.3.7.3 Biobased polyethylene terephthalate

There are numerous biobased polyesters that are mainly produced from the sources of sugarcane. One of the most common polyester is polyethylene terephthalate (PET) produced from 70% terephthalic acid and 30% monoethylene glycol (MEG) by weight. For biobased PET, the MEG is produced from renewable raw materials using ethanol from sugarcane instead of fossil raw materials. So far, two major manufacturing companies have produced biobased PET. Teijin has a product mainly for use in textile and automotive fibres under the tradename of ECO Circle PlantFiber. Since Teijin's biobased PET uses MEG from renewable sources and the production still uses the existing reaction path, the biobased PET has similar properties to fossil-based PET. On the other hand, a US company, Anellotech, developed the Bio-TCat technology to obtain high-purity biobased paraxylene, which is a key component for making 100% biobased PET bottles. The biobased paraxylene is sourced from biomass which requires a reforming and gasification process to produce hydrogen gases to react with the biomass to produce benzene, toluene and xylene (BTX). With the high purity of paraxylene, it can be converted to terephthalic acid and later react with biobased MEG to produce biobased PET, as shown in Fig. 7.15.

Recently, Suntory Group, which is one of the largest beverage manufacturing companies, introduced 100% biobased PET bottles. They are marketing the Orangina brand of beverage in Europe and the Suntory Tennensui mineral water brand in Japan using these 100% biobased PET bottles (Packaging Europe, 2021)

7.4 General environmental concerns about biopolymers

One of the most common environmental concerns about biopolymers is whether they are really more eco-friendly compared to the fossil-type polymers. In order to answer this concern, several factors need to be considered to determine the impacts on the environment. First, Papong et al. (2014) has made a comparison of drinking water bottles made of PLA and PET. Generally, PLA overall has fewer impacts as compared to PET bottle production because it is produced from plant sources. However, growing crops to the harvesting stage with the use of fertilizers and pesticides and herbicides has led to a higher eutrophication and acidification potential of PLA. The leaching of fertilizers by rainwater can pollute lakes and rivers. On the other hand, the emissions are limited to hydrocarbons, chemicals, catalysts and electricity as the main energy source in a factory for PET production. The pollutants can be contained within the factory boundary with only a small possibility of involving eutrophication and leaching. One of the merits of PLA production is the abundance of agricultural residues suitable for combustion to generate renewable energy.

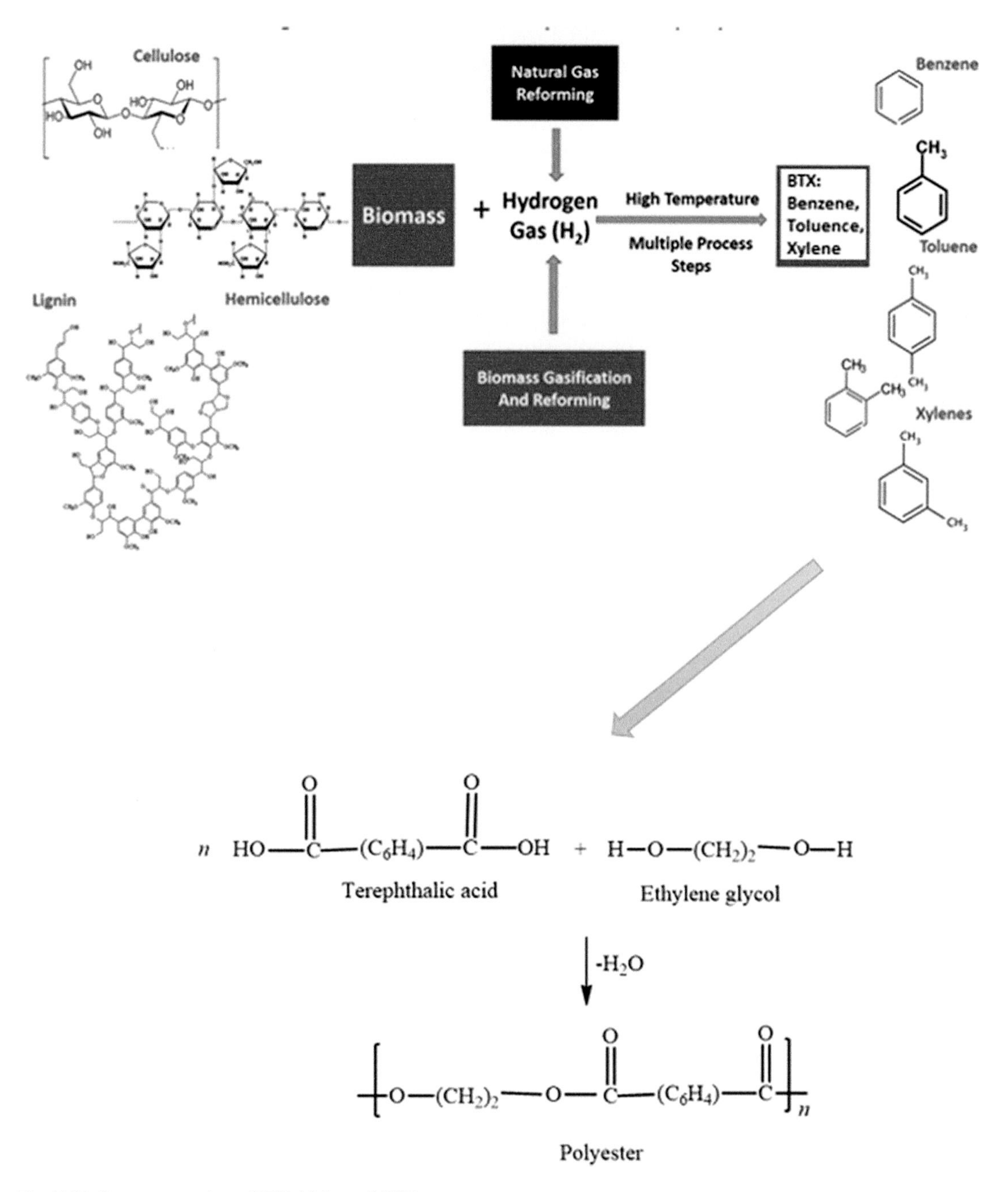

Fig. 7.15 Route to produce 100% biobased PET.

Moreover, wind turbines can be installed over the farm area to harvest wind energy so that the dependency on fossil energy can be further reduced. In the near future, the introduction of electric types of harvesters with rechargeable capability could further reduce the diesel dependency. Papong et al. (2014) summarized details of the production systems of both PLA and PET, as shown in Figs 7.16 and 7.17. Obviously, PLA production is more complicated with a wide spectrum of elements needed such as fertilizers, herbicides and enzymes as the inputs for PLA, whereas both PLA and PET have similar inputs of fuel, electricity, variety of chemicals, water and catalysts. As a result of the complexity of process involvement, the sources of pollution can be more difficult to manage for PLA as compared to PET.

Secondly, while people are looking for more eco-friendly materials, recycled materials are the more environmentally friendly compared to biopolymers, as reported by Simon et al. (2016) who compared aluminium, polyethylene terephthalate (PET), PLA, carton and glass beverage bottles. The postconsumer bottles that generate the least amount of greenhouse gases (GHGs) are produced from recycling, inasmuch as the bottles are recycled into secondary materials and added into virgin materials. The difference in the GHG emissions among recycling, incineration and landfills can be as large as 7.64 times, especially for aluminium bottles. Meanwhile, Simon et al. (2016) also reported that PLA has the lowest GHG of 66 kg CO_2-eq, followed by the large 1.5 L PET bottle with 85 kg CO_2-eq and the carton at 88 kg CO_2-eq. PLA seems to be the most favourable eco-friendly product. However, when PLA bottles undergo incineration and landfilling, the GHG can increase severalfold to 498 CO_2-eq and 500 CO_2-eq, respectively. This result clearly shows that recycling is the ultimate method to keep the environment greener, while incineration and landfills should only be considered after the end-of-life of the materials. Nonetheless, the tremendous increment in GHG was also found similar to that of other materials such as glass, PET, aluminium and cartons when undergoing incineration and landfill.

Besides comparisons on the emissions, the water consumption of production of plastics also needs to be considered. Cheroennet et al. (2017) conducted an informative analysis on the water impact of biobased box production of polystyrene, PLA from sugarcane derivation (PLA-S), PLA from sugarcane derivation with starch blended (PLA-S/starch) and polybutylene succinate (PBS). In this study, water footprint (WF) assessment was further broken down into green WF, blue WF and grey WF. The green WF is the ratio of rainwater to crop yield from the field during the growing period, while the blue WF is the ratio of irrigation water used for crop yield from the field during the growing period. The grey

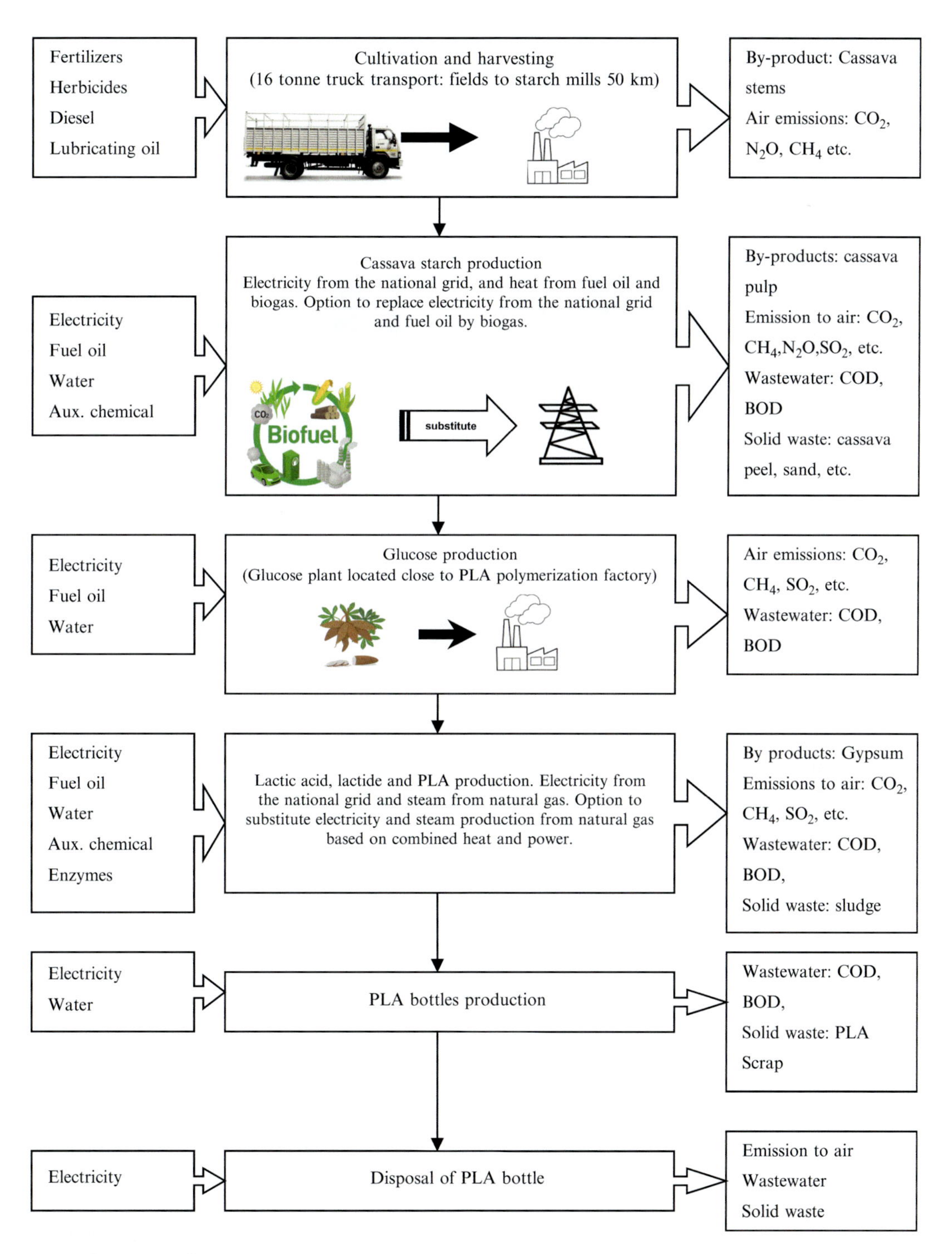

Fig. 7.16 PLA bottle production—inputs, process and emissions. From Sin, L. T., & Tueen, B.-T. (2019). Polylactic acid. A practical guide for the processing, manufacturing, and application of PLA. United States: Elsevier, with permission from Elsevier.

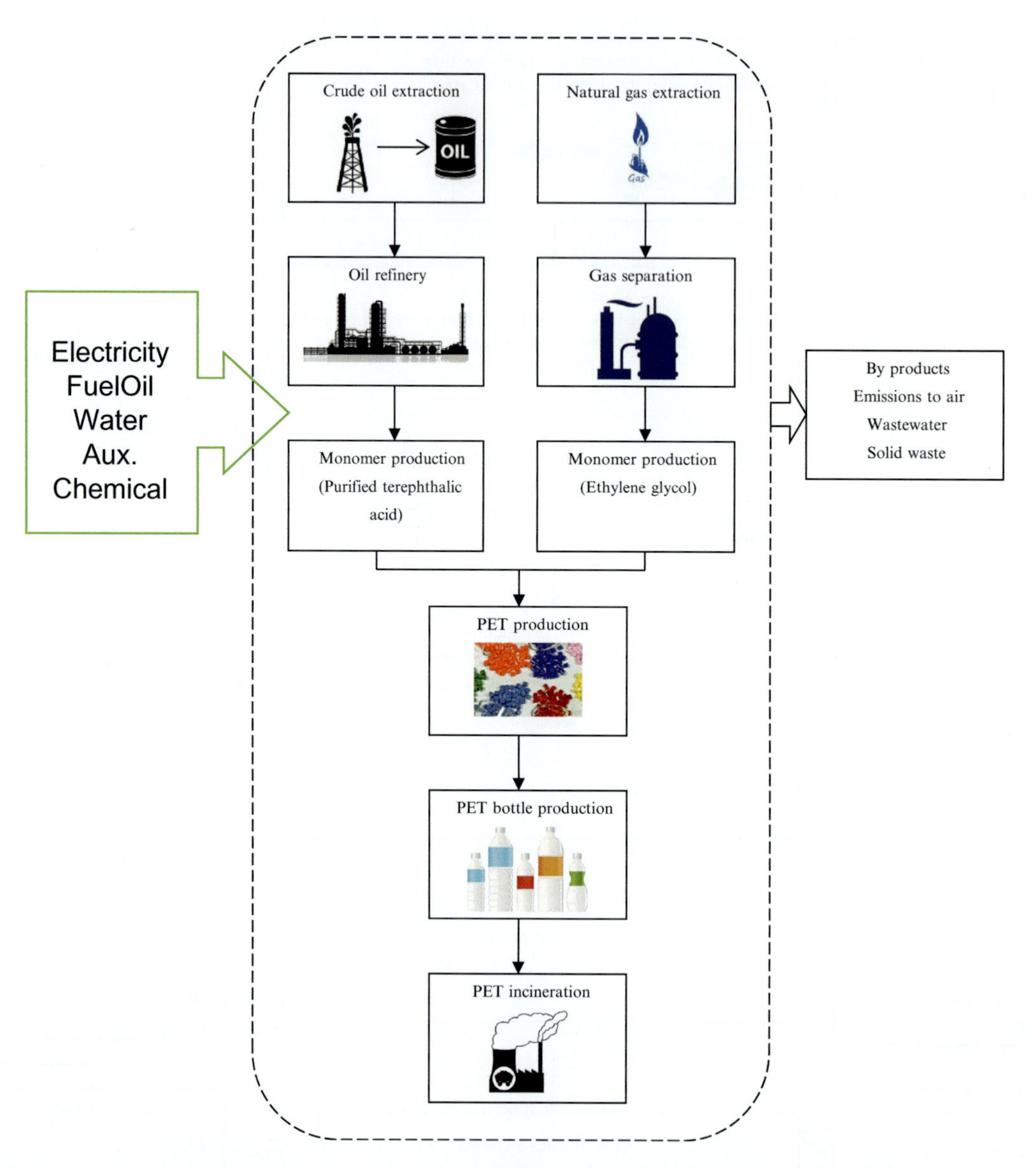

Fig. 7.17 PET bottle production—Inputs, process and emissions. From Sin, L. T., & Tueen, B.-T. (2019). Polylactic acid. A practical guide for the processing, manufacturing, and application of PLA. United States: Elsevier, with permission from Elsevier.

WF is the volume of water required to dilute the pollutant concentration. The analysis shows that PLA-S actually consumes more water to produce biobased boxes due to the fact that growing sugarcane requires a large volume of water from rainwater and irrigation water. As a result, PLA-S has a water footprint of 1.11 m^3 with 36.14% from green water contribution, 49.82% from blue water contribution, while grey water contributes 14.05%. PS as the non-renewable polymer material has 0.70 m^3 with 100% contribution from blue water. PLA-S/starch contains starch blending with PLA-S, which required 0.55 m^3 water to produce per box and the smallest WF is PBS at 0.38 m^3. Finally, the addition of starch into PLA-S is able to reduce the WF of PLA-S by 50.42%, mainly because of the large quantity of water required to grow sugarcane as compared to cassava starch. Cheroennet et al. (2017) has further confirmed that irrigation water for planting of sugarcane is significantly higher than for cassava, resulting in the blue WF for PLA-S and PLA-S/starch being 0.55 and 0.27 m^3. The grey water for PLA-S is also higher than for PLA-S/starch because PLA-S requires 4.06 kg to produce per PLA-S box, whereas the sugarcane and cassava are 1.75 and 0.11 kg to produce per PLA-S/starch box. This can be explained by the fact that the blending of starch is a direct process where the raw starch is added as filler in the PLA-S. However, production of PLA-S involves a reaction during which the multiple stages of conversion can cause significant mass losses. Hence, PLA-S/starch is more favourable in terms of less material consumption to produce the biobased box.

In addition, Cheroennet et al. (2017) found that the carbon footprint (CF) of the PLAS (0.675 kg CO_2 equivalent) was the highest among PS (0.05 kg CO_2 equivalent) and PLAS-starch (0.303 kg CO_2 equivalent). These data were obtained from the observations of Cheroennet et al. (2017) that box formation is also highly dependent on the amount of use of plastic material to make the boxes. For instance, PS boxes can be produced using a very small amount of resin because PS foamed boxes are very light (~0.053 kg PS/box), whereas PLAS requires ~0.243 kg PLAS to produce a PLAS box, and 0.105 kg PLAS pellet and 0.032 kg of cassava starch are required to produce a PLAS-starch box. This also indicates the environmental friendliness of plastic boxes not only depends on the selection of materials, but at the same time factors such as (1) amount of material used, (2) water source, (3) complexity of production process, (4) transportation of the raw material to factory, (5) delivery distance to consumer and (6) recyclability and reusability can affect the environmental footprint of the plastic products. The factor of transportation indeed is very important when PLA produced in Nebraska, United States is sent to Europe, or from Thailand to Europe, as a great deal of fuel is needed compared to localized production, which can greatly save on fuel consumption. Overall, the issue of transportation needs to be thoroughly examined including the route of delivery to the consumer so that the environmental friendliness of the plastic products can be justified.

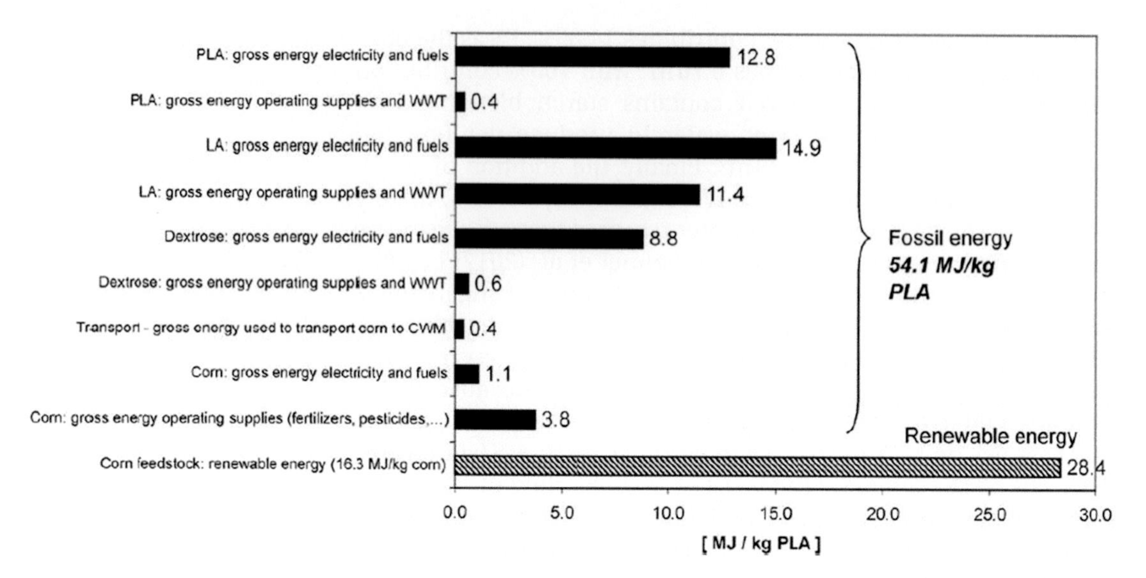

Fig. 7.18 Gross energy consumption for production of PLA. *LA*, lactide; *WWT*, wastewater treatment. From Vink, E.T.H. Rábago, K.R. Glassner, D.A. Gruber, P.R. (2003). Applications of life cycle assessment to NatureWorks™ polylactide (PLA) production. *Polymer Degradation and Stability, 80*, 403–419.

The major US-based producer of PLA, NatureWorks, has revealed a life cycle analysis (LCA) using corn as the feedstock by Vink et al. (2003). In the early days, the PLA production used major energy input from fossil fuels before employing studies on shifting to wind energy and biomass energy. At that time, PLA lacked attractiveness in terms of emissions. As shown in Fig. 7.18, the consumption of fossil energy can be 54.1 MJ/kg for PLA, whereas with renewable energy it is 28.4 MJ/kg. Nonetheless, in comparison with other petroleum-based polymers, PLA still remains outstanding in consuming fewer fossil inputs, as shown in Fig. 7.19.

NatureWorks has spent years of efforts to reduce dependency on fossil inputs by examining various alternatives. When selecting the approaches to be adopted, NatureWorks found that the PLA plant in Nebraska was not a promising place to harvest wind energy. Thus the dependency of NatureWorks on power generated other than fossil sources is not possible unless solar energy is used. In order to overcome the difficulty, NatureWorks decided to subscribe to a Renewable Energy Certificate (REC) to reduce (1) indirect emissions from electricity production at 1.561 CO_2 eq (kg/kg PLA) and (2) fuel, material, corn production and reclamation at 1.244 CO_2 eq (kg/kg PLA). In fact, the REC is a sort of

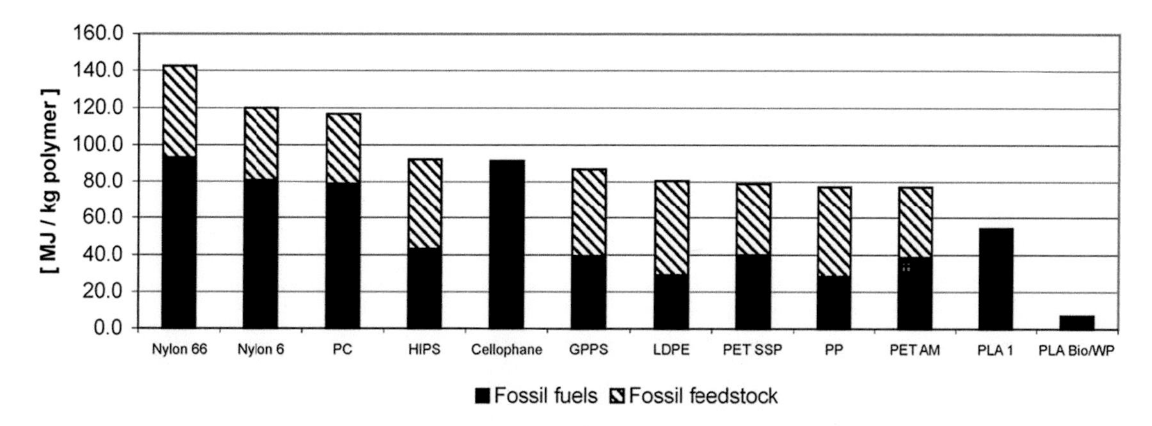

Fig. 7.19 Comparison of fossil fuel energy consumption for petroleum-based polymers and PLA. The cross-hashed part of the bars represents the fossil energy used as chemical feedstock (the fossil resource to build polymer chains). The solid part of each bar represents the gross fossil energy used for fuels and operation supplies used to drive the production processes. *PC*, polycarbonate; *HIPS*, high impact polystyrene; *GPPS*, general purpose polystyrene; *LDPE*, low-density polyethylene; *PET SSP*, polyethylene terephthalate, solid state polymerization (bottle grade); *PP*, polypropylene; *PET AM*, polyethylene terephthalate, amorphous (fibres and film grade); *PLA1*, PLA without adoption of biomass and wind power; *PLA B/WP*, PLA with adoption of biomass wind power. Reproduced with Elsevier permission from Vink, E.T.H. Rábago, K.R. Glassner, D.A. Gruber, P.R. (2003). Applications of life cycle assessment to NatureWorks™ polylactide (PLA) production. *Polymer Degradation and Stability, 80*, 403–419.

carbon credit trade which encourages the companies that produce renewable energy to trade off in the voluntary market to foster green energy development. The carbon credit can be traded to other businesses that are unable to produce renewable energy efficiently but have interest in participating in the renewable energy industry to promote lower emissions in their business activities. As a result, Vink, Glassner, Kolstad, Wooley, & O'Connor (2007) reported a 90% reduction of carbon emissions after the purchase of RECs, as shown in Table 7.18.

In addition to the preceding study, Groot and Borén (2010) in a life cycle assessment of lactide and PLA production from sugarcane in Thailand reported that every ton of PLA emitted 500 kg CO_2. Although alternative renewable energy through burning of sugarcane bagasse can be obtained in the range of 17–95 kWh/tonne of sugarcane, Groot and Borén (2010) highlighted that the environmental credit varies on the type of by-products, combustion technology and the mix of energy in the application. In other words, every source of PLA has a unique eco-profile, with the selection of an eco-friendly process being of the utmost importance to develop a preferable green PLA. This can be evidenced in Fig. 7.20, showing the environmental impact of PLA and some of the petrochemical polymers. Certain

Table 7.18 Scenario of PLA emissions with the purchasing of Renewable Energy Certificate (REC).

Process	CO_2 eq (PLA) (kg/kg PLA)	
	Before purchase REC	*After purchase REC*
(1) NatureWorks/Cargill site, direct emissions	1.038	1.038
(2) Indirect emissions from electricity production	1.561	1.561
(3) Fuel, material, corn production, reclamation	1.244	1.244
Corn feedstock-CO_2 uptake	−1.820	−1.820
REC purchased to offset electricity emission from (1)	–	−1.553
REC purchase to offset electricity emissions from (2)	–	−0.197
Total	2.023	0.272

Reproduced with Elsevier permission from Sin, L. T., & Tueen, B.-T. (2019). Polylactic acid. A practical guide for the processing, manufacturing, and application of PLA. United States: Elsevier, with permission from Elsevier.

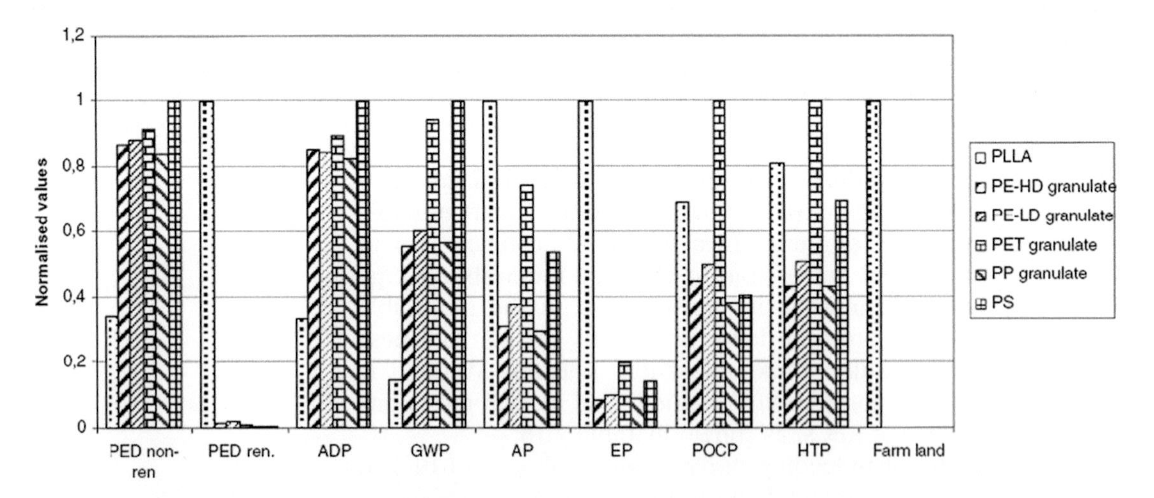

Fig. 7.20 Comparison of the most relevant ecological factors involved in the production of PLA and fossil-based derived polymers. *PED*, primary renewable energy; *PED nonren*, primary nonrenewable energy; *GWP*, global warming potential; *AP*, acidification potential; *EP*, eutrophication potential; *POCP*, photochemical ozone creation potential; *ADP*, abiotic resource depletion potential; *HTP*, human toxicity potential. Based on Groot, W.J., Borén, T. (2010). Life cycle assessment of the manufacture of lactide and PLA biopolymers from sugarcane in Thailand. *International Journal of Life Cycle Assessment, 15*, 970–984.

ecological aspects of PLA still require further improvement for greener production. Similar to any crop-growing activities, the farming of sugarcane contributes significantly to the environmental impact, including eutrophication, acidification, and photochemical ozone creation due to the nitrogen emission of ammonia-based fertilizer. Meanwhile, the combustion of agricultural residues for co-generation operations can release greenhouse gases such as NO_x, SO_x and CO. Some of the related soil activity by microorganisms can cause emissions of NO_x and methane as well. So far, PLA is the mass-produced biopolymer that has an impact on farmland due to the continuous replanting causing soil erosion and loss of natural nutrients.

Besides the comparison based on the emissions of polymers corresponding to their production process, another important issue to be addressed concerns the ecotoxicological effects on the surrounding species. Such a study has been conducted by several researchers on marine species and the effect has been summarized by Manfra et al. (2021) in Table 7.19. Although the biopolymers are known to be biodegradable, when they are being exposed to living organisms, the effects are still pronounced. In many scenarios, exposure to biodegradable plastic can affect the assemblage structures of different species. Overall, although biopolymers are generally known to be more eco-friendly, a wide spectrum of examination of their benefits and impacts still remains worthy of study. Consumers should always look into ways to reduce, reuse and recycle plastic materials to minimize the impacts on the environment.

7.5 Conclusion

Biopolymers are commonly known to be more eco-friendly compared to existing petroleum-based commodity polymers. They are either biodegradable or derived from renewable plant sources. Although biopolymers have properties comparable to petroleum-based commodity polymers, the use of biopolymers remains low due to their high cost and limited production. Besides, biopolymers derived from plant sources require a wide range of activities including planting, irrigation, pest control, harvesting, segregation, fermentation and polymerization, and finally they undergo plastic processing to produce the end products. As the result of more production steps, the environmental impact of biopolymers is actually almost equal to that of the plastic types on the market. Another reason for the limited use of biopolymers is the fact that adoption of production technology requires advanced knowledge. For instance, the processing of biopolymers into the end products requires tuning of processing parameters and modifications of the system.

In conclusion, biopolymers have mainly been introduced to reduce the environmental impacts of fossil-type nondegradable polymers. Nonetheless, the main approach to mitigate the

Table 7.19 A summary of biopolymer effects on marine organisms.

Type of biopolymer	Concentration and exposure	Organism species	Effects	Reference
PLA (0.6–363 μm) powdered form	0.8–80 μg/L (seawater)@ 60 days	*Ostrea edulis* (oyster) and *Scrobicularia plana* (peppery furrow shell clam)	Effects on the species were minimal, but benthic assemblage structures differed and species richness and the total number of organisms were ~1.2 and 1.5 times greater in control mesocosms than in those exposed to high doses of microplastics	Green (2016)
PLA (1.4–707 μm) powdered form	0.02–0.2–2% (wet sediment weight)@ 31 days	*Arenicola marina* (commonly known as lugworms)	Species produced fewer casts in sediments containing microplastics. Metabolic rates of species increased, while microalgal biomass decreased at high concentrations, compared to sediments with low concentrations or without microplastics	Green et al. (2016)
PLA (200–500 μm) shredded particles	0.1–0.2–0.4 mg /mL@24 h	*Microcosmus exasperates* (ascidian)	Submerged PET plates accumulated a richer invertebrate community than PLA plates. PET and PLA microparticles lowered fertilization rates of species	Anderson and Shenkar (2021)
PHB powdered form	0.5–1–2–4–8%@ 30–60 days diet	*Liza haemotocheila* (fish)	Dietary PHB supplementation could increase antioxidant enzyme activity, including total antioxidant capacity, catalase and superoxide dismutase	Qiao et al. (2019)

PHB (10–90 μm) suspended in acetone	1000 MPs/mL@96 h	*Mytilus edulis* (blue mussel)	Activities of superoxide dismutase (SOD), catalase (CAT), glutathione peroxidases (GPx), glutathione S-transferase (GST), and glutathione reductase (GR) were found to be significantly susceptible to fluoranthene and plastics in both tissues. Interestingly, a single exposure to PHB microplastics led to decreased activity levels of CAT and GST in gills, SOD in digestive glands and SeGPx in both tissues	Magara et al. (2019)
PHB dissolved in chloroform	1–3–5%@35 days diet	*Litopenaeus vannamei* (crustacean)	Dietary PHB altered the composition and diversity of intestine microbiota, and the microbiota diversity decreased with the increasing doses of PHB.	Duan, Zhang, Dong, Wang, and Zhang (2019)
Novamont biobags (vegetable oils and corn starch) (14 cm × 14 cm, 20 μm thickness) bag piece of the same size horizontally inserted into the sediment	Bags@180 days	*Cymodocea nodosa* and *Zostera noltei* (seagrasses)	*C. nodosa* root spread and vegetative recruitment increased compared to controls, both intra- and interspecific interactions shifted from neutral to competitive, and the growth form changed from guerrilla (loosely arranged group of widely spaced ramets) to phalanx form (compact structure of closed spaced ramets) but only with *Z. noltei.*	Balestri et al. (2017)

environmental impacts of plastics is still the basic reduce, reuse and recycle (3R). The successful implementation of 3R is highly dependent on community education, particularly the younger generation who will take over the care of Mother Earth in the future.

References

Anderson, G., & Shenkar, N. (2021). Potential effects of biodegradable single-use items in the sea: Polylactic acid (PLA) and solitary ascidians. *Environmental Pollution, 208,* 115364.

Balestri, E., Menicagli, V., Vallerini, F., & Lardicci, C. (2017). Biodegradable plastic bags on the seafloor: A future threat for seagrass meadows? *Science and The Total Environment., 605-606,* 755–763.

CCM International Limited. (2010). Corn products. *China News, 3*(1), 99–100.

Cheroennet, N., Pongpinyopap, S., Leejarkpai, T., & Suwanmanee, U. (2017). A trade-off between carbon and water impacts in bio-based box production chains in Thailand: A case study of PS, PLAS, PLAS/starch, and PBS. *Journal of Cleaner Production, 167,* 987–1001.

Duan, Y., Zhang, Y., Dong, H., Wang, Y., & Zhang, J. (2019). Effects of dietary poly-β-hydroxybutyrate (PHB) on microbiota composition and the mTOR signaling pathway in the intestines of litopenaeus vannamei. *Journal of Microbiology, 55*(946), 954.

European Bioplastics. (2009). *Fact sheet Nov 2009 industrial composting.* Available at www.european-bioplastics.org.

Fujimaki, T. (1998). Processability and properties of aliphatic polyesters, 'BIONOLLE', synthesized by polycondensation reaction. *Polymer Degradation and Stability, 59,* 209–214.

Green, D. S. (2016). Effects of microplastics on European flat oysters, Ostrea edulis and their associated benthic communities. *Environmental Pollution, 216,* 95–103.

Green, D. S., Boots, B., Sigwart, J., Jiang, S., & Rocha, C. (2016). Effects of conventional and biodegradable microplastics on a marine ecosystem engineering (Arenicola marina) and sediment nutrient recycling. *Environmental Pollution, 208,* 426–434.

Groot, W. J., & Borén, T. (2010). Life cycle assessment of the manufacture of lactide and PLA biopolymers from sugarcane in Thailand. *International Journal of Life Cycle Assessment, 15,* 970–984.

Hartmanna, H. (1998). High molecular weight polylactic acid polymers. In D. L. Kalan (Ed.), *Biopolymer from renewable resources* (pp. 367–411). Berlin: Springer-Verlag.

Li, Z., Yang, J., & Loh, X. J. (2016). Polyhydroxyalkanoates: Opening doors for a sustainable future. *NPG Asia Materials, 265.*

Magara, G., Khan, F. R., Pinti, M., Syberg, K., Inzirillo, A., & Elia, A. C. (2019). Effects of combined exposures of fluoranthene and polyethylene or polyhydroxybutyrate microplastics on oxidative stress biomarkers in the blue mussel (Mytilus edulis). *Journal of Toxicology and Environmental Health, Part A, 82,* 616–625.

Manfra, L., Marengo, V., Libralato, G., Costantini, M., De Falco, F., & Cocca, M. (2021). Biodegradable polymers: A real opportunity to solve marine plastic pollution? *Journal of Hazardous Materials., 416,* 125763.

Packaging Europe. (2021). *Suntory to introduce 100% plant-based PET bottle prototypes [online]*. Available at https://packagingeurope.com/news/suntory-to-introduce-100-plant-based-pet-bottle-prototypes/7665.article. (Accessed 8 March 2022).

Papong, S., Malakul, P., Trungkavashirakun, R., Wenunun, P., Chom-in, T., Nithitanakul, M., & Saronol. (2014). Comparative assessment of the environmental profile of PLA and PET drinking water bottles from a life cycle perspective. *Journal of Cleaner Production, 65*, 539–550.

Patten, A. M., Vassao, D. G., Wolcott, M. P., Davin, L. B., & Lewis, N. G. (2010). 3.27 Trees: A remarkable biochemical bounty. In H.-W. Ben, & L. Mander (Eds.), *Chemistry and biology: Vol. 3. Comprehensive natural products II* (pp. 1173–1296).

Pellegrio, C., Faleschini, F., & Meyer, C. (2019). 2. Recycled materials in concrete. In S. Mindess (Ed.), *Developments in the formulation and reinforcement of concrete* (pp. 19–54).

Perstorp Trade Brochure. Available at: https://www.pcimag.com/ext/resources/PCI/Home/Files/PDFs/Virtual_Supplier_Brochures/Perstorp.pdf. n.d.

PTT MCC Biochem BioPBSTM trade catalog, Thailand, n.d.

Qiao, G., Su, Q., Zhang, M., Xu, C., Lv, T., Qi, Z., Yang, W., & Li, Q. (2019). Antioxidant system of soiny mullet (Liza haematocheila) is responsive to dietary poly-β-hydroxybutyrate (PHB) supplementation based on immune-related enzyme activity and de novo transcriptome analysis. *Fish & Shellfish Immunology, 95*(314), 327.

Showa Denko. (2016). *SDK to terminate production and sale of biodegradable plastic [online]*. Available at: https://sdk.co.jp/english/new/2016.16250.htm (Accessed 10 October 2022).

Siegenthaler, K. O., Künkel, A., Skupin, G., & Tamamoto, M. (2011). Ecoflex® and Ecovio®: Biodegradable, performance-enabling plastics. *Advances in Polymer Science, 245*, 91–135.

Simon, B., Amor, M. B., & Földényi. (2016). Life cycle impact assessment of beverage packaging system: Focus on the collection of post-consumer bottles. *Journal of Cleaner Production, 112*, 238–248.

Sin, L. T., & Tueen, B.-T. (2019). *Polylactic acid. A practical guide for the processing, manufacturing, and application of PLA*. United States: Elsevier.

Sudesh, K., Abe, H., & Doi, Y. (2000). Synthesis, structure and properties of polyhydroxyalkanoates: Biological polyesters. *Progress in Polymer Science, 25*, 1503–1555.

Sun, H., Mei, L., Song, C., Cui, X., & Wang, P. (2006). The in vivo degradation, absorption and excretion of PCL-based implant. *Biomaterials, 27*, 1735–1740.

Vink, E. T. H., Glassner, D. A., Kolstad, J. J., Wooley, R. J., & O'Connor, R. P. (2007). The eco-profiles for current and near future NatueWorks polylactide (PLA) production. *Industrial Biotechnology, 3*, 58–81.

Vink, E. T. H., Rábago, K. R., Glassner, D. A., & Gruber, P. R. (2003). Applications of life cycle assessment to NatureWorks™ polylactide (PLA) production. *Polymer Degradation and Stability, 80*, 403–419.

Woodruff, M. A., & Hutmaacher, D. W. (2010). The return of a forgotten polymer—Polycaprolactone in the 21st century. *Progress in Polymer Science, 35*, 1217–1256.

8

International policies of plastic use and consumption

8.1 Introduction

Over the past decades, plastics consumption has been a tremendous burden to solid waste treatment facilities. The low prices and abundance of single-use plastics have worsened the problem, as people tend to dispose of plastic rubbish everywhere, leading to drain blockage, breeding of mosquitoes, a cluttered environment and life-threatening pollution. Thus the authorities of many countries have taken actions to implement regulations, policies, and guidelines to govern the use of plastics, especially related to domestic applications. Indeed, most of the authorities are focusing on reducing single-use plastics while embracing reduce, reuse and recycle (3R) in daily life to minimize the impacts of plastic wastes. This chapter discusses the approaches that have been engaged by countries to tackle plastic wastes.

8.2 Policies in European Union

8.2.1 Italy

In April 2019, the Italian Council of Ministers unanimously approved 'SalvaMare', which allows fishermen to bring back plastics caught in their nets during fishing and deposit them at designated recycling areas located in ports without paying for disposal. This policy was first piloted in the region of Tuscany and over the course of 5 months, more than 60 tons of plastic wastes were collected. It was deemed successful and was then carried out along major Italian coastlines from the Tyrrhenian to the Adriatic Sea. The Italian government rewards these fishermen with awards and incentives as a token of appreciation of their efforts to minimize plastic pollution (The Local, 2019). In addition, Italy has

Plastics and Sustainability. https://doi.org/10.1016/B978-0-12-824489-0.00009-X
Copyright © 2023 Elsevier Inc. All rights reserved.

banned retail shops from selling or providing plastic bags or, more specifically, plastic bags made from polyethylene. The bill took full effect in January 2011, and retailers are only allowed to provide bags made from cloth, paper and biodegradable plastic. A proportional tax on the manufacturing of single-use plastic products was introduced through the Italian Budget Law 2020 in efforts to reduce the production and consumption of plastics. Under the proposed tax, the amount was fixed at 0.45€/kg. The products that fall under the said tax include bottles, bags, food containers made from polyethylene, tetrapack containers, packaging made from extended polystyrene, rolls of pluri-ball plastics and caps made from plastic. The mentioned tax is applicable to all single-use plastic items made either fully or partially of organic polymers of synthetic origins meant to be used for the purpose of containment, protection and delivery of goods. To enforce this tax, failing to pay it results in an amplification of the amount from 2 to 10 times the unpaid tax. Both manufacturers and consumers are eligible to be taxed under this law (EY Global, 2020).

8.2.2 Germany

The European Waste Directive (2008/98/EC) was infused into German Law through the Kreislaufwirtschaftsgesetz (KrWG), which came into effect in 2012 and has led the way to the five-step waste hierarchy, which clearly gives priority to recycling over energy recovery and disposal of wastes in landfills or incineration. The KrWG's main objective is to protect the environment and public health through better waste management. KrWG introduced extended producer responsibility (EPR), giving major responsibility to the producers to ensure that their products are disposed of in a manner that does not harm the environment or the public. Another policy that has been passed by the Germans is Verpackungsverordnung (VerpackV). VerpackV stipulates that a minimum of 60% materials from plastic packaging must be covered, of which 60% must be recycled. To keep it simple, 36% is the bare minimum for recycling quotas for plastic packaging wastes in Germany. The VerpackV puts EPR into play so that producers and distributors of packaging products are obligated to recover packaging wastes of any material and to provide a collection and return system (Gandenberger et al., 2014). Starting from July 2021, Germany has banned the sale and distribution of single-use plastic products in order to boost the efforts to reduce the generation of plastic wastes. The first act to amend the Packing Act was entered into force in Germany in February of 2021. The amendment banned the overall sale and distribution of lightweight plastic bags starting from January of 2022. Plastic bags with a thickness below 50 microns are considered to be lightweight. However, plastic bags with a thickness below 15 microns are exempt from the ban. Under the same act, any merchants or retailers caught violating the ban may face fines up to 100,000 euros (Gesley, 2021).

8.2.3 Britain

In 2018, the government announced that a new type of tax would be introduced in England, on the production and import of plastic packaging starting from April 2022. The tax is applied to all plastic packaging produced in the country and also imported plastics. The tax only applies to plastic packaging that contains less than 30% recycled materials. The tax has a flat rate of £200 per ton of plastic packaging. Since October 2015, the law required all large retailers to charge 5p for all types of single-use carrier bags made from plastic. Only retailers with above 250 employees are obliged to carry out the tax; however, smaller businesses can participate on a voluntary basis if they wish. There are some cases in which consumers are not charged. Retailers aren't required to charge consumers for single-use plastic carriers on airports, trains, aeroplanes or ships. Products that contain items with a risk of contamination such as raw meat, medicines, health products and several others are not charged under the new law. Since then, there has been an 80% decrease in the number of plastic bags sold to consumers. It has been estimated that over the next 10 years, the tax on single-use plastic carriers will inject almost £780 million into the economy and save up to £60 million in litter clean-up costs.

In order to tackle the issue of rising plastic wastes generated from drinks, in March 2018 the government confirmed that it would introduce a deposit return scheme (DRS) in the country for single-use drink containers under the Environmental Bill (Smith, 2021). Major retailers such as Iceland are voluntarily installing machines in their stores in support of the government's decision. These machines work fairly simply: the consumers are rewarded with coupons that are worth 10p for every bottle that is deposited into the machine. The trial by the retailer so far has showed significant promise, as they managed to collect a daily average of 2583 bottles per day (Smithers, 2019).

A ban on microbeads in the manufacturing and import of cosmetics and care products was implemented in January 2018 and took effect starting from late June 2018 based on the notification of regulations to the World Trade Organization (WTO). In the United Kingdom alone, 680 tons of microbeads are present in cosmetic products sold in the country annually. The draft of the ban was published the year before, and by the end of the year more than 70% of major cosmetic companies halted the sale of products containing microbeads (Chemicalwatch, 2018).

8.2.4 Spain

The Spanish government came to a decision in 2020 to approve a draft law which introduces a new tax on plastic wastes. The purpose of the levy is to indirectly tax the import and manufacturing of nonreusable plastic packaging from other member states of the European Union. The tax is set at €0.45/kg of plastic packaging and, based on the figures from the year 2017, the tax is estimated to bring in an annual revenue of €724 million to the Spanish economy. In accordance with the same law, single-use plastic products such as cotton swabs, plastic cutlery and plastic stirrers are prohibited from being distributed or sold in Spain. Also, in July of 2018, the Spanish Government introduced a new law aimed at completely replacing the plastic bags and containers provided by retailers with compostable or recyclable substitutes by January 2021. The law requires merchants to charge customers for the purchase of plastic bags that have a thickness of 15 microns or more, with exemptions made for bags having 70% or more recycled materials. In the latest report (InSpain News, 2022), EU has criticized Spain for negligent of waste policy by violating the European Direction 2019/904 for plastic waste management. Spain has the new law in discussion in the Senate about tax landfill and waste incineration across Spain to encourage waste recycling.

8.2.5 Portugal

In late 2015, the government of Portugal introduced a tax on plastic bags that stipulated that retailers were to charge 10 cents for each plastic bag requested by the consumer (Theportugalnews, 2016). As of June 2020, the Portuguese Parliament introduced a bill to ban the use of lightweight plastic bags and disposable styrofoam packaging for fruits, vegetables and bread. Retailers are no longer allowed to distribute plastic bags below 15 μm in thickness for the sale of the mentioned items (Freshplaza, 2019). The Portuguese government has adopted a new law in 2022 of Deposit Return Scheme (DRS). Many member states of the European Union have already implemented such a system and it has been proven to increase the collection rate of single-use plastics. For instances, Scotland (2022), England Wales and Northern Ireland (2023–2024), Portugal (2022), Slovakia

(2022), Romania (2022), Turkey (2023), Latvia (2022), Malta (2022), New Zealand (2022), Singapore (2022), Victoria Australia (2023). In Portugal, the system is fairly simple: students and staff simply pay a €0.15 deposit when purchasing any beverages within the campus store. For retrieving the deposit, beverage packaging such as glass, plastics and cans are simply returned to the TOMRA reverse vending machines located within the campus (TOMRA, 2020). In addition, Portugal introduced a landfill tax in 2007 at £2 per ton, which was hiked to £4 per ton in 2011. The main objective for introducing the tax was to encourage waste producers to seek more sustainable and environmentally friendly alternatives such as recycling and incineration with energy recovery. The recycling rate has been noted to steadily rise over the years since the introduction of the tax; however, the landfill tax barely had any effect on landfilling (Bakas, 2013).

8.2.6 Belgium

Since 2020, in Flanders, Belgium, organizers of events are prohibited from providing disposable cups along with any single-use plastic products unless 90% of them can be collected back for the purpose of recycling (OECD, 2021). In 2007, Belgium introduced several environmental taxes, one of which is an environmental charge that covers single-use plastic bags along with several other environmentally damaging products. Its effectiveness was questioned initially; however, the implementation was deemed successful as it has achieved the goal of reducing the usage of single-use plastic bags. It has been observed that from the years 2008 to 2009, the distribution of single-use plastic bags dropped by a staggering 60%. The tax rate for plastic bags is as follows: single-use carrier bags (EUR 3/kg), single-use plastic (EUR 2.70/kg) and disposable plastic cutlery (EUR 3.60/kg), while biodegradable bags were exempted from the tax.

Policies such as landfill taxes have existed since 1997 and the rate has increased over time. In Flanders, the average price for landfilling is EUR 50 per ton for household waste and EUR 43 per ton for industrial wastes, while in Wallonia, the average price for landfilling is between EUR 40–80 per ton depending on the type of waste; the Belgium Capital Region has banned landfills. There are three main collection procedures: via recycling stations, retailers and kerbside collection. Residents pay about EUR 1.25 per bag of 60 L of residual wastes, while more billing options exist for the kerbside collection system. In 2010, fines for the residents were introduced in BCR if they did not separate their wastes accordingly. This initiative was mainly to increase the public awareness and to improve the quality of recycling waste streams. The organization responsible for the disposal and recycling of household packaging wastes is known as Fost Plus. Fost Plus has concentrated more on the collection of bottles (PET) and flasks (HDPE) compared to other types of plastic wastes. These two types of plastic account for 43% of the total plastic packaging in Belgium and about 71% is recycled. The organization collects from the producer about EUR 36 per household and a packaging recycling cost of EUR 0.5 per inhabitant annually. Residents do not have to pay a dime as the fees are already included in the product's price (Plastic-Zero, 2012).

8.2.7 France

Several policies were introduced by French policymakers in 2019 to combat the rising issue of plastic pollution. The French national roadmap of a circular economy (FREC) has announced a goal of 100% plastic recycling by the year 2025, a major leap from 22% in 2016. This is to be accomplished by the reduction of a value added tax (VAT) on recycling activities and the implementation of plastic disposal mechanisms. The Law on the Recovery and Protection of Biodiversity (2016-1087) placed a complete ban on cosmetics that contain microbeads and the sale of cotton buds by the year 2020. The Law on Energy Transition for Green Growth (LTECV) (2015-992) banned most single-use plastic products such as certain plastic bags, plastic packaging of mail advertising, plastic cups and plates and many more. Under the same law, the number of existing landfills was planned to be reduced through a phased tax increase, along with strengthening of existing extended producer responsibility (EPR) systems such as CITEO. Other policies such as Grenelle II (2010), which aims to bring down the amount of household wastes produced per capita through local prevention programs, and NOTRe (2015-991), which gives higher authority to regional and communal action plans in the department of waste reduction and management (A Plastic System Guidebook for France, 2019). In order to make waste prevention and recovery less costly, the VAT is reduced to 5.5% for prevention, separate collection, sorting and material recovery of wastes.

8.2.8 Sweden

The sharp drop in dependency on landfilling can be attributed to the introduction of a landfill tax in the country, approved by the government in the first quarter of 2000. As of 2015, the tax rate per ton of waste landfilled is at approximately €47.9. The law states that every material that enters the landfill facilities must be taxed, while those removing materials from landfills may request a reduction. The tax is paid by whoever owns the landfill on the basis of weight. It is clear that the law has had a significant effect on the rates of landfilling, as the rate decreased from 22% in the year 2001 to a mere 0.6% in 2013. Although the reduction in the use of landfills can't be attributed to the landfill tax alone, it definitely played a huge role in the decline (Municipal Waste Management-Sweden, 2016).

The Swedish Parliament passed a new tax on plastic bags in January 2020. The law took effect in March 2020, and under the new law plastics of thickness below 15 μm are taxed at a rate of US$0.03 a piece, while plastic bags above the mentioned thickness are taxed at a rate of US$0.31 a piece (Hofverberg, 2020). Sweden also announced a ban on the distribution and sale of cosmetics containing microbeads, a type of microplastics, at the start of June 2017. However, the ban is only limited to rinse-off products such as exfoliant scrubs (Abbing, 2017).

Sweden's deposit return scheme (DRS), commonly known as pant, has one of the highest return rates globally, capturing 85% of material. The system is run by an organization known as Returpack, first launched in late 1984. These reverse vending machines are able to accept one-way aluminium cans and one-way PET bottles.

8.2.9 Greece

The Greek administration drafted a bill to ban the use of single-use plastics by July 2021. The ban includes plastics such as straws, cotton buds, drink stirrers, styrofoam containers and more. One of the objectives of the bill is to bring down the consumption of plastic cups by the public by 60% by the year 2026 and to promote the circular economy among its citizens. Simultaneously, the bill lays the foundation for the nation's transition into a greener economy by offering incentives for the support of environmentally friendly actions. It is also set to include a levy for the use of plastic cups and food containers that are used from January 2022 and onwards. The penalties include a percentage of turnover from the previous year for manufacturers of these banned plastics, and food and beverage service providers such as restaurants and hotels will be charged €5 per item and a minimum fine of €1000 (GTP Headlines, 2020). In January of 2018, the Greek government introduced a new law which applied a levy on the consumption of plastic bags. The public was first taxed at a rate of €0.04 for each plastic bag in retail stores, and a year later the tax was raised to €0.09.

In 2012, the Greek government implemented a law that stated that as of January 2014 any organization that used landfills to dispose of untreated wastes must pay a tax between €35 and €60 per ton of waste. However, if these taxes are paid by end consumers, the tax will amount to an extra cost of €50–€150 per household annually. The law, however, was postponed to December 2016, as it was thought this law could lead to higher rates of illegal dumping as Greece has been suffering from a poor economy (Zachariadis, 2016).

The extended producer responsibility (EPR) was initiated by the government of Greece on legislation passed in 2001. The main packaging compliance scheme is under HERRCo (Hellenic Recovery Recycling Corporation). Just like any other existing system in Europe, HERRCo adopted the green dot system and by the year 2010, more than 1650 companies joined HERRCo; the number of municipalities under HERRCo by the end of 2010 was estimated to be 679. HERRCo introduced a new system called Blue Bins, where these blue bins are meant to be used to dispose of recyclable items. Participating municipalities were provided with a total of 110,000 blue bins with 1000 L capacities. The system is fairly simple: households dispose of recyclable wastes into the blue bins, which are then collected by local municipalities, which then deliver them to facilities known as Recycling Sorting Centres (RSCs); in 2011, there were a total of 28 RSCs in the country (Letsrecycle, 2012).

8.2.10 Poland

The Polish government unvealed a policy to take effect starting on 1 September 2019 in which plastic bags sold in the country were charged a fee of EUR 0.06. However, the levy was only applied on plastic bags sold that are over 15 μm in thickness. The Polish government had been considering the implementation of a Single-Use Plastics Directive into the Polish legal system, which was the main subject of a consultation meeting held on 17 October 2019 at the Ministry of the Environment. The main content of the directive was the complete ban of some single-use plastic products such as cotton buds, plastic cutlery, plates, straws, containers made from polystyrene and many more. The directive was scheduled to be introduced in the second quarter of 2021 and in addition to the ban on single-use plastics, the directive only allows the sale of plastic products with plastic caps and lids which are permanently attached to the container; by the year 2025, all plastic bottles must be produced using a minimum of 25% recycled materials and the threshold will be increased to 30% in the year 2030. The directive also includes a collection target for plastic bottles, which is a minimum of 77% by the year

2025. This means that a minimum of 77% of the total mass of plastic bottles disposed of needs to be collected back (Poland Government, 2019).

8.2.11 Lithuania

In February 2016, the Lithuanian government implemented the deposit return system (DRS) in efforts to further increase the collection rates of wastes generated from glass, plastic and metal beverage containers. The DRS operates in a very simple manner: consumers pay about €0.10 as a deposit when purchasing drinks and are paid back once the emptied container is brought back for recycling. The Lithuanian Parliament amended the law on the Management of Packaging and Packaging Waste which prohibited the distribution of free plastic bags with thickness of 15–50 μm at retail stores as of January 2019. Under the law, retailers found that violating the new rule would lead to facing fines up to EUR 3000 and up to a maximum of EUR 5000 for repeated violations (LRT, 2020).

8.2.12 Austria

Austria's cabinet approved a ban on single-use plastic bags from 1 January 2020. Companies and retailers are no longer allowed to distribute plastic bags to consumers that are in the range of 15–50 μm in thickness and as of 15 March 2020, manufacturers are no longer allowed to manufacture and import plastic bags which fall in that category. Moreover, retailers are to report the correct numbers of all types of plastic bags that were delivered to the country in the previous year. The main purpose of the ban is to reduce public littering and to encourage citizens to rely on reusable bags which are more environmentally friendly. Based on the government's estimates, the new ban would eliminate about 5000–7000 tons of plastic wastes each year (Deutsche Recycling Service, 2020).

8.3 Policies in American countries

8.3.1 Antigua and Barbuda

In 2016, Antigua and Barbuda, an island country, introduced a ban on plastic shopping bags, an initiative taken by the government in order to reduce plastic pollution. The government of Antigua and Barbuda introduced a plastic ban law known as 'The External Trade (Shopping Plastic Bags Prohibition) Order, 2017' (Department of Environment Antigua and Barbuda, 2016). In this policy, 'shopping plastic bags' refers to polyethylene or petroleum-based plastic bags. After 30 June 2016, the shopping plastic bags were strictly barred in terms of importation, distribution, trade and usage. However, six categories of plastic bags were not barred under this policy. The banning of the plastic shopping bags began at the big grocery stores and was followed by the smaller convenience stores and shops. As the plastic bags are banned, the supermarkets are compelled to supply only paper bags that are made from recycled paper. The supermarkets also can provide reusable bags for the customers to purchase. These banning activities are separated into Phase 1, Phase 2 and Phase 3. Phase 1 is about stopping plastic bag imports, which takes around 6 months. Phase 2 is mainly targeted at elimination of plastic bags in the big grocery stores, while Phase 3 is a target on the smaller convenience stores. Phase 3 will start once Phase 2 is completed. The smaller shops were given a longer period for this transition due to their small business scale and lack of flexibility.

Antigua also established External Trade (Expanded Polystyrene) (Prohibition) Order, 2018. This styrofoam banning policy was added as a result of the positive results from the banning of shopping bags, and it encourages use of biodegradable and reusable materials. This styrofoam ban is classified into three stages (Antigua Nice Ltd, 2017). In the first stage, the importation and use of food containers were banned. The examples of food containers are clamshell and hinge containers, bowls, plates, hot dog containers and hot and cold beverage cups. The period for this stage was 1 July to 31 July 2017. In the second stage, plastic utensils, straws, fruit trays, meat trays, vegetable trays and egg cartons were banned in terms of importation and usage. The period for this stage was 1 January to 30 June 2018. In the third stage, 'naked' styrofoam coolers were banned by ceasing importation and use. The period for this stage was 1 July 2018 to 1 January 2019. This styrofoam ban will be broadened to every business within the food service industry including supermarkets, grocery stores and the catering sector. This ban does not apply to airline carriers, private charters and large cruise liners. When styrofoam was banned, the government listed the alternatives to styrofoam.

8.3.2 Bahamas

The Bahamas government has established Environmental Protection (Control of Plastic Pollution) Act 2019 to tackle plastic pollution problems. This act mainly emphasizes the following three actions (The Government of Bahamas, 2019):

(i) Forbids the use of single-use plastic dishware and nonbiodegradable, oxo-biodegradable, and biodegradable single-use plastic bags.

(ii) Forbids the release of balloons.
(iii) Controls the use of compostable single-use plastic bags and related matters.

It was approved on 19 December 2019 by the Parliament of the Bahamas and was enforced on 1 January 2020. However, those who manufacture and export expanded polystyrene in Bahamas were exempted from this act.

8.3.3 Canada

In 2007, the town of Leaf Rapids in Manitoba, Canada introduced a plastic bag ban law and it was the first community in North America to do so. The Leaf Rapids authorities had noticed that plastic bags not only caused harm to the environment but also caused them to allocate almost $5000 every year to clean up the plastic bags that were scattered around the town. Hence, the Municipal Council legalized By-Law 462, a plastic bag ban by-law, on 22 March 2007 (Bernhardt, 2022).

The elements in the By-Law 462 are:

(1) The banning of plastic shopping bags will take effect starting from 2 April 2007.
(2) The merchants in Leaf Rapids are not allowed to sell or give single-use plastic shopping bags.
(3) Any person who breaches this By-Law is considered as guilty of an offence and will be given a fine of not more than $1000.
(4) If a violation persists for more than one day, the offender is guilty of a second offence for each day the violation persists.

Prince Edward Island, a province in eastern Canada, has implemented a bill known as the Plastic Bag Reduction Act. This Act was enforced on 1 July 2019 (Government of Prince Edward Island, 2019). Through this Act, no business shall refuse or dissuade a consumer from using his or her own reusable bag to transport things purchased or received from the business. In addition, no business shall sell or give a plastic bag to a client or give a client a free paper or reusable checkout bag.

In 2016, Montreal, the second largest city in Canada, announced the enaction of By-Law 16-051. The objective of implementing this by-law was to forbid the giving of certain shopping bags in the retail outlets. These certain shopping bags refer to those bags made from traditional, oxo-degradable or biodegradable plastic. The city council hoped to urge a shift in mindset when it comes to using plastic bags and reduce the environmental impact through the implementation of this by-law (Nicholas Institute for Environmental Policy Solutions, 2016). Under this By-Law 16-051, the following actions are prohibited:

(a) Supplying customers in retail outlets with traditional plastic shopping bags fewer than 50 microns thick regardless of whether free of charge or charged for.
(b) Supplying any thickness of oxo-degradable, oxo-fragmentable or biodegradable plastic bags.

8.3.4 California, United States

In 2016, Proposition 67 was passed in California after a referendum that made Senate Bill 270 begin to take effect. The introduction of this bill was mainly due to the worsening of plastic pollution, because Californians use approximately 15 billion single-use plastic bags every year, which is around 400 bags for each Californian. Senate Bill 270 forbids stores from offering single-use plastic bags to the customers (Legislative Analyst's Office, 2016). The two vital provisions in the bill are:

(1) Prohibition of single-use plastic bags

Supermarkets, convenience stores, large pharmacies and liquor stores in California are not allowed to offer single-use plastic bags to customers. However, the stores still can provide plastic bags for specific purposes such as unwrapped produce (fresh meat/fish).

(2) Creates new standards for reusable plastic carryout bags

This provision is mainly to set a standard for the material composition and durability of reusable bags or recycled paper bags. The bag manufacturers are monitored by the California Department of Resources Recovery and Recycling (CalRecycle) to ensure the bags produced meet the requirements.

8.3.5 Hawaii, United States

8.3.5.1 Disposable Foodware Ordinance (Ordinance 19-30/Bill 40)

The government of Hawaii has been implementing a plastic bag ban as early as 2011. For the past few years, some ordinances have been introduced for regulating the plastic bag ban. Regarding the Disposable Foodware Ordinance (Ordinance 19-30/Bill 40), the Mayor of Honolulu in Oahu, Hawaii approved this bill and signed it into law in 2019. The objective of this bill/ordinance is to clarify the provision of plastic bags and single-use plastic products. There are several sections in this bill and the details of the bill follow (City and County of Honolulu, 2019).

SECTION 3: A new Article 27 with title (Polystyrene Foam and Disposable Food Service Ware) is introduced to replace the old one.

(1) Regulation on polystyrene foam dish ware, disposable plastic service ware and plastic food ware.
  (i) Polystyrene foam food ware, disposable plastic service ware and disposable plastic food ware are not allowed to be sold to clients or used to contain prepared food.

(2) Exemptions apply on:
  (i) Grant exemption upon application which comes with reasonable evidence.
  (ii) Industry exemption.
  (iii) Disposable plastic straws can be given if the client's physical or medical conditions are not suitable to use a nonfossil fuel-based straw.
  (iv) Packaging used to contain unprepared produce such as raw meat, seafood and poultry.
  (v) Prepackaged food packaging, shelf stable food packaging, and catered food packaging.
  (vi) Packaging in whichever circumstance presumed by the city to be an emergency demanding urgent action to protect life, health, property, safety, or essential public infrastructure.

(3) Selling of polystyrene foam food ware, disposable plastic service ware, and disposable plastic food ware are prohibited.
  (i) Polystyrene foam food ware, disposable plastic service ware, and disposable plastic food ware are not allowed to be sold in any stores in Honolulu, but exclusively for the following:
    (a) Packaging used to contain unprepared produce such as raw meat, seafood and poultry.
    (b) Prepackaged food and shelf stable food packaging.
    (c) Products that are not compliant and are sold to a food vendor who has been granted an exemption.
  (ii) The stores can apply for an exemption to comply with the prohibition.

(4) Upon request, disposable service ware is available.
  (i) Only upon the request or definite response of a client or person receiving the prepared food or beverage, or in a self-service area or dispenser, the food vendor can give or distribute disposable service ware for prepared food or beverages.

8.3.6 Jamaica

In 2018, the Jamaican government introduced a plastic ban policy called 'The Trade Act' also known as the Plastic Packaging Materials Prohibition Order, 2018. Under this policy, single-use plastic is not allowed to be imported and distributed in large quantities, starting from 1 January 2019. There are two guidelines on the dimensions of the single-use plastic where each takes effect in a different timeline. Plastic with a size of 610 mm × 610 mm and a thickness of 0.03 mm is prohibited for the duration before 1 January 2021. On the other hand, single-use plastic with a size of 610 mm × 610 mm and a thickness of 0.06 mm is forbidden after 1 January 2021. The prohibition of single-use plastic does not take effect on the exemptions mentioned here (Government of Jamaica, 2018).

Exemptions:

(a) Single-use plastics that are brought into Jamaica prior to 1 January 2019.

(b) Single-use plastics used to contain eggs, flour, rice, baked goods and fresh produce such as raw meat for hygiene and safety purposes.

(c) Single-use plastics that are required for medical use and are distributed by the Health Ministry.

(d) Single-use plastics used when entering or leaving Jamaica for carrying personal belongings.

(e) Drinking straws that are brought by an organization responsible for disabled persons from overseas.

(f) Drinking straws made entirely or partially of polyethylene or polypropylene, manufactured for single use, and attached to or constituting part of the packaging of juice boxes or drink pouches, until 1 January 2021.

This Act prohibits all types of single-use plastics including those plastic bags defined as oxo-degradable, photodegradable, biodegradable, degradable or compostable. For those who violate this Act, the person will be subjected to a fine of not more than two million dollars or an incarceration of maximum 2 years.

Jamaica also has The Natural Resources Conservation Authority (Plastic Packaging Materials Prohibition) Order, 2018. This Order is actually similar to the Trade Act. The difference is that this Order prohibits the commercial use and manufacture of single-use plastic bags (Government of Jamaica, 2018). Other than banning single-use plastic, the government has also been taking some measures to ban styrofoam and plastic straws. There are two phases, with the first phase starting on 1 January 2019 and the second phase starting on 1 January 2020. In the first phase, expanded polystyrene foam is restricted from being imported into Jamaica. In the second phase, products made from polystyrene foam are prohibited from manufacture and distribution. For the banning of plastic straws, the measures are divided into phase 1 and phase 3. In phase 1, plastic straws are not allowed to be brought into

Jamaica from other countries and the manufacturing of plastic straws is banned. In phase 3, those straws that are affixed on a juice box are restricted from importing. The businesses are given a period of 6 months for the transition from plastic straws to other alternative material straws.

8.3.7 Chicago

In April 2014, the aldermen in Chicago, Illinois authorized the ordinance Amendment of Municipal Code Chapter 11-4 by adding Article XXIII as an effort to curb the plastic pollution problem. Through this ordinance, stores are not allowed to offer plastic bags to their clients to contain the goods that are bought by them. This ordinance took effect starting from 1 August 2015 but this was only for the stores that occupy a land area of more than 10,000 ft^2. For those stores that occupy 10,000 ft^2 or less, this ordinance took effect starting from 1 August 2016 (Chicago City Council, 2014). In addition, Chicago implemented the Chicago Checkout Bag Tax Ordinance and this levy is called the Checkout Bag Tax. This tax is intended for the distribution of checkout bags in Chicago citywide. Customers are charged $0.07 upon the purchase of a checkout bag. However, this tax should not be considered as a tax inflicted to the business or stores (Chicago City Council, 2016).

8.3.8 Saint Vincent and Grenadines

In 2017, the government of Saint Vincent and Grenadines (SVG) introduced a policy known as the Environmental Health (Expanded Polystyrene Ban) Regulations 2017. This policy was implemented to help mitigate the plastic pollution problem in the country and it took effect from 1 May 2017. No one is permitted to import, produce or sell dishware products that are made of expanded polystyrene. For the person who breaches this policy, that person will be subjected to a penalty of $5000 or incarceration of 12 months or both. The restaurants are also not allowed to serve or offer foods using dishware made of expanded polystyrene (styrofoam). Other than the Regulation mentioned here, the government also waived the Value Added Tax (VAT) in 2017. The waiver of the VAT targets biodegradable packaging and dishware. Through the repeal of VAT, the costs of manufacturing the plastic substitutes can be reduced. This repeal took effect starting from 1 May 2017 (Jamaica Observer Ltd, 2017). Another regulation is known as Environmental Health Control of Disposable Plastics Regulations 2019. The purpose of these Regulations is to tackle the usage of single-use plastic bags (SearchLight, 2020). Any person breaching this regulation will be subjected to a penalty of $5000 and if the person still continues to violate, an additional penalty of $500 will be issued every day until the violation stops. Through the implementation of this regulation with a heavy penalty, the Minister expected to ease the plastic problem in a 5-month period.

8.4 Policies in Asian countries

8.4.1 Japan

Japan is the Asian country which has the most established legislative system and policies to achieve a recycling-oriented society. The Resource Recycling Promotion Law entered into effect in 1991 (later amended to Law for the Promotion of Utilization of Recyclable Resources). According to the legislative framework establishment and enforcement by the Ministry of Economy, Trade and Industry, Japan (METI), as shown in Fig. 8.1, METI Recycling Guidelines are comprehensive, covering a wide spectrum of industries and every type of waste.

8.4.1.1 Law for the promotion of effective utilization of resources

This law entered into force in April 2001, to promote integrated initiatives for 3Rs so that Japanese society can be more sustainable. The law designates 10 industries and 69 product categories, and actions stipulated include 3R policies at the product manufacturing stage, 3R consideration at the design stage, product identification for fulfilment of separated waste collection, and the creation of voluntary collection and recycling systems by manufacturers and others. This law pertains to businesses engaged in manufacturing requirements to meet the Standards of Judgement concerning Reduction in Generation of Used Products, Recycled Resources or Reusables Parts for personal computers, unit-type air conditioners, television sets, microwave ovens, clothes driers, refrigerators, washing machines and copying machines.

8.4.1.2 Containers and packaging recycling law

This law entered into force in April 2000. In general, it is aimed at reducing the amount of garbage generated, by promoting effective utilization of resources in containers and packaging waste, applying to about 60% of household waste (by volume). All the players should take responsibility for recycling, with consumers separating their waste according to category, municipalities

Life cycle of product	Legislative System	Role of Parties Concerned
Production	Law for Promotion of Effective Utilization of Resources	**Business** -3R oriented design (resource-saving, long-life) -Reuse, promoting recycling -Labeling of materials
Consumption/Use	Green Purchasing Law	**National government, national organizations, and local authorities.** -Taking the lead in purchasing environmentally friendly products
Collection/Recycling	Law for Promotion Effective Utilization of Resources	**Consumers** -Proper disposal -Proper payment of costs **Business** -Self-collection
	Containers and Packaging Recycling Law	
	Home Appliance Recycling Law	
	Food Recycling Law	
	Construction Material Recycling Law	
	End-of-Life Vehicle Recycling Law	
Waste	Waste Management Law	**Businesses/Municipalities** **-Proper disposal of waste**

Fig. 8.1 Japan's legislative framework for creating a sustainable society based on the 3Rs (METI, 2022).

collecting the separated waste, and businesses recycling what has been collected into new products.

8.4.1.3 Act on recycling of specified home appliances

This law entered into force in April 2001. All the stakeholders are required to work together to meet completion of the recycling process in a responsible approach involving home appliances

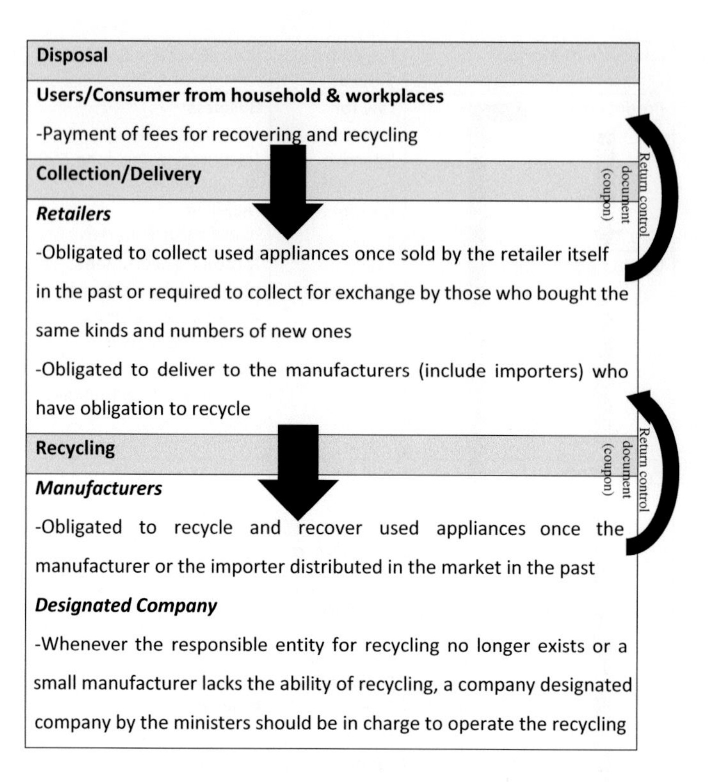

Fig. 8.2 Procedure of handling unwanted home appliances.

such as air conditioners, television sets, refrigerators, washing machines and drying machines. The details of the roles are listed in Fig. 8.2. Every stage involves a coupon system to assure proper collecting and recycling of home appliances waste. The Act requires a coupon to be issued at each set of home appliance waste and on which the information of the parties concerned is to be written.

8.4.1.4 End-of-life vehicle recycling law

This law will be enforced in January 2025. The main purpose of this law is to share the roles among car owners, end-of-life vehicle collecting businesses, car manufacturers and car importers. All the stakeholders are working together to reduce waste and resources used for building a recycling-oriented society. This law covers all types of four-wheel vehicles (including commercial vehicles like trucks and buses). The car owners are required to pay the recycling fee upon purchasing a new car. When a vehicle is at end of life, the car owner is required to send the car to the end-of-life car collecting operators that are registered with local authorities. The end-of-life car is delivered to fluorocarbon (CFC) collection operators, auto dismantling operators and shredding operators. The collected and recycled CFC, airbags, spare parts and shredder dust from the end-of-life vehicles are delivered to the car manufacturers/importers for further reusing and recycling market, while all the reports between the stages as described are recorded through an Information Management Center which is managed by Japan Automobile Recycling Promotion Center.

In the latest Japan development, the Cabinet of Japan approved the Bill for the Act on Promotion of Resource Circulation for Plastics on 9 March 2021.

It was also passed by the House of Councilors on 3 June. The Act shall come into effect on the date specified by a Cabinet Order within a period not exceeding one year from the date of promulgation (Shotaro, 2021).

8.4.2 India

Since plastic has been widely employed as a necessity in everything from toothbrushes to debit cards, India is struggling to dispose of its growing volume of plastic waste. Kerala has outlawed the manufacturing, transportation, storage and sale of single-use plastics (Government of Kerala, 2019). This implementation is the result of the Indian government's unsuccessful attempt to legislate plastic waste management rules in 2016. In order to ease the separation and recycling of plastic waste, the state government of Kerala banned the use and sale of plastic carry bags having a thickness of less than 50 microns. However, this implementation was ineffectual due to irresponsible plastic consumption, which caused an increase in environmental issues as well as health risks among individuals. As a result, the chief minister of Kerala implemented a blanket ban on the use of plastic (Government of Kerala, 2019).

On the other hand, Tamil Nadu, a southern Indian state with a large population, is the country's second largest source of plastic garbage. On 1 January 2019, the Tamil Nadu government outlawed throwaway and single-use plastic without regard for the thickness of the material (Preetha, 2020). Plastic carry bags used solely for export, plastic used to store milk, and carry bags made from biodegradable polymers were excluded from this implementation. According to a Tamil Nadu Pollution Control Board officer in Coimbatore, the previously granted exemption to use plastic bags for packaging products has been abolished by releasing a notification. As per the notification, plastic bags that contain or form an important element of the packaging in which items are sealed before using at manufacturing or processing units are prohibited (Preetha, 2020).

In addition, the Maharashtra state government released a notice on limiting the use of plastic in order to combat pollution. The notification forbids the manufacture, use, transport, distribution, wholesale, retail sale, and storage of plastic bags with and without handles, as well as the import of disposable items made of plastic and thermocol (polystyrene), such as single-use disposable dishes, cups, plates, glasses, forks, bowls, containers and disposable plates or bowls used for food containers in hotels and cutleries (Environmental Department, 2018). These restrictions apply to any person, group, government or nongovernment organization, educational institution, or other organization or institution in Maharashtra that utilizes plastics (Environmental Department, 2018).

8.4.3 China

The Chinese government has made attempts to tackle pollution and environmental degradation, since the nation has been afflicted by smog and contaminated soil as a result of decades of rapid urbanization (Xinhua, 2017). In order to address this issue, China issued a new ban on foreign waste imports on 27 July 2017, known as the Solid Waste Import Management Reform Plan, also known as the 'China ban', which prohibits the importation of 24 types of solid waste in four broad categories, including plastic waste, paper waste, metal waste and textile waste. Five federal agencies announced the ban, which went into effect on 31 December 2017 (Zhang & Laney, 2017). Another 16 kinds of solid rubbish, including waste hardware and industrial waste plastics, were added to the list of prohibited products by the Ministry of Ecology and Environment in April 2018.

In order to address waste management issues, China adopted a household garbage categorization policy in July 2019. The Regulation is a stringent new policy that requires rubbish to be sorted into one of four categories: dry, moist, recyclable or hazardous, and placed in colour-coded containers (Wang & Jiang, 2020). China began its initiative in the early 2000s, focusing largely on eight cities. However, this plan was poorly handled, and its goal was not met. In 2017, two trillion tons of household solid garbage were generated in 202 Chinese cities, demonstrating the failure of implementation (Wang & Jiang, 2020). In March 2017, China unveiled a more ambitious national strategy to combat rising domestic waste mountains, requiring 46 pilot cities, including Shanghai, to pass local decrees or local laws mandating rubbish classification by the end of 2020 (Wang & Jiang, 2020). As a result, China began a period of mandatory rubbish sorting.

The National Development and Reform Commission of China unveiled a 5-year plan to phase out single-use plastic, which is divided into two phases to gradually reduce consumption, manufacturing and trash output. Nonbiodegradable plastic bags were prohibited in large supermarkets and shopping malls in major cities by the end of 2020, and in all cities and municipalities by 2022 (Zhang & Laney, 2017). Aside from that, China intended to switch from plastic straws to paper straws or polylactic acid straws at restaurants, food stalls and coffee shops across the country. Single-use straws were banned entirely by 2020, and single-use cutlery will be prohibited entirely by 2022 (Joe, 2021). Furthermore, one-time plastic consumption in food delivery packaging is planned to be decreased to 30% by 2025, and hotels are recommended not to sell single-use plastics goods (Joe, 2021). During phase 2 in 2022, all cities and municipalities are prohibited from using nonbiodegradable plastic bags (Joe, 2021). The fresh products market merchants will still have some flexibility, since a blanket bag ban is scheduled to take effect in 2025.

8.4.4 Malaysia

Malaysia is a global participant in the plastic business, with over 1300 plastic manufacturers today (METI, 2022). Over the years, environmental issues related to plastic waste have become a major issue. As a result, the No Plastic Bag Campaign Day was launched in 2011 by the Ministry of Domestic Trade, Cooperatives, and Consumerism (MDTCC) with the purpose of reducing plastic bag consumption. The campaign was launched at the consumer level in selected supermarkets or hypermarkets, big stores, and significant shopping malls around the country. Every weekend, the programme charged MYR 0.20 cents for each plastic bag (Zen et al., 2013). The funds raised by the campaign were used to sponsor environmental initiatives. Later, the Malaysian Ministry of Energy, Science, Technology, Environment, and Climate Change published a Roadmap to Zero Single-Use Plastics (Chen et al., 2021; MESTECC, 2018). Pollution taxes are imposed on consumers and manufacturers as a part of this Roadmap. In 2019, the governments of Selangor and all three federal regions in Malaysia outlawed the use of plastic straws (Chen et al., 2021). A nationwide ban on the use of straws was enacted in 2020, but there has been no enforcement so far.

Malaysia has also implemented the 'National Strategic Plan for Solid Waste Management (2000–20)' in response to public concern about waste reduction. The Sustainable Waste Management

Hierarchy's purpose is to decrease the amount of waste that enters landfills and disposal facilities (Hashim et al., 2011). Consumer environmental awareness and preference for more environmentally friendly products appears to be expanding across the industrialized globe (Rashid, 2009). Malaysia's government has launched a Malaysian national ecolabelling programme which approves products based on environmental standards such as ecologically compostable and nonhazardous plastic packaging material (Suki, 2013). The maker, retailer, or marketer makes self-declared claims about a product, which might be based on a single attribute or an overall evaluation of the product.

8.4.5 Singapore

8.4.5.1 Singapore packaging agreement

The Singapore Packaging Agreement (SPA) was a collaborative effort by the government, industry and nongovernmental organizations (NGOs) to minimize packaging waste, which accounts for around one-third of Singapore's domestic trash by weight (Kojima, 2010). The Agreement was voluntary in order to provide the sector the freedom to develop cost-effective waste-reduction measures. During the review of the Singapore Green Plan (SGP) 2012, the Clean Land Focus Group, which was appointed to review waste management, advised that Singapore should consider adopting the concept of Extended Producer Responsibility (EPR) to reduce waste, including waste generated from packaging, which was proven to be an effective way to reduce plastic wastes in other countries. This was consistent with an online poll done in connection with the assessment of the SGP 2012, in which 94% of respondents agreed that producers should take steps to minimize the amount of packaging they use. Following that, the National Environment Agency (NEA) investigated various packaging practises in a number of nations, including Australia, Japan and New Zealand. The NEA also met with industry representatives to learn about their concerns. The industry believed that enacting laws would raise costs, which would eventually be passed on to consumers. Furthermore, laws would not allow industry to be flexible in order to innovate.

Therefore, the parties agreed to start a voluntary programme instead. It was to be modelled after New Zealand's Packaging Accord, with aspects from Australia's National Packaging Covenant included. This programme was to be based on product stewardship, in which everyone involved in the product's lifespan would be held accountable for minimizing the product's environmental effect. It was expected that such a programme would foster government-industry collaboration and promote government-industry-community interaction, engage the entire packaging supply chain, and provide industry with opportunities to assume greater corporate responsibility and shift the focus from compliance to continuous improvement.

In 2007, the NEA announced the first SPA to minimize packaging waste. On World Environment Day, 5 June 2007, it was signed by numerous parties including industry associations, individual enterprises, nongovernmental organizations, the Garbage Management & Recycling Association of Singapore, public waste collectors, and the NEA. Initiatives like the Packaging Partnership Programme and Mandatory Packaging Reporting Regulations have continued to support sustainable packaging waste management in Singapore when the SPA expired on 30 June 2020. The SPA's major goals are to reduce waste from product packaging by optimizing production processes, improving packaging and increasing packaging waste reuse and recycling. Also, raising consumer awareness and education about waste reduction is critical since customers' behaviours, for example, choosing items with less packaging and participating in recycling, have a direct influence on the program's performance.

The focus was on food and beverage packaging, which made up a significant portion of household packaging trash. When the Agreement was first signed, it had 32 signatories involved. The Waste Management andRecycling Association of Singapore, the four public waste collectors, and the National Environmental Agency were among the five industry associations representing more than 500 companies, 19 individual companies, two nongovernmental organizations, and the Waste Management & Recycling Association of Singapore. The initiative was expanded in October 2009 to include additional product packaging, such as detergents, home items, toiletries, and personal care products. By the end of 2009, the number of signatories had risen to 95, with additional signatories from a variety of businesses, including hotel and shopping centre owners or managers.

The SPA's implementation is supervised by a Governing Board. The Governing Board provides guidance to the signatories, assists them in resolving challenges, and ensures that they continue to actively engage in the SPA in order to fulfil their promises. Members of the Governing Board are strong supporters of the SPA and frequently seek out chances to engage with other industry players and partners in order to get them to join. The Governing Board, for example, has been holding monthly CEO luncheons and inviting both signatories and nonsignatories to these luncheons. These luncheons allow signatories to exchange best practises for reducing packaging waste, as well as the financial savings they experience when they implement more ecologically sustainable methods. At the same time, nonsignatories will learn about the SPA and be urged to join the programme through such activities. Since the signing of the Agreement, the signatories have achieved great progress in decreasing waste. Over the last 4 years of the Agreement, from 1 July 2007 to 30 June 2011, signatories decreased about 7100 tons of packaging waste, saving a total of $14.9 million in packaging-related expenditures.

8.4.5.2 National recycling program

Under this programme, which began in April 2001, all Housing and Development Board estates, private landed properties, and condominiums/private flats that have opted into the public garbage collection scheme are required to offer recycling containers and recycling collection services (National Environmental Agency, 2022). Paper, plastic, glass and metal recyclables are all dumped in the same blue recycling containers for collection by waste collectors under the National recycle programme. The plan was mainly allocated for houses, schools and workplaces. The waste minimization and recycling at these places are planned to be done using the 3R concept, as shown in Table 8.1.

Upon disposal in the designated area, specialized recycling trucks collect the discarded garbage and transport it to Materials Recovery Facilities (MRFs). Plastics are sent to a recycling plant and sorted by plastic type. Each type of plastic is crushed into smaller fragments and mixed together to create a homogeneous slurry. In order to form strands, the combined substance is heated and forced over a screen. Extrusion is the term for this process. The plastic strands are then chopped into pellets and utilized as material for new products after cooling.

8.4.5.3 Extended producer responsibility (EPR)

EPR is a policy strategy that holds producers responsible for the end-of-life consequences of their goods. EPR enables proper collection and disposal of items after use by assuring responsible manufacturing methods, and it extends beyond that to encourage waste reduction and environmental impact in general. With plastics being the most common type of trash produced in Singapore and a poor recycling rate, it is past time for upstream initiatives like the EPR to be considered. Despite the fact that many multinational corporations in Singapore have introduced their own 2025 targets aimed at reducing their overall plastic footprint, these

Table 8.1 Singapore's 3R strategies plan.

Place	Plan
At home	*Reduce* • Bring a reusable shopping bag to reduce usage of plastic or paper bags • Purchase items with less packaging *Reuse* • Reuse medicine containers for travel toiletries • Use unwanted plastic bags to contain household waste before disposal into refuse bins or chutes. Plastic bags are then incinerated with other waste at the waste-to-energy incineration plants which are designed to Incinerate waste safely, recovering energy from incineration process as well as meeting air quality standards through good combustion and fuel gas emission control • Use reusable bags when doing grocery shopping *Recycle* • Paper, glass, plastic and metal are advised to be recycled under certain conditions • Steps to recycle at home: (1) **Segregate recyclables from waste** (2) **Empty food or liquid residue, if any; rinse if needed** (3) **Place recyclables in a household recycling receptacle, e.g. used plastic bag or cardboard box** (4) **Deposit recyclables into recycling bin**
At school	*Reduce* • Avoid overcatering food and drinks for school activities and events • Use refillable soap and drinks dispensers instead of providing individual packets • Purchase devices and stationery that are durable • Purchase refillable stationery *Reuse* • Reuse plastic and glass containers to water plants or as pots for planting • Unwanted items in good/usable condition to charitable organizations *Recycle* • Paper, plastic and metal are advised to be recycled under certain conditions • Steps to recycle at home: (1) **Put an unwanted box/container and label it 'Recycling Box' to collect recyclable items. Empty food or liquid residue, if any; rinse if needed** (2) **Ensure recyclable container (if any) is empty, i.e. no food or liquid waste in container** (3) **Put recyclables (e.g. old newspapers/magazines, torn plastic files, etc.) into Recycling Box** (4) **Transfer the recyclables from the box to the school's recycling bin**

Continued

Table 8.1 Singapore's 3R strategies plan—cont'd

Place	Plan
At work (Office)	*Reduce* • Avoid the use of disposable items • Do not ask for plastic carriers if packet food can be held in hand • Bring own reusable lunchbox and cutlery for takeaways • Bring own mug to meetings to reduce use of plastic or styrofoam cups *Recycle* • Appoint an officer to implement the recycling programme • Engage a recyclables collector. • Provide recycling bins with appropriate labelling at strategic locations • Instill awareness and engage staff regularly through circulars, newsletters, talks and activities • Educate the cleaners to keep the recyclables separate from general waste during collection to avoid contamination • Monitor the scheme regularly, and keep staff informed by charting progress of the recycling programme
At work (Industry)	*Reuse* • Proper management of raw materials, intermediate products, final products and the associated waste streams • Companies can join the Packaging Partnership Programme to reduce packaging waste from consumer products • Purchase only the amount of raw materials needed for a production run or a set period of time • Purchase the material in the proper sized container/package to reduce packaging waste *Reuse* • Look for ways to use a waste as a raw material in another process • Look for ways to avoid contamination of a waste so that it can be put back into the originating process as a substitute for a raw material *Recycle* • A Recycling Programme for JTC Industrial Estates was launched in November 2003. Under this programme, designated corners were set up in bin centres to collect wood waste for reuse. Recycling bins were also provided in strategic locations for the collection of paper, aluminium cans, plastic, metal and glass

efforts may not translate into meaningful societal shifts in the means of consumption and disposal of plastic, especially if these are one-time strategies such as switching to greener materials (Qiyun, 2020). Due to physical or budgetary restrictions, businesses are frequently unable to undertake substantial supply chain modifications. As a result, EPRs are critical for tying legal imperatives to financial levers, compelling enterprises to make essential transformations, and strengthening municipal waste management.

Singapore's Resource Sustainability Act (RSA), which focuses on packaging waste, is likely to be the first EPR in Southeast Asia in 2025. This is a significant step forward in Singapore's efforts to manage packaging waste, especially after more than two decades of the Singapore Packaging Agreement (SPA), which intended to raise packaging waste awareness and reduce waste. The SPA has been criticized as ineffective in promoting circular economy transitions. The RSA, on the other hand, was passed in 2019 and outlines regulatory actions aimed at waste streams that include packaging with high generation and low recycling rates. Large manufacturers of packaging and packaged items are presently required to declare the amount and type of packaging they are placing on the market. By 2022, producers are required to outline how they intend to reduce, reuse, and recycle these packaging materials. This information is most likely to be utilized as a starting point for the EPR's development.

8.4.6 Thailand

On 25 May 2021, the Pollution Control Department (PCD) issued a 'Notification of Pollution Control Department Regarding the Definitions of Plastic Wastes and Plastic Scraps B.E. 2564 (2021)' (Leungsakul, 2019). Plastic waste and plastic scrap have distinct definitions according to the PCD. Plastic waste is defined as any plastic object or portion of a plastic object that has been dumped after use or after its intended use, degraded and useless plastic, or plastic that has been contaminated with other types of garbage. Plastic scrap, on the other hand, is a piece of plastic or useless plastic residue. Thailand imports only 'plastic scraps' as raw materials in the industrial sector, according to the customs tariff regulation on plastic importation, and does not import 'plastic wastes'. In 2020, 150,807 tons of plastic wastes were imported, while 44,307 tons were imported between January and April 2021. Thailand is currently in a period of adjustment to avoid affecting operators, and will progressively restrict the importation of plastic scrap, with the goal of '100% plastic scrap import ban within 5 years'.

In order to support the efficient recycling process of plastic waste within the country and recycle as a high-quality raw material that is in growing demand in industrial sectors, collaboration between the government and private sectors has been developed to increase people's awareness of plastic waste separation. The Minister of Natural Resources and Environment, as chairman of the National Environment Board's Plastic and Electronic Waste Management Subcommittee, stated emphatically that the

Ministry of Natural Resources and Environment's policy is to promote plastic waste utilization as much as possible within the country. Within the next 5 years, the government will prohibit the import of all foreign plastic scraps, and the government has never had a policy of importing plastic garbage. They are unlawfully smuggled if they are imported. In this case, the Customs Department and related agencies have been assigned to expedite the delivery of the containers back to the country of origin immediately.

On 17 April 2018, the Prime Minister of Thailand issued an order to the Ministry of Natural Resources and Environment at a cabinet meeting to collaborate with all sectors and develop a series of action policies to avoid and eliminate the problem of plastic waste throughout its life cycle (Jangprajak, 2021). In line with this development, the National Environment Board established a subcommittee on plastic waste management. As the subcommittee's secretary, a working group on the development of plastic management mechanisms was created. The working group submitted a draught plastic waste management strategy from 2018 to 2030, which would act as a framework and guidance for avoiding and resolving Thailand's plastic waste problem.

When developing the roadmap, the working group on the development of plastic management mechanisms invited participation from government agencies, the private sector, nongovernmental organizations, international organizations, educational institutions, and other relevant parties through three meetings with the working group, three meetings with plastics industry entrepreneurs, and five meetings with the plastic waste management subcommittee. The roadmap's draught version was presented to and approved by the National Environment Board.

The cabinet eventually acknowledged the strategy during its meeting on 4 January 2019. Months later, the roadmap had been altered in response to comments. On 17 April 2019, 'the roadmap on plastic waste management, 2018–30' was presented to and approved by the cabinet. The plan will be utilized as a policy framework to address Thailand's plastic waste problem. The roadmap's goal is to limit and eliminate the use of plastic and replace it with other ecologically friendly alternatives. This roadmap is divided into three sections. The first phase began in 2019, with the prohibition of three plastic products, which are cap seal (plastic that covers the bottle top), oxo-degradable plastic, and microbeads. During the second phase, the use of four more categories of single-use plastic were to be prohibited by 2022, including thin plastic bags with a thickness of less than 36 microns, styrofoam food boxes, plastic straws, and single-use plastic cups.

The guide pamphlet has been shared widely in the media. The third phase of the roadmap has to be done by 2027, where 100% of plastic waste shall be reusable. Three action plan measures were also implemented, as stated in Table 8.2.

Campaigns to encourage public engagement in reducing and eliminating plastic usage were started using traditional and online medias at the end of 2019. The Ministry of Natural Resources and Environment anticipates that this regulation will reduce the volume of plastic garbage by 0.78 million tons per year and save 3.9 billion baht in waste management costs per year. The strategy will also contribute to a reduction in greenhouse gas emissions of 1.2 million tons of CO_2 eq and 1000 acres of landfill area.

Table 8.2 Thailand action plan's measures to tackle plastic waste.

Action plan	Measures
Reduction of plastic waste at sources	• Reducing single-use plastics • Ecodesign of packaging • Use alternatives to replace single-use plastics • Set plastic product standard • Green procurement • Support for ecoinvestment • Creating a plastic database • Tax incentives to promote biodegradable plastic packaging
Reduction of use of single-use plastic at consumption process	• Educate and outreach to promote green consumption • Cooperate among stakeholders to reduce single-use plastics • Set rules/regulations/procedures in preventing marine littering • Establish policy on plastic waste management under international cooperation
Postconsumption plastic waste management	• Issue rules and regulations for waste separation according to the 3R principle by the local government • Develop and promote the Circular Economy • Promote waste-to-energy • Capacity building informal sector and waste buyer • Develop a law to prevent/solve the problem of marine plastic litter • Control the import of plastic scraps from abroad • Reward those who have contributed to good plastic management

8.4.7 South Korea

8.4.7.1 Extended producer responsibility (EPR)

Korea has been one of the Asian countries to adopt the Extended Producer Responsibility (EPR) System in the waste management industry. The EPR system, which was implemented in 2003, requires producers and importers to recycle a specified percentage of their products. At the time of its inception, only 15 products were covered by the policy. By 2008, the list of things covered had grown to include 24 items, including four packaging materials. The scheme included 10 electronic products, as defined by Article 8 of the presidential decree 'Act on Resource Recirculation of Electrical and Electronic Waste and End-of-Life Vehicles', as well as styrofoam floats and packaging materials used to pack food and beverages and livestock products, cleansers, medicines, and cosmetics, as of 2013.

The Korea Environment Corporation oversees the EPR system, ensuring that manufacturers and importers comply with reporting requirements for sales and imports, as well as waste collection and recycling data (Kengo, 2021). The federal government is in charge of developing and enforcing EPR rules, while local governments are in charge of ensuring effective, responsible trash collection and increasing recycling and reuse rates. Private recycling collectors are hired by apartment buildings to collect trash and sell it to recyclers. A number of labelling solutions for items covered by the EPR system improve monitoring, including information on packaging recyclability and how it should be disposed of. Importers and manufacturers generate these labels.

The Ministry of Environment of Korea (MOE) issued a legislative notice on 21 July 2021, proposing a partial amendment to the Presidential Decree on the Act on the Promotion of Saving and Recycling of Resources. This would broaden the scope of items subject to the EPR System; they invited public comment by 30 August 2021. Business companies that manufacture 17 products, including pallets, industrial films, and polyvinyl chloride pipes, were required to collect and recycle the relevant items in quantities that are computed and publicized by MOE every year, according to the proposed amendment. From fiscal year (FY) 2022 to FY2023, the proposed amendment will apply to four items, which are industrial films, agricultural films, 20 types of domestic products, and replacement water purifier filters, followed by the remaining 13 items from FY2023 (Kengo, 2021). Pallets, safety nets, fishing nets, ropes, industrial films, agricultural films, polyethylene (PE) pipes, artificial turf, 20 types of household goods (e.g. sealed and storage containers for kitchens), plastic transport boxes, profiles, polyvinyl chloride (PVC) pipes, floor materials, insulation materials for construction, power and communication lines, replacement water purifier filters, and a variety of other items have been added to the list of products subjected to mandatory recycling. The total number of items subject to mandatory recycling under the Act increases to 29 with the addition of the preceding 17 items.

The Voluntary Agreement System for the Collection and Recycling of Plastic Waste was used to handle the goods that were suggested to be added to the list. During the term of the Agreement, manufacturers of target items must build and manage an effective collection and recycling system. The Voluntary Agreement of Plastic Waste Collection and Recycling is a system in which manufacturers of plastic products subject to the waste Charge can receive a reduction or exemption from the waste Charge if they sign a voluntary agreement on waste collection and recycling with the Minister of Environment and meet the collection and recycling targets set in the agreement. In the case of pallets, for example, the Korea Pallet Container Association has signed a voluntary agreement with the Minister of Environment as a representative organization to fulfil the requirement of collecting and recycling pallets at the end of each year.

8.4.7.2 Framework act on resource circulation (FARC)

The FARC was implemented in an attempt to turn Korea into a resource-circulating society that maximizes resource efficiency, consequently addressing resource energy and environmental issues. This law was set to take effect in 2018. A resource-circulating society is one in which individuals and businesses collaborate to reduce waste and increase reuse or recycling of materials into new products or energy. By encouraging innovative ideas and technologies for reusing and recycling waste rather than simply sending it to incinerators or landfill sites, the FARC aims to transform the mass production-oriented and mass waste-producing economic structure into a much more sustainable and efficient resource-circulating one at an ideal fundamental level.

The FARC defines a waste hierarchy, assigns duties to key players, and mandates the establishment of measurable objectives and methods to encourage resource circulation, such as waste disposal levies. The FARC outlines waste management prioritizing based on the waste hierarchy, which includes waste generation minimization through efficient resource use, circular waste usage, and waste disposal in the priority sequence of reuse, reclamation, energy recovery from waste, and suitable disposal.

When it comes to garbage disposal, the influence and harm to the environment and human health should be taken into account. The FARC, in particular, is concerned about hazardous chemicals, stating that the Ministry of Environment may assess hazards and circular utilization for any products suspected of being hazardous to human health and the environment due to containing any harmful substances, such as air pollutants or toxic chemicals, and suspected of not being circularly used when they become wastes, under specific regulations, such as the Clean Air Conservation or Chemicals Control Acts.

The two main provisions of FARC are the 'Recyclable Resource Recognition Program (RRRP)' and the 'Resource Circulation Performance Management Program (RCPMP)'. First and foremost is RRRP, which recognizes waste products that satisfy specified criteria as 'recyclable resources' and exempts them from waste regulations in general. The term 'recyclable resource' refers to substances that are generated or utilized in the recycling of waste and that fulfil the MOE environmental, economic and technological standards. Even after being recycled, waste is now subject to waste regulations. However, it is no longer subjected to the burdensome waste control rules if formally recognized as a recyclable resource under the RRRP. Businesses that collect, transport, recycle and distribute recyclable resources will encounter fewer regulatory constraints in their economic operations, allowing them to better engage in the market for recycled products. Waste products that do not meet the safety criteria may have their recyclable resource designation revoked at any time by the MOE. Waste restrictions will apply to items that lose their recyclable resource classification.

According to the FARC, the Minister of Environment has the responsibility to limit waste generation and promote resource circulation by establishing mid- to long-term and phase-by-phase national resource circulation goals for final waste disposal, resource circulation, and energy recovery rates, and taking appropriate actions based on Article 14. Municipal and provincial governments, as well as enterprises, must define their own targets for final waste disposal and resource rotation rates. The key to these provisions is the implementation of RCPMP on businesses. Under the RCPMP, the MOE is required to encourage businesses to strengthen and improve their recycling performance by establishing resource circulation goals for various types of businesses and providing financial and technological incentives for those businesses to strive to meet those goals. RCPMP establishes mandatory recycling quotas to enterprises and organizations that generate large volumes of waste and evaluates their recycling performance against those quotas.

8.4.7.3 National marine litter management plan

The South Korean Ministry of Oceans and Fisheries (MOF) announced the formation of the 1st Framework on Marine Debris Management from 2021 to 2030 in May 2021, which lays out systematic policies on marine debris management as well as implementation initiatives for the next decade (NOWPAP CEARAC, 2020). Previously, the framework for dealing with marine debris was governed by the principles of the Marine Environment Management Act, which primarily dealt with its collection and disposal. The 3rd National Marine Litter Management Plan 2019–23 was authorized in 2019 under this act.

The 3rd Management Plan intends to analyse the preceding 2nd National Marine Litter Management Plan (2nd Management Plan) (2014–18) by examining global and domestic trends in marine litter management and calculating the level of domestic marine litter. The evaluation of the second management plan will aid in the development of objectives and strategies for the third management plan, as well as in the implementation and execution of the suggested plans. The framework includes 5 key implementation methods as well as 16 initiatives with the purpose of improving the life cycle of marine debris creation, collection and disposal, as well as increased collaborations with appropriate agencies. The strategy seeks to minimize 60% of total marine plastics trash influx by 2030 and zero by 2050. Strategy 1 is 'preventative'. This strategy implements a deposit system on buoys and fishing gear to prevent the root cause of debris production, broadens the distribution scope of eco-friendly buoys, prevents the influx of land debris into the oceans, and establishes a preventive mechanism against overseas marine debris through participation in international organizations and bilateral partnerships. Meanwhile, an improved system for collection and delivery was set as Strategy 2, to avoid debris collection blind spots, where the Ministry intends to deploy seven cleaning vessels in island areas, secure enough drop-off stations for collected marine debris, and implement a responsive mechanism to deal with massive influxes of debris caused by natural disasters such as typhoons and floods. The MOF also runs consultation groups with relevant authorities for cooperative debris collection and works to enhance the overall collection method. Furthermore, a prediction system based on the active use of ICT, artificial satellites, and drones will be created to anticipate the production and movement of maritime trash. This will also lead to the continued improvement of effective collection mechanisms, including suitable equipment, in distant places where marine plastics and debris are inaccessible.

In Strategy 3 the government aims to encourage the recycling concept. In this target, the management infrastructure will be developed and supported further by improving the installation of the life cycle of marine trash and the capability of relevant authorities. In addition, an energized system tailored to islands and fishing villages will be designed, and distributable models will be tested, all of which will contribute to the construction of a circular economy town. Strategy 4 is to enhance the management basis. The Ministry will further establish and operate an interministerial committee to handle the life cycle of marine debris at the government level, with a specialized governing authority with significantly expanded experience in marine debris management. Furthermore, frequent study will be carried out in order to alleviate worries about the effects of microplastic waste on marine ecosystems, and countermeasures will be devised that take into account all potential consequences and the type of microplastics. Finally, Strategy 5 promotes public awareness towards marine litter. In order to boost public participation, the government, local governments, and relevant agencies will work together, and different public engagement initiatives, such as the Adopt-a-Beach Movement, will be implemented. Meanwhile, instructional programmes will be designed and distributed, as well as promotional content for various social groups such as citizens, fishermen and students (Ministry of Oceans and Fisheries, 2021).

8.4.8 Indonesia

8.4.8.1 National Plastic Action Partnership (NPAP)

As part of the Global Plastic Action Partnership (GPAP), Indonesia's Coordinating Minister of Maritime Affairs and Investment, Minister Luhut B. Pandjaitan, businesses, civil society groups, and local stakeholders launched the Indonesia National Plastic Action Partnership (NPAP) in March of 2019. The Indonesia NPAP is the country's primary platform for public-private partnership, working closely with leading change-makers from all sectors to develop a collaborative path towards a plastic pollution-free Indonesia. It has more than 150 member organizations (Zorp, 2019).

GPAP intends to accelerate the country's transition to a circular economy by choosing feasible investment ideas that may be recognized worldwide and replicated and implemented in other nations. It is the most recent move in Indonesia's ambitious national effort to eliminate marine plastic litter by 70% and address solid waste. The World Economic Forum hosts GPAP, which is supported internationally by the governments of Canada and the United Kingdom, as well as The Coca-Cola Company,

Dow, PepsiCo, and Nestlé, and also the World Resources Institute, the World Bank, Pew Charitable Trusts, SYSTEMIQ, the Platform for Accelerating the Circular Economy, Friends of Ocean Action and others.

Through NPAP, the Indonesian government aspires to achieve three goals. To begin, the NPAP collaborates with Indonesia's National Action Plan on Marine Debris, the Indonesian National Waste Management Policy and Strategy (Jakstranas), and other initiatives to reduce marine plastic leakage by 70% by 2025. Furthermore, it is anticipated that Indonesia will prevent 16 million tons of plastic trash from entering the ocean by 2040 by following the NPAP Multistakeholder Action Plan. Moreover, the NPAP's suggested system reform initiatives are predicted to increase Indonesia's economic development by more than 150,000 direct jobs, improving quality of life.

8.4.8.2 Ecolabelling scheme

On World Environment Day, 5 June 2004, the Indonesian Eco-label Logo and Scheme of Eco-label Accreditation and Certification were launched. The word 'Ekolabel' was originally used by the Indonesian Ministry of Environment in a regulation that discusses environmental management, ecolabels, clean manufacturing, and environmentally based technology. Indonesia has a national Ecolabel Scheme that allows items to be given a Type I Ecolabel known as 'Ramah Lingkungan', which means 'environmentally friendly'. The Indonesian Eco-label, also known as Eko-label Indonesia, aspires to be a powerful tool for protecting the environment and human lives while simultaneously improving product efficiency and competitiveness.

Indonesia's Environmental Protection and Management Act No 32/2009 aims to improve the quality of the environment by implementing mechanisms to alleviate pollution and degradation, as well as shifting the paradigm from 'end-of-pipe' to 'preventive'. As a result, Indonesia's Ecolabel Program is being implemented with a few goals in mind, which are synergy of negative environmental impact mitigation throughout the product life cycle to the environment, encouraging demand and supply of environmentally friendly products, providing proactive guidance to industry to proactively improve their products, and educating and assisting consumers in understanding and identifying environmentally friendly products.

The ecolabel programme employs ISO 14020 General Principles of Environmental Labels and Declarations, ISO 14024 Guideline of Ecolabel Type I, and ISO 14021 Environmental Label and Declarations—Self-Declared Environmental Claims as technical references for implementation (Type II Environmental Labeling). In Indonesia, this ecolabel may be found on retail items, as shown in Fig. 8.3. The ecolabel certification criteria are based on rigorous technical examinations of the environmental elements of the items throughout their life cycles. As of October 2013, product eco-label standards for 12 product groupings, including plastic shopping bags, were created.

The verification agencies, which are recognized by the Ministry of the Environment, have the power to verify the claims made by corporations based on their proenvironmental accomplishments. To comply with the standards for affixing the symbol to the product, the manufacturing business must fulfil the necessary ecolabel certification requirements. The required certification encompasses various environmental management requirements, as well as the implementation of an environmental management system, the adoption of quality and standardized products, and the use of environmentally friendly packaging. In the certification procedure, all of the required features will be employed as assessment materials for the ecolabel.

Fig. 8.3 Ecolabel used by Indonesia.

8.4.8.3 Waste management act (Law number 18 of 2008)

People's consumption habits are leading to the production of a growing variety of waste, particularly packaging waste that is hazardous and difficult to break down naturally. President Doctor Haji Susilo Bambang Yudhoyono approved Law Number 18 of 2008 on Waste Management on 7 May 2008 in Jakarta. After being proclaimed on 7 May 2008 by the Minister of Law and Human Rights Andi Mattalatta in Jakarta, Law 18 of 2008 concerning Waste Management was published in the State Gazette of the Republic of Indonesia, Number 69, so that everyone would be aware of it. The Supplement to the State Gazette of the Republic of Indonesia Number 4851 contains the explanation of Law Number 18 of 2008 on Waste Management.

The principal policy under Law Number 18 of 2008 on Waste Management controls the execution of integrated and comprehensive waste management, as well as the community's rights and obligations and the Government's and regional governments' duties and authority to provide public services. The principles of responsibility, sustainability, benefit, justice, awareness, togetherness, safety, security and economic value underpin the legal regulation of waste management in Law 18 of 2008 concerning Waste Management. Everyone has the right to a pleasant and healthy living environment under Article 28H paragraph (1) of the Republic of Indonesia's 1945 Constitution. As a result of the Constitution's mandate, the government is required to offer public waste management services. This has legal consequences in that the government is the authorized and accountable entity in the sector of waste management, despite the fact that its administration can collaborate with businesses. Furthermore, waste management operations can be carried out by solid waste organizations and community groups operating in the solid waste sector.

8.5 Conclusion

In conclusion, many efforts, policies and regulations have been made to minimize the use of plastic materials and control disposal behaviour for plastics. It can be seen that most of the countries have adopted a soft-landing strategy by reducing the dependency of the society on substantial usages and transforming to responsible types of uses by promoting reduce, reuse and recycle. This is because plastics are very convenient for humans, and a drastic ban of plastics could have a domino effect on the entire economic chain. In other words, the main purpose is not to penalize the users, but to control the unnecessary consumption of plastic materials and ensure disposal in an appropriate manner, so that the negative impacts can be progressively mitigated. More important, many of the policies made are placing the responsibility not only on the users but also on the producers, with the extended producer responsibility (EPR) program where the producers are responsible for managing the end-of-life products as part of their social-business responsibility. With the current trend, it is believed that soon our global societies will manage plastics usage and disposal proactively for the betterment of future generations and healthier lives.

Acknowledgements

The authors would like to express their sincere appreciation to Saravanan Raj A/L Pariyasamy, Mishaliny A/P Supramaniam and Teoh Shi Xuan for their support in the preparation of this chapter.

References

A Plastic System Guidebook for France. (2019). *World Wildlife fund for Nature* (pp. 1–19). [ebook]. Available at: https://wwfeu.awsassets.panda.org/downloads/05062019_wwf_france_guidebook.pdf>. (Accessed 25 March 2021).

Abbing, M. (2017). *Sweden leads European ban on microbeads.* [online]. Beat the Microbead. Available at: https://www.beatthemicrobead.org/sweden-leads-european-ban-on-microbeads/. (Accessed 26 March 2021).

Antigua Nice Ltd. (2017). *Antigua news: Styrofoam ban in Antigua: Stages and implementation.* [online]. Available at: https://www.antiguanice.com/v2/client.php?id=806&news=10298>. (Accessed 1 September 2021).

Bakas, I. (2013). *Municipal waste management in Portugal* (pp. 1–17). European Environmental Commission. [ebook] Available at: https://ec.europa.eu.ds_resolveuid. (Accessed 25 March 2021).

Bernhardt, D. (2022). *Leaf Rapids lead the way: Manitoba town was 1st in North America to ban plastic bags 15 years ago. CBC New.* Available at: https://cbc.ca/news/canada/manitoba/plastic-bag-ban-leaf-rapids-manitoba-1.6405046. (Accessed 9 October 2022).

Chemicalwatch. (2018). *UK microbeads ban enters into force.* [online]. Available at: https://chemicalwatch.com/62944/uk-microbeads-ban-enters-into-force>. (Accessed 24 March 2021).

Chen, H. L., Nath, T. K., Chong, S., Foo, V., Gibbins, C., & Lechner, A. M. (2021). The plastic waste problem in Malaysia: Management, recycling and disposal of local and global plastic waste. *SN Applied Sciences, 3*(4).

Chicago City Council. (2014). *Office of the City Clerk-Record #: SO2014-1521.* [online]. Available at: https://chicago.legistar.com/LegislationDetail.aspx?ID=1676473&GUID=E704F6F5-3960-4327-BC60-07B8766C35FA&Options=Advanced&Search=.>. (Accessed 1 September 2021).

Chicago City Council. (2016). *Chicago checkout bag tax.* [online]. Available at: https://www.chicago.gov/content/dam/city/depts/rev/supp_info/TaxPublicationsandReports/3-50ChicagoCheckoutBagTaxOrdinance.pdf>. (Accessed 1 September 2021).

City and County of Honolulu. (2019). *A bill for an ordinance.* [online]. Available at: http://www4.honolulu.gov/docushare/dsweb/Get/Document-248953/ORD19-030.pdf>. (Accessed 1 September 2021).

Department of Environment Antigua and Barbuda. (2016). *The external trade (shopping plastic bags prohibition) order.* [online]. Available at: https://elaw.org/system/files/attachments/publicresource/External_Trade_Prohibition_of_Plastic_Bags_Order_2017.pdf>. (Accessed 1 September 2021).

Deutsche Recycling Service. (2020). *Ban on plastic bags in Austria: What has changed?.* [online]. Available at: https://deutsche-recycling.de/en/blog/ban-on-plastic-bags-in-austria-what-has-changed/>. (Accessed 28 June 2021).

Environmental Department. (2018). *Notification.* [online]. MPCB gov India. Available at: https://www.mpcb.gov.in/sites/default/files/plastic-waste/rules/plasticwasteenglish119102020.pdf. (Accessed 1 August 2021).

EY Global. (2020). *Italy introduces proportional tax on plastic items.* [online]. Available at: https://www.ey.com/en_gl/tax-alerts/ey-italy-introduces-proportional-tax-on-plastic-items>. (Accessed 23 March 2021).

Freshplaza. (2019). *Portugal to ban the use of plastic bags for fruit.* [online]. Available at: https://www.freshplaza.com/article/9093709/portugal-to-ban-the-use-of-plastic-bags-for-fruit/>. (Accessed 25 March 2021).

Gandenberger, C., Orzanna, R., Klingenfuß, S., & Sartorius, C. (2014). *The impact of policy interactions on the recycling of plastic packaging waste in Germany* (pp. 7–15). Econstor. [ebook]. Available at: https://www.econstor.eu/bitstream/10419/100033/1/791618277.pdf. (Accessed 24 March 2021).

Gesley, J. (2021). *Germany: Lightweight plastic bag ban to take effect January 1, 2022|Global Legal Monitor.* [online] Loc.gov. [online]. Available at: https://www.loc.gov/law/foreign-news/article/germany-lightweight-plastic-bag-ban-to-take-effect-january-1-2022/>. (Accessed 24 March 2021).

Government of Jamaica. (2018). *The trade act.* [online]. Available at: https://www.nepa.gov.jm/sites/default/files/2019-11/Proc_1_Trade_Act.pdf>. (Accessed 1 September 2021).

Government of Kerala. (2019). *Sanitation.kerala.gov.in.* [online]. Available at: http://sanitation.kerala.gov.in/wp-content/uploads/2019/12/output.pdf>. (Accessed 13 July 2021).

Government of Prince Edward Island. (2019). *RSPEI 1988, c P-9.2|Plastic Bag Reduction Act|CanLII.* [online]. Available at: https://www.canlii.org/en/pe/laws/stat/rspei-1988-c-p-9.2/latest/rspei-1988-c-p-9.2.html>. (Accessed 1 September 2021).

GTP Headlines. (2020). *Greece to ban single-use plastics by 2021.* [online]. Available at: https://news.gtp.gr/2020/10/05/greece-ban-single-use-plastics-2021/>. (Accessed 26 March 2021).

Hashim, K. H., Mohammed, A. H. B., & Redza, H. Z. S. (2011). *Developing conceptual waste minimization awareness model through community based movement: A case study of green team, International Islamic University Malaysia.* [online]. *intechopen,* Malaysia: Persidangan Kebangsaan Masyarakat, Ruang dan Alam Sekitar (MATRA 2011) pp. 1–11. Available at: http://irep.iium.edu.my/8655/1/FULL_PAPER_DEVELOPING_CONCEPTUAL_WASTE_MINIMIZATION_AWARENESS_MODEL_THROUGH_COMMUNITY_BASED_MOVEMENT.pdf. (Accessed 5 March 2022).

Hofverberg, E. (2020). *Sweden: Parliament votes to adopt tax on plastic bags.* Loc.gov. [online]. Available at: https://www.loc.gov/law/foreign-news/article/sweden-parliament-votes-to-adopt-tax-on-plastic-bags/>. (Accessed 26 March 2021).

InSpain News. (2022). *European Commission wastes Spain to ban disposable plastic products.* Available at: https://inspain.news/european-commission-wants-spain-to-ban-disposable-plastic-products/. (Accessed 9 October 2022).

Jamaica Observer Ltd. (2017). *St Vincent bans styrofoam products.* [online]. Available at: https://www.jamaicaobserver.com/news/St-Vincent-bans-styrofoam-products&template=MobileArticle>. (Accessed 1 September 2021).

Jangprajak, W. (2021). *National action plan on Plastic waste management in Thailand.* [Online]. Available at: https://www.iges.or.jp/sites/default/files/inline-files/S1-5_PPT_Thailand%20Plastic%20Action%20Plan.pdf. (Accessed 3 March 2022).

Joe, T. (2021). *China single-use plastic bag & straw ban now in effect across major cities.* [online]. Green Queen. Available at: https://www.greenqueen.com.hk/china-single-use-plastic-bag-straw-ban-now-in-effect-across-major-cities/. (Accessed 6 July 2021).

Kengo, W. (2021). *Korea announces Legislative Notice of Amendment to add 17 items to EPF to be started from FY 2022.* [online]. Available at: https://enviliance.com/regions/east-asia/kr/report_4036. (Accessed 4 March 2022).

Kojima, M. (2010). *3R policies for Southeast and East Asia.* ERIA Research Project Report 2009, No. 10.

Legislative Analyst's Office. (2016). *Referendum to overturn ban on single-use plastic bags.* [online]. Available at: https://lao.ca.gov/ballot/2016/Prop67-110816.pdf>. (Accessed 1 September 2021).

Letsrecycle. (2012). *Recycling initiatives in Greece.* letsrecycle.com. [online]. Available at: https://www.letsrecycle.com/news/latest-news/recycling-initiatives-in-greece/>. (Accessed 26 March 2021).

Leungsakul, S. (2019). *Situation of plastic scrap & I-waste import into Thailand.* [Online]. Available at: https://www.env.go.jp/en/recycle/asian_net/Annual_Workshops/2019_PDF/Session1/S1_11_Thailand_ANWS2019.pdf. (Accessed 3 March 2022).

LRT. (2020). *Supermarkets in Lithuania to face fines for free plastic bags.* [online]. Available at: https://www.lrt.lt/en/news-in-english/19/1137448/supermarkets-in-lithuania-to-face-fines-for-free-plastic-bags>. (Accessed 27 June 2021).

MESTECC-Ministry of Energy, Science, Technology, Environment & Climate Change. (2018). *Malaysia's roadmap towards zero single-use plastics 2018-2030 (pp. 2–10).*, [online]. https://www.moe.gov.my/images/KPM/UKK/2019/06_Jun/Malaysia-Roadmap-Towards-Zero-Single-Use-Plastics-2018-2030.pdf. (Accessed 5 August 2021).

METI Ministry of Economy, Trade and Industry. (2022). (online) https://www.meti.go.jp/policy/recycle/main/english/law/legislation.html. (Accessed 3 March 2022).

Ministry of Oceans and Fisheries. (2021). *Key components of the 1st Framework on Marine debris Management (2021∼2030)—Achieve 60% Reduction in Marine Plastic Debris inflow by 2030, Zero by 2050.* [online]. Available at https://www.mof.go.kr/en/page.do?menuIdx=1480>. (Accessed 4 March 2022).

Municipal Waste Management-Sweden. (2016). *European Environmental Agency* (pp. 4–10). [ebook]. Available at: https://www.eionet.europa.eu/etcs/etc-wmge/products/other-products/docs/sweden_msw_2016.pdf>. (Accessed 26 March 2021).

National Environmental Agency. (2022). *National Recycling Program.* [Online]. Available at https://www.nea.gov.sg/our-services/waste-management/3r-programmes-and-resources/national-recycling-programme>. (Accessed 3 March 2022).

Nicholas Institute for Environmental Policy Solutions. (2016). *Montreal Bylaw 16-051 prohibiting the distribution of certain shopping bags in retail stores.*

[online]. Available at: http://ville.montreal.qc.ca/sel/sypre-consultation/afficherpdf?idDoc=27530&typeDoc=1>. (Accessed 1 September 2021).

NOWPAP CEARAC. (2020). *National actions on marine microplastics in the NOWPAP Region.* [Online]. Available at http://www.cearac-project.org/RAP_MALI/National_actions_on_MPs.pdf>. (Accessed 4 March 2022).

OECD. (2021). *OECD ocean—Preventing single-use plastic waste.* [online]. Available at: https://www.oecd.org/stories/ocean/preventing-single-use-plastic-waste-d18c8d38>. (Accessed 6 February 2021).

Plastic-Zero. (2012). *Review on plastic waste in the municipal waste streams, Belgium* (pp. 3–9). ebook]. Plastic-Zero, Available at: http://www.plastic-zero.com/media/62436/annex_d20b_-_action_1.3_-_review_on_plastic_waste_in_the_municipal_waste_stream_-_belgium_final.pdf>. (Accessed 6 February 2021).

Poland Government. (2019). *Website of the Republic of Poland. 2019. Consultation meeting on the single-use plastics directive.* [online]. Available at: https://www.gov.pl/web/climate/consultation-meeting-on-the-single-use-plastics-directive. (Accessed 24 June 2021).

Preetha, M. S. (2020). State bans plastic for primary packaging too. *The Hindu.* [online]. 9 Jun. Available at: https://www.thehindu.com/news/national/tamil-nadu/tn-bans-plastic-for-primary-packaging-too/article31790446.ece. (Accessed 18 July 2021).

Qiyun, W. (2020). Beyond plastic recycling: A look at extended producer responsibility in Singapore. *Singapore Policy Journal.* [Online]. Available at https://spj.hkspublications.org/2020/09/08/beyond-plastic-recycling-a-look-at-extended-producer-responsibility-in-singapore/. (Accessed 3 March 2022).

Rashid, N. R. N. A. (2009). Awareness of eco-label in Malaysia's green marketing initiative. *International Journal of Business and Management, 4*(8).

SearchLight. (2020). *Ban on use of disposable plastic bags suspended—PM—Searchlight.* [online]. Available at: https://searchlight.vc/searchlight/breaking-news/2020/08/15/ban-on-use-of-disposable-plastic-bags-suspended-pm/>. (Accessed 1 September 2021).

Shotaro, N. (2021). *Japan cabinet approves bill for plastic resource management* (Online) https://enviliance.com/regions/east-asia/jp/report_2667. (Accessed 3 March 2022).

Smith, L. (2021). *Plastic waste.* [ebook] House of Commons Library, pp. 1–7, 25–26. Available at: https://researchbriefings.files.parliament.uk/documents/CBP-8515/CBP-8515.pdf. (Accessed 24 March 2021).

Smithers, R. (2019). Plastic bottle deposit scheme in UK proving hit with shoppers. *The Guardian.* [online]. Available at: https://www.theguardian.com/environment/2019/jan/02/plastic-bottle-deposit-scheme-in-uk-proving-a-hit-with-shoppers>. (Accessed 24 March 2021).

Suki, N. M. (2013). Green awareness effects on consumers' purchasing decision: Some insights from Malaysia. *Green Awareness Effects, 9*(2), 50–63. [online]. Available at: http://ijaps.usm.my/wp-content/uploads/2013/07/Art3.pdf. (Accessed 20 August 2021).

The Government of Bahamas. (2019). *Environmental protection (control of plastic pollution) Act, 2019—Government—Notices.* [online]. Available at: https://www.bahamas.gov.bs/wps/portal/public/gov/government/notices/environmentalprotection(control of plastic pollution)act2019/. (Accessed 1 September 2021).

The local. (2019). *What is Italy doing about the shocking level of plastic pollution on its coastline?.* Available at: https://www.thelocal.it/20190521/what-is-italy-doing-about-the-shocking-level-of-plastic-pollution-on-its-coastline/. (Accessed 23 March 2021).

Theportugalnews. (2016). *Plastic bag use plummets a year after tax introduction.* [online]. Available at: https://www.theportugalnews.com/news/plastic-bag-use-plummets-a-year-after-tax-introduction/37473>. (Accessed 25 March 2021).

TOMRA. (2020). *Deposit return system kicks off in Portugal with TOMRA.* [online]. Available at: https://newsroom.tomra.com/nova-tomra-portugal/>. (Accessed 25 March 2021).

Wang, H., & Jiang, C. (2020). Local nuances of authoritarian environmentalism: A legislative study on household solid waste sorting in China. *Sustainability, 12*(6), 2522.

Xinhua. (2017). *China announces import ban on 24 types of solid waste—China—Chinadaily.com.cn.* [online]. Available at: https://www.chinadaily.com.cn/china/2017-07/21/content_30194081.htm. (Accessed 10 August 2021).

Zachariadis, P. (2016). *Landfill tax in Greece* (pp. 1–4). Cyprus University of Technology. [ebook]. Available at: https://ieep.eu/uploads/articles/attachments/8192ea44-2204-4756-b71c-d70c8558730e/EL%20Landfill%20Tax%20final.pdf?v=63680923242>. (Accessed 26 March 2021).

Zen, I. S., Ahamad, R., & Omar, W. (2013). No plastic bag campaign day in Malaysia and the policy implication. *Environment, Development and Sustainability, 15*(5), 1259–1269.

Zhang, & Laney. (2017). *China: National plan on banning "foreign garbage" and reducing solid waste imports.* Washington, DC: Library of Congress. [online]. Available at: https://www.loc.gov/item/global-legal-monitor/2017-08-08/china-national-plan-on-banning-foreign-garbage-and-reducing-solid-waste-imports/>. (Accessed 15 August 2021).

Zorp, Y. (2019). *Indonesian government and partners announce next steps to tackle plastic pollution.* [online]. Available at: https://www.weforum.org/press/2019/03/indonesian-government-and-partners-announce-next-steps-to-tackle-plastic-pollution/>. (Accessed 4 March 2022).

Index

Note: Page numbers followed by *f* indicate figures and *t* indicate tables.

R

Printed in the United States
by Baker & Taylor Publisher Services